COMPUTATIONAL AND STATISTICAL APPROACHES TO GENOMICS

COMPUTATIONAL AND STATISTICAL APPROACHES TO GENOMICS

edited by

Wei Zhang
and
Ilya Shmulevich

University of Texas M. D. Anderson Cancer Center

KLUWER ACADEMIC PUBLISHERS
Boston / Dordrecht / London

Distributors for North, Central and South America:
Kluwer Academic Publishers
101 Philip Drive
Assinippi Park
Norwell, Massachusetts 02061 USA
Telephone (781) 871-6600
Fax (781) 681-9045
E-Mail < kluwer@wkap.com>

Distributors for all other countries:
Kluwer Academic Publishers Group
Post Office Box 322
3300 AH Dordrecht, THE NETHERLANDS
Telephone 31 786 576 000
Fax 31 786 576 254
E-Mail < services@wkap.nl>

 Electronic Services < http://www.wkap.nl>

Library of Congress Cataloging-in-Publication Data

A C.I.P. Catalogue record for this book is available
from the Library of Congress.

Printed on acid-free paper.
Printed in the United States of America

**The Publisher offers discounts on this book for course use and bulk purchases.
For further information, send email to <joanne.tracy@wkap.com>.**

Contents

This book is dedicated
to our families and
colleagues

Foreword

Looking back at the last century, some of the most magnificent discoveries that have had everlasting impact on society and people's lives were made in the areas of life science and computer science. One does not have to go very far to witness the result of such advances hand-in-hand—the completion of the draft of the human genome—the genetic blueprint for all human beings. This huge map of the sequence, over 3 billion nucleotides long, allows us to "see," with the help of computers, how highly complex we really are. The revelation of these complexities leads to the further realization that the single-gene approaches we have used in biology and medicine are so deficient, and that the sciences, molecular biology and genetics, which we know so well and with which we grew comfortable, are being surpassed by a new breed of disciplines. Now, genomics, proteomics, microarrays, and bioinformatics become the new buzz-words for symposiums, news releases, top magazines, and new curriculum development in undergraduate and graduate schools. The major biomedical institutions are investing millions to establish state-of-the-art laboratories with robotic systems and to recruit engineers, physicists, statisticians, computer scientists, and mathematicians to live and work among the biologists. We now seek well beyond tens of genes or proteins to thousands of genes and proteins to satisfy researchers' inquiry into the secrets of life—be it normal cells at different stages of development or diseased cells with different phenotypes. Genomics and informatics have become the new dream subjects for graduate, college, and even high school students. At the State Science Fair of Texas in 2000, the top prize went to a high school senior who presented his volunteer work in the area of cancer genomics at the Cancer Genomics Core Lab of The University of Texas M. D. Anderson Cancer Center. Genomics era is indeed here to stay and to move forward.

Although new, genomics has already made a major impact in biomedicine. Using microarrays and related high throughput technologies, scientists have already identified "signature" gene expression patterns for different cancer cells, cancer specific vessel cells, and marker genes for important processes such as cancer metastasis. The gene expression patterns also allow identification of "diseases within a disease." An expanded effort is being made to search for genes that have predictive values for response to therapy. That way, based on the gene expression information, those patients who are resistant to certain treatments, such as radiation, may not have to go thorough the debilitating treatment. Efforts are also being made to identify the subtle genetic differences called Single Nucleotide Polymorphisms that may tell us something about certain propensities, be it tolerance to alcohol or likelihood of lung cancer development. Knowledge in this area may eventually impact disease prevention and counseling to improve people's lives.

This sounds like a dream too good to come true any time soon; but dreams have been the driving force for so many marvelous things that happened to us. How, then, do we make this genomic dream come true? We will have to develop new tools to generate accurate information and process it reliably. We will have to approach biology in a more quantitative manner. We will have to educate a new generation of genomic biologists who are also well-versed in statistics, computer sciences, and engineering. As a cancer researcher for the last thirty years, I am fortunate to be part of this revolution. The mission of M. D. Anderson Cancer Center is to eradicate cancer. I see the Cancer Genomics Program, headed by Dr. Stanley Hamilton, as a major vehicle to facilitate our mission. We have and will continue to invest in cancer genomics operations and education.

It is timely that Drs. Zhang and Shmulevich from the Cancer Genomics Core Laboratory at The University of Texas M. D. Anderson Cancer Center have conceived and achieved a volume with contributions spanning a broad range of topics in genomics research. These chapters should be invaluable to researchers who are establishing high throughput array capacity in their institution, typically with inadequate means to build a strong cadre of specialists in informatics, statistics and computation. This book will also be welcomed by investigators from a variety of disciplines, who work together in multidisciplinary teams to advance our understanding of the complex reality of the genome and proteome.

John Mendelsohn, M.D.
President and Professor of Cancer Medicine
The University of Texas
M. D. Anderson Cancer Center

Preface

From Nature's chain whatever link you strike,
Tenth or ten thousandth, breaks the chain alike.

—Alexander Pope

One of the most exciting and at the same time challenging aspects of genomics is its multifaceted nature. Like a mosaic, this relatively new and multidisciplinary field of study integrates many tightly intertwined components, which may at first appear to be unrelated, into a unified whole. Successful research in genomics *ipso facto* requires every participating specialist to be able to see this 'whole picture.' This last requirement suggests an interesting implication: existing methods or approaches in the specialist's field of expertise may not be adequate or appropriate for successful integration and implementation in genomics. This makes it necessary for the researcher to develop new methodologies and techniques or 'customize' existing ones. Such a scenario often has the potential to advance the state of the art not only in genomics, but also in the specialist's own area of expertise.

In this book, we have tried to collect contributions that cover a wide range of topics in genomics and span a variety of disciplines, such as image analysis, statistics, machine learning, pattern recognition, time-frequency and nonlinear signal processing, parallel computing, and molecular biology to name but a few. The organization of the chapters roughly follows the structure *data* $\implies$ *analysis* $\implies$ *synthesis* $\implies$ *application*. Each of these stages is equally important and crucial for tackling the complex problems one encounters in genomics.

The first step is to collect *data* and prepare it for further analysis. If this link in the chain is weak, all further work is compromised. This could not be more true than in the case of microarrays. Chapter 1 deals with the important issues and challenges of microarray image analysis as well as statistics of microarray data and quality assessment. Chapter 2 addresses the data acquisition process with cDNA microarray technology. Special emphasis is placed on amplification methods and ensuing statistical analysis. Chapters 3 and 4 go even deeper into the statistical nature and challenges in microarray experiments and discuss the sources of variation, error models, experimental design issues, and the statistical analysis of replicate measurements.

The next step involves the *analysis* of collected and possibly preprocessed data in order to extract and visualize meaningful information for further study. Chapter 5 describes exploratory clustering and visualization for analyzing gene expression data, using the Self-Organizing Map. The authors illustrate their methods with mutated yeast strain data. Chapters 6 and 7 focus on classification using gene expression data. Chapter 6 is concerned with the selection of informative genes for cancer classification, using newly developed algorithms and Boolean threshold functions. Chapter 7 addresses the fundamental classifier design issues in microarray experiments, such as error estimation, classifier complexity, and sample sizes as well as provides an illustration of various classification techniques. Chapter 8 is an excellent example of the use of powerful signal processing methods in biology. The authors employ time-frequency analysis of signals using wavelets and the Wigner-Ville distribution to analyze protein sequences and provide visualization techniques. Chapter 9 takes up the challenging goal of characterizing the statistical distributions appearing in various genome-related phenomena. Exciting results are presented about the distributions for transcript copy numbers in a single eukaryotic cell, sizes of protein clusters, protein domain occurrence values in bacteria and eukaryotic proteomes, and for numbers of regulatory DNA-binding proteins in the prokaryotic genome. A new methodology is also developed for removing experimental errors from SAGE gene expression data.

The next stage (*synthesis*) involves model building in order to understand the enormous complexity of molecular interactions that take place in our cells. At some level or another, living organisms are systems and it takes formal methods and computational modeling to be able to make non-trivial assertions and predictions about their complex dynamical behavior. Chapter 10 starts out by addressing the problem of multivariate gene expression prediction and classification. This approach is in stark contrast to traditional 'single-gene' approaches in biology and harnesses the true power of modeling. The authors look for flexible classes of models with good predictive power, but also consider the complexity of the models, developing effective methodologies based on normalized maximum likelihood models for Boolean regression. Chapter 11 addresses the problem of inferring the structure of a genetic regulatory network from data, using the Boolean network model class. Using a so-called Best-Fit Extension strategy in order to deal with errors and inconsistencies in the data, the computational complexity as well as the performance accuracy of the inference algorithm are studied. Chapter 12 continues the investigation of genetic regulatory network models by studying methods for imposing robustness. The authors utilize the powerful concept of noise-injection to improve the performance of genetic network models and tackle the so-called dimensionality problem. Chapter 13 is concerned with the implementation of multivariate code-termination analysis of gene expression relations on a parallel computing environment. The authors discuss various parallelization methods, provide performance results, and describe a flexible tool for visualization of gene expression.

The ultimate hope is that the massive amounts of quantitative data, combined with powerful analysis tools and simulations of models, will result in significant advances in the life sciences (*application*). One cannot overrate the enormous benefit of objectively diagnosing and classifying cancers, predicting potential drug targets for clinical intervention, and gaining a deeper understanding of biological systems and the disruption of their normal functioning in diseases. Chapter 14 addresses the importance of high-throughput technologies, such as cDNA microarrays and tissue arrays, for cancer diagnosis and classification. The authors focus on human glioma diagnosis and discuss various statistical and computational approaches, such as linear discriminant analysis, multidimensional scaling, and k-nearest neighbor classification. Chapter 15 focuses on breast cancer. The authors investigate the important problem of the effect of tissue heterogeneity on gene expression measurements and interpretations. Methods such as hierarchical clustering and multidimensional scaling are used to demonstrate the impact of gene selection on clustering outcomes.

Chapter 16 explores the role of alternative RNA splicing in cancer. It is crucial to understand this important phenomenon in order to appreciate the enormous complexity of genetic regulation. The chapter overviews computational tools for gene finding and alternative splicing analysis and reviews useful bioinformatic tools, such as databases and annotation programs, available on the Internet. It goes on to review protein diversity from alternative splicing, microarray technology for exon prediction, and the dysregulation of alternative splicing in cancer. Chapter 17 discusses another important phenomenon – Single-Nucleotide Polymorphisms (SNP). The authors review the relationships between

COMPUTATIONAL GENOMICS

SNPs and DNA damage and repair, SNP detection using data-mining strategies, known associations with cancer, several SNP databases, and methods for analyzing SNP data using machine-learning approaches.

We hope you will find this book not only interesting, but also useful in your work or research, whether you are working in biology, medicine, computer science, engineering, mathematics, or statistics.

Acknowledgments

We are of course deeply indebted to all the contributors (67 in total) without whom this book would not have existed. We are also glad to thank Joanne Tracy at Kluwer Academic Publishers for her professionalism and encouragement. We are grateful to Drs. John Mendelsohn, Stanley Hamilton, Joe Gray, Daniel Von Hoff, Albert Deisseroth, and Raphael Pollock for their support of this book from its conception. Finally, we are pleased to acknowledge the support of The University of Texas M. D. Anderson Cancer Center.

I.S. thanks Drs. Gonzalo Arce and Kenneth Barner for giving him the opportunity to organize a special session on Signal Processing in Biocomputing at the 2001 IEEE–EURASIP Workshop on Nonlinear Signal and Image Processing (NSIP), where he had the pleasure of meeting some of the contributors to this book.

W.Z. thanks Beth Notzon for her professional editing assistance on several chapters.

WEI ZHANG AND ILYA SHMULEVICH

Chapter 1

MICROARRAY IMAGE ANALYSIS AND GENE EXPRESSION RATIO STATISTICS

Yidong Chen[1], Edward R. Dougherty[2], Michael L. Bittner[1],
Paul Meltzer[1], and Jeffery Trent[1]

[1] *Cancer Genetics Branch, National Human Genome Research Institute, National Institutes of Health, Bethesda, Maryland, USA*

[2] *Department of Electrical Engineering, Texas A&M University, College Station, Texas, USA*

1.   Introduction

A cell relies on its protein components for a wide variety of its functions, including energy production, biosynthesis of component macromolecules, maintenance of cellular architecture, and the ability to act upon intra- and extra-cellular stimuli. Each cell in an organism contains the information necessary to produce the entire repertoire of proteins the organism can specify. Since a cell's specific functionality is largely determined by the genes it is expressing, it is logical that transcription, the first step in the process of converting the genetic information stored in an organism's genome into protein, would be highly regulated by the control network that coordinates and directs cellular activity. A primary means for regulating cellular activity is the control of protein production via the amounts of mRNA expressed by individual genes. The tools required to build an understanding of genomic regulation of expression reveal the probability characteristics of these expression levels.

Complementary DNA microarray technology provides a powerful analytical tool for human genetic research (Schena *et al.*, 1995; Schena *et al.*, 1996; DeRisi *et al.*, 1996; DeRisi *et al.*, 1997; Duggan *et al.*, 1999). It combines robotic spotting of small amounts of individual, pure nucleic acid species on a glass surface, hybridization to this array with multiple fluorescently labeled nucleic acids, and detection and quantitation of the resulting fluor-tagged hybrids with a scanning confocal microscope (Fig. 1.1). A basic application is quantitative analysis of fluorescence signals representing the relative abundance of mRNA from distinct tissue samples. cDNA microarrays are prepared by printing thousands of cDNAs in an array format on glass microscope slides, which provide gene-specific hybridization targets. Distinct mRNA samples can be labeled with different fluors and then co-hybridized on to each arrayed gene. Ratios of gene-expression levels between the samples can be used to detect meaningfully different expression levels between the samples for a given gene. Given an experiment design with multiple tissue samples, microarray data can be used to cluster genes based on expression profiles (Eisen *et al.*, 1998; Khan *et al.*, 1998), to characterize and classify disease based the expression levels of gene sets (Golub *et al.*, 1999; Ben-Dor *et al.*, 2000; Bittner *et al.*, 2000; Hedenfalk *et al.*, 2001; Khan *et al.*, 2001), and for the many statistical methods presented in this book. When using cDNA microarrays, the signal must be extracted from the background. This requires image processing to extract signals arising from tagged mRNA hybridized to arrayed cDNA locations (Chen *et al.*, 1997; Schadt *et al.*, 2000; Kim *et al.*, 2001), and variability analysis and measurement quality control assessment (Bittner *et al.*, 2001; Newton *et al.*, 2001; Wang *et al.*, 2001).

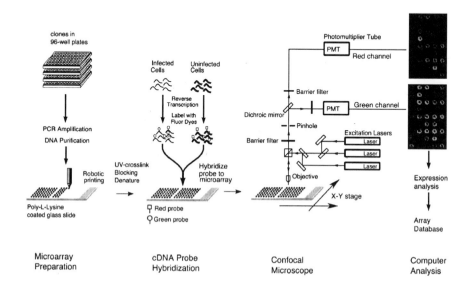

Figure 1.1. Illustration of a microarray system.

This chapter discusses an image processing environment whose components have been specially designed for cDNA microarrays. It measures signals and ratio statistics to determine whether a ratio is significantly high or low in order to conclude whether the gene is up- or down-regulated, and provides related tools such as those for quality assessment.

2.   Microarray Image Analysis

A typical glass-substrate and fluorescent-based cDNA microarray detection system is based on a scanning confocal microscope, where two monochrome images are obtained from laser excitations at two different wavelengths. Monochrome images of the fluorescent intensity for each fluor are combined by placing each image in the appropriate color channel of an RGB image (Fig. 1.2). In this composite image, one can visualize the differential expression of genes in the two cell types: test sample typically placed in red channel, while the reference sample in green channel. Intense red fluorescence at a spot indicates a high level of expression of that gene in the test sample with little expression in the reference sample. Conversely, intense green fluorescence at spot indicates relatively low expression of that gene in the test sample compared to the reference. When both test and reference samples express a gene at similar levels, the observed array spot is yellow. We generally assume that specific DNA products from two samples have an equal probability of hybridizing to the specific target. Thus, the fluorescent intensity measurement is a function of the amount of specific RNA available within each sample, provided samples are well-mixed and there is sufficiently abundant cDNA deposited at each target location.

The objective of the microarray image analysis is to extract probe intensities or ratios at each cDNA target location, and then cross-link printed clone information so that biologists can easily interpret the outcomes and perform further high-level analysis. The block diagram of the image analysis system is shown in Fig. 1.3. A microarray image is first segmented into individual cDNA targets, either by manual interaction or an automated algorithm. For each target, the surrounding background fluorescent intensity is estimated, along with the exact target location, fluorescent intensity and expression ratios. Microarray image sources are not from a single print mode (i.e., different printing tip arrangement or different arrayers (Bowtell, 1999)) or one hybridization method (i.e., fluorescent, radioactive probe, and others); nevertheless, to simplify the presentation, we model the microarray images from two fluorescent probes as our main processing input images.

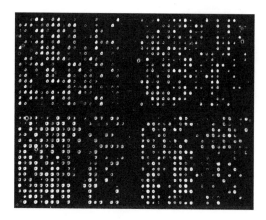

Figure 1.2. An example of microarray image. (see color plates)

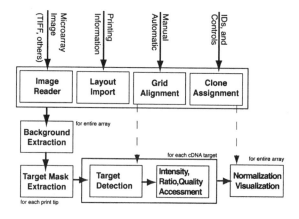

Figure 1.3. Block diagram of cDNA microarray image analysis.

2.1   Target Segmentation and Clone Information Assignment

Since each element of an array is printed automatically to a pre-defined pattern and position, we assume the final probe signals from a regular array to be automatically aligned to a predefined grid-overlay. The initial position of the grid (Fig. 1.4) can be manually determined if no particular orientation markers are printed or no visible signals that can be used as orientation markers, or automatically determined if the orientation markers are presented in the final image and the entire array has no obvious missing row or column signals (Bittner *et al.*, 2001). Due to the complication of customized print procedures and various hybridization protocols, we assume the initial grid-overlay is manually determined. Usually, the initial target segmentation achieved by grid overlaying does not need to be precise. An automatic refinement of the grid position is preferable following the manual grid overlaying. For each sub-array, and after the initial alignment of the grid pattern, we utilize the following procedure:

1 Adjust the four corners' horizontal coordinates with [-1, 0, 1].

2 Calculate the column average along the vertical edge of the bonding boxes for all columns in the sub-array, and then sum the average intensities I, as illustrated in Fig. 1.5 by red open circles.

3 Calculate the center of gravities between two valleys, as marked in Fig. 1.5 by dark "+". Evaluate the average difference between the calculated center to the center of gravities, D.

4 Select the corner coordinates that produce the minimum value of $I \times D$, or overlaying grid pattern between the cDNA targets. The minimum value $I \times D$ ensures that the grid will be mostly overlaid within the background region, and when the background space between columns of the target is large, the center of gravity shall overlay with the centers of the each grids.

5 Repeat Steps 1-4 for the horizontal direction of the grid pattern.

6 Repeat steps 1-5 once, to limit the adjustment to each corner to no more than 2 pixels (more repeats are optional to allow a larger adjustment).

This is a semi-automatic grid alignment procedure that depends on the initial user interaction. A post-refinement of the grid alignment may be activated if the image needs to be processed again. Statistics derived from the prior detection procedure, such as spot measurement quality, intensity or size, are available to further qualify the spots for precise grid alignment.

After segmentation, clone information is linked to array locations before further processing. Information regarding non-printed, duplicated, negative-control, or housekeeping-genes is useful for normalization and other calibration purposes.

2.2   Background Detection

Typically, the background of a microarray image is not uniform over the entire array and extraction of local background intensity is necessary.

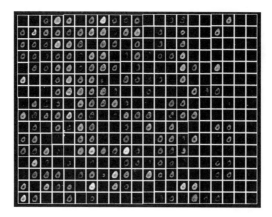

Figure 1.4.   Manual grid alignment. (see color plates)

Changes of fluorescent background across an array are usually gradual and smooth, and due to many technical reasons. Abrupt changes are rare; when they happen, the actual signal intensities of array elements near these changes are not reliable. Conventionally, pixels near the bounding box edge are taken to be background pixels, and thus the mean gray-level of these pixels provides an estimation of the local background intensity (Schena *et al.*, 1995; Schena *et al.*, 1996; Eisen *et al.*, 1998). This method becomes inaccurate when the bounding box size is close to 10×10 pixels or the target fills entire bounding box. Alternative methods require some knowledge of the target location, such that the corner regions between targets may be chosen to be the candidate region for background estimation.

Fluorescent background is typically modeled by a Gaussian process (Wang and Herman, 1996). If we choose a large area that covers the center of its neighboring targets, then the gray-level histogram within the box is usually unimodal, since the majority of the background pixel values are similar while the target pixel values spread up to very high gray levels. The histogram mode provides the mean local background intensity μ_b and the left tail of the histogram provides the spread (standard deviation σ_b) of the background intensity. A unimodal histogram is shown in Fig. 1.6 that has been derived from a region (the small insert at right-hand side) with no effort made to eliminate the target region, even though it contains a bright target. The background statistics derived here provide essential information for target detection.

In practice, the histogram is not always unimodal. Thus, care must be taken. We first use the minimal value from average intensities from

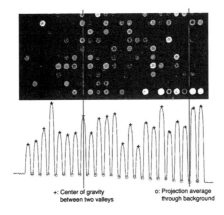

Figure 1.5. Semi-automatic grid alignment. (see color plates)

the four sides of the bounding box of the target to estimate of the mean background level μ_b. We then take the gray level, g, at 1 percentile within the bounding box with the intention being to eliminate possible dark pixels due to scratches (random or in the center of the target). We consider an initial estimate $\sigma_b = (v - g)/3$. We then refine the estimate μ_b and σ_b from within the initial range $(\mu_b - 3\sigma_b, \mu_b + 3\sigma_b)$, and then estimate the mode within the range $(\mu_b - \sigma_b, \mu_b + \sigma_b)$.

Finding exact background intensities is problematic. Some hybridizations yield much higher or lower intensities than those from the peripheral region. It is highly recommend that microarray design include randomly distributed negative control locations. The background intensity for a cDNA target can be assessed by its nearest negative control locations. Background subtraction can produce negative intensities. We set a negative intensity to 1.0 to avoid an unattainable ratio quotient. Alternatives include global background subtraction, conservative estimation of background level by the morphological opening operator, or no subtraction at all.

2.3   Target Detection

Identifying the target region within a bounding box is a difficult image-processing task. Targets are somewhat annular (Fig. 1.7) owing to how the robot print-tip places the cDNA on the slide and how the slide is treated; however, the final target image may be a collection of subregions within the nominal circular target region owing to variability caused by cDNA deposition or the hybridization process. To reduce the noise-to-signal interaction, the final signal intensity should be mea-

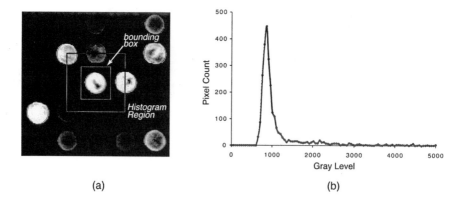

(a)   (b)

Figure 1.6.   Background estimation.

sured over regions corresponding to the probe-hybridized-to-target area. Conventional adaptive thresholding segmentation techniques are unsatisfactory when the signal is weak because there is no marked transition between foreground and background. Standard morphological methods also fail because for weak signals there is no consistent shape information for the target area. Given the knowledge that each sub-array in a microarray slide is produced by a single print-tip, resulting in a consistent spot morphology, the target detection algorithm has two steps: (1) target shape (mask) extraction; and (1) target detection within each optimal matched target mask region.

Target Mask Detection. For a given sub-array and local background statistics (μ_b, σ_b), we can easily identify strong targets, for which $I_{avg} > \mu_b + 4\sigma_b$, where I_{avg} is the mean intensity over the entire bounding box. Perform auto-correlation to the first detected strong target, S, and then add a detected target to the first detected target. After summing all strong targets to S, the target mask is obtained by thresholding S at a value, t, determined by Otsu's histogram method (see Eq. 1.1 below) (Otsu, 1979), where t is limited by $(\mu + 4\sigma) \le t \le 2\mu$. If the pixel count in the bounding box is less than 200, the threshold value is $\mu + 4\sigma$, where μ and σ are the average mean background and average standard deviation over all detected strong targets, respectively. We keep largest component in the mask to be the final mask for each print-tip, M.

Target Detection – Mann-Whitney Method   (Chen   *et   al.*, 1997).   For this method, a target site is segmented from the target patch by the following procedure. A predefined target mask M is used to identify a portion of the target patch that contains the target site.   Eight sample pixels are randomly selected from the known background (outside the target mask) as $Y_1, Y_2, ..., Y_8$. The eight lowest samples from within the target mask are selected as $X_1, X_2, ..., X_8$. The rank-sum statistic W is calculated and, for a given significance level α, compared to $w_{\alpha,8,8}$. Eight samples are chosen because the Mann-Whitney statistic is approximately normal when $m = n \geq 8$ (Gibbons and Chakraborti, 1996). If the null hypothesis is not rejected, then we discard some predetermined number (perhaps only 1) of the 8 samples from the potential target region and select the lowest 8 remaining samples from the region. The Mann-Whitney test is repeated until the null hypothesis is rejected. When H_0 is rejected, the target site is taken to be the 8 pixels causing the rejection together with all pixels in the target mask whose values are greater than or equal to the minimum value of the eight. The resulting site is said to be a target site of significance level α. If the null hypothesis is never rejected, then it is concluded that there is no appreciable probe at the target site. One can require that the Mann-Whitney target site contain some minimum number of pixels for the target site to be considered valid and measured for fluor intensity. Figures 1.8 and 1.9 show the detection results of target sites at $\alpha = 0.0001$ and $\alpha = 0.05$, respectively, where detected site boundaries are superimposed with original images.

Target Detection – Fixed Threshold Method. First, match the mask M to the target by auto-correlation (this step also performed in Mann-Whitney method). The threshold value is the maximum

$$\max_{all \ t}\{q_1(t)q(t)[\mu_1(t) - \mu_2(t)]^2\} \tag{1.1}$$

where $q_1(t)$ and $q_2(t)$ are the number of pixels whose gray levels are below or above the threshold value t, respectively, and $\mu_1(t)$ and $\mu_2(t)$ are the mean gray levels from pixels below or above t, respectively. We limit the threshold by $(\mu_b + 2.3\sigma_b) \leq t \leq (\mu_b + 4\sigma_b)$ in case the histogram method fails to find an appropriate threshold value. Finally, we union the detected target pixels from both fluorescent channels as T_i. Delete the target if the number of pixels within T_i is too few (default to 10 pixels).

2.4 Intensity Measurement and Ratio Calculation

Intensity measurements are carried out after target regions are determined. For a two-color system, the target regions detected from the red and green channels are combined. Both probes have been hybridized to the same target, so that if we observe either one of them, the underlying

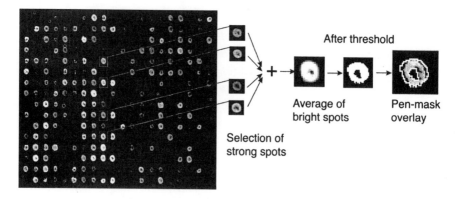

Figure 1.7.   Pen-mask extraction illustration. (see color plates)

region must belong to original target unless the scanner has an obvious mechanical channel alignment problem. Upon obtaining the correct target site, the measurements of fluorescent intensities from both channels for a target region are obtained in the following manner:

1 Trim 5% of the pixels from the top and 5% from the bottom of the intensity-sorted lists for both red channel and green channels.

2 Select pixels not trimmed from either channel. This step may effectively trim up to 10% of the pixels from top and up to 10% from the bottom.

3 Calculate mean intensities, standard deviations and other measurement statistics from both fluorescent channels and corresponding expression ratios based on the trimmed pixel collection.

4 Perform background intensity subtraction, and if the background subtracted intensity is less than 1, set it to 1.

5 The ratio is calculated by $T = R/G$, where R and G are the background subtracted intensities from the red and green channels, respectively.

The trimmed mean is used for estimating the average intensity to lessen the interference of spike noises or possible mis-registration between the two channels.. Trimming suppresses most noise spikes while maintaining the robustness of the averaging process and compatibility with the background level estimation. We take the background subtracted mean intensity, which has a normal distribution when large numbers are presented (central limit theorem), to estimate the signal intensity.

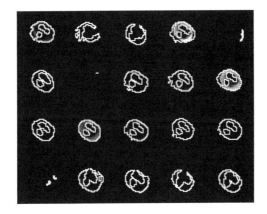

Figure 1.8. Detection result at $\alpha = 0.0001$. (see color plates)

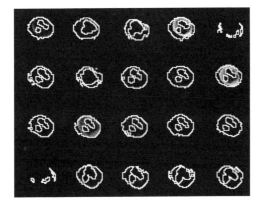

Figure 1.9. Detection result at $\alpha = 0.05$. (see color plates)

3.   Ratio Statistics

In many practical microarray analyses, the ratio statistic has been chosen to quantitate the relative expression levels differentially expressed in two biological samples. It is well-known that working with ratio distributions can be problematic (Feldman and Fox, 1991; Sokal and Rohlf, 1995) and recent research on the matter is generally confined to normality study of the ratio distribution and numerical calculations (Shanmugalingam, 1982; Schneeberger and Fleischer, 1993; Korhonen and Narula, 1989). However, as we now discuss, a special situation arises for gene expression that permits a more detailed statistical analysis, as well as hypothesis tests and confidence intervals based on a single microarray. The unique conditions in gene expression analysis are (1) the level of a transcriptor depends roughly on the concentration of the related factors which, in turn, govern the rate of production and degeneration of the transcript; (2) the random fluctuation for any particular transcript is normally distributed; and (3) as a fraction of abundance, the variation of any transcript is constant relative to most of the other transcripts in the genome, which means that the coefficient of variation, cv, can be taken as constant across the genome. These observations provide the statistical foundation for many of our routine analyses.

3.1   Constant Coefficient of Variation

Consider a microarray having n genes, with red and green expression values $R_1, R_2, \ldots, R_n$ and $G_1, G_2, \ldots, G_n$, respectively. Regarding the expression ratios, $T_i = R_i/G_i$, even if red and green measurements are identically distributed, the mean of the ratio distribution will not be 1. Moreover, a hypothesis test needs to be performed on expression levels from a single microarray. Letting μ_{R_k} and σ_{R_k} denote the mean and standard deviation of R_k (and similarly for G_k), the equal-cv assumption means that

$$\sigma_{R_k} = c\mu_{R_k}$$
$$\sigma_{G_k} = c\mu_{G_k}$$

(1.2)

where c denotes the common coefficient of variation (cv). Previously, we have proposed a ratio-based hypothesis test for determining whether R_k is over- or under-expressed relative to G_k, assuming constant cv for all genes in the microarray (Chen *et al.*, 1997). This assumption facilitates the pooling of statistics on gene expression ratios across the microarray. The desired hypothesis test using the ratio test statistic $T_k = R_k/G_k$ is

$$H_0 : \quad \mu_{R_k} = \mu_{G_k}$$
$$H_1 : \quad \mu_{R_k} \neq \mu_{G_k} \tag{1.3}$$

Under the null hypothesis H_0, Eq. 1.3 implies that $\sigma_{R_k} = \sigma_{G_k}$. Assuming R_k and G_k to be normally and identically distributed, T_k has the density function

$$f_{T_k}(t; c) = \frac{(1+t)\sqrt{1+t^2}}{c(1+t^2)^2\sqrt{2\pi}} \exp\left[\frac{-(t-1)^2}{2c^2(1+t)}\right]. \tag{1.4}$$

Since the subscript k does not appear on the right-hand side, the density applies to all genes, which enables us to pool expression ratios across the entire microarray to perform statistical analyses, such as normalization and construction of confidence intervals. Figure 1.10 depicts the probability density given in Eq. 1.4 for $c = 0.05$, 0.1, and 0.2. The density function of Eq. 1.4 is an asymmetric function and its peak is close to 1 under the null hypothesis. In most practical microarray applications, the evidence of equally distributed genes, regardless of their expression levels, around the 45° diagonal line in a log-log scatter plot shows that the constant-coefficient-of-variation assumption is not overly violated. And it has been demonstrated that some variation of cv is tolerable (Chen *et al.*, 1997). However, even if the assumption intrinsically holds for expression levels, a detected expression level may not satisfy the assumption, particularly when the gene expresses weakly. Expression-level variation increases when the levels approach the background fluorescence, even though the aforementioned image processing techniques are capable of detecting the cDNA target reliably. This problem is addressed in Section 3.4.

3.2   Ratio Confidence Interval

If a set of internal control genes whose expression levels are assumed to be stable between the reference and test samples (satisfying the null hypothesis in Eq. 1.3), or housekeeping genes, can be identified, then according to Eq. 1.4, then we can pool to estimate the parameter of Eq. 1.4 by using maximum likelihood. We select a parameter c to maximize the likelihood function

$$L(c) = \prod_{i=1}^{n} \frac{(1+t_i)\sqrt{1+t_i^2}}{c(1+t_i^2)^2\sqrt{2\pi}} e^{-\frac{(t_i-1)^2}{2c^2(1+t_i^2)}} \tag{1.5}$$

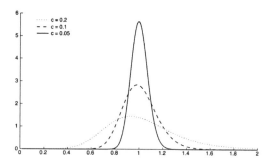

Figure 1.10. Ratio density function.

where $t_1, t_2, ..., t_n$ are ratio samples taken from a single collection of expression values, for example, all housekeeping-genes ratios in a microarray. The maximum-likelihood criterion requires that $d[\log L(c)]/dc = 0$. Hence, the estimator for c is

$$\hat{c} = \sqrt{\frac{1}{n} \sum_{i=1}^{n} \frac{(t_i - 1)^2}{(t_i^2 + 1)}} \qquad (1.6)$$

The 99% confidence interval for a given c can be evaluated by Eq. 1.4 and used to identify over- or under-expressed genes, relatively to their internal control genes' variation. A set of confidence intervals for different cv values is shown in Fig. 1.11.

For two identical mRNA samples co-hybridized on a slide (self-self experiment), c (cv of the fluorescent intensity) provides the *variation of assay*. However, it is not possible to guarantee the null hypothesis condition implied in Eq. 1.5 for an arbitrary experiment. One alternative is to duplicate some or all clones where the same expression ratio is expected. The ratio of expression ratios, $T = t/t'$, satisfies the null hypothesis required by Eq. 1.3. It can be shown (Chen *et al.*, 2001) that $c = \sigma_{\log T}/2$, where $\log T$ is the natural log-transform of T, and with the assumption that the measurement of the log-transformed expression level is approximately normally distributed. For any given experiment with some duplicated clones, $\sigma_{\log T}^2$ is easily calculated.

For a given microarray experiment, one can either use the parameter derived from pre-selected housekeeping genes (Eq. 1.6), or a set of duplicated genes, $c = \sigma_{\log T}/2$, if they are available in the array. The former confidence intervals contain some levels of variation from the fluctuation of the biological system that also affects the housekeeping genes,

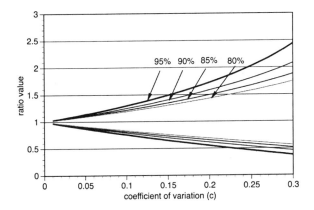

Figure 1.11. Confidence intervals for different *cv*.

while the latter contains no variation of the biological fluctuation, but contains possible spot-to-spot variation. This spot-to-spot variation is not avoidable if one wishes to repeat the experiment. Thus, the confidence interval derived from the duplicated genes is termed the *confidence interval of the assay*.

3.3   Ratio Normalization

In practical application, a constant amplification gain m may apply to one signal channel, in which case the null hypothesis may become $\mu_{R_k} = m\mu_{G_k}$. Under this uncalibrated signal detection setting, the ratio density can be modified by

$$f_T(t; c, m) = \frac{1}{m} f_T(t/m; c, 1) \tag{1.7}$$

where $f_T(\cdot; c, 1)$ is given by Eq. 1.5. In Ref. (Chen *et al.*, 1997), an estimation procedure for the parameter m is proposed that has proven to be very efficient.

Other studies utilize the approximate normality of the distribution of the log-transformed ratio parameter. Therefore, the normalization procedure is to move the mean of the log-normal distribution to 0 (or ratio of 1) (Schena *et al.*, 1995; Schena *et al.*, 1996; DeRisi *et al.*, 1996; DeRisi *et al.*, 1997). More recent proposals include non-linear regression (Bittner *et al.*, 2001) and rank-order-based methods (Schadt *et al.*, 2000).

3.4   Ratio Statistics for Low Signal-to-Noise Ratio

The condition of constant cv is made under the assumption that R_k and G_k are the expression levels detected from two fluorescent channels. The assumption is based on image processing that suppresses the background noise relative to the true signal. It is quite accurate for strong signals, but problems arise for weak signals. Even with image processing, the actual expression intensity measurement is of the form

$$R_k = (SR_k + BR_k) - \mu_{BR_k} \tag{1.8}$$

where SR_k is the expression intensity measurement of gene k, BR_k the fluorescent background level, and $\mu_{BR_{*k}}$ the mean background level. The null hypothesis of interest is $\mu_{BR_k} = \mu_{SG_k}$. The constant cv assumption applies to SR_k and SG_k, not R_k and G_k, and the density of Eq. 1.5 is not applicable. The situation is treated in detail in (Chen *et al.*, 2001). Here we provide a brief summary.

The matter is quantified by defining an appropriate signal-to-noise ratio. Assuming SR_k and BR_k, are independent,

$$c_{R_k}^2 = c^2 + SNR_{R_k}^{-2} \tag{1.9}$$

where the signal-to-noise ratio is defined as $SNR = \mu_{SR}/\sigma_{BR}$. If $SNR \gg 1$ for gene k, meaning the expression signal is strong, then the measured cv is close to a constant, namely $c_{R_k} \approx c$, but if the signal is weak, the constant cv condition is violated. Figure 1.12 shows how in most practical applications the weaker expression signals (at the lower left corner of the scatter plot) produce a larger spread of gene placement. Given the complication introduced by Eq. 1.9 into Eq. 1.5, we can only evaluate the ratio distribution by a Monte Carlo method. By using the numerical method, the 99% confidence interval at different SNRs is shown in Fig. 1.13.

3.5   Measurement Quality Assessment

Before submitting expression ratios from an experiment for further analysis, it is useful to forward a quality measurement with each ratio. For instance, if a cDNA target has very low SNR, then we may discard it. A quality metric must be designed at an early processing stage, because information is lost through image processing and when ratios are extracted and forwarded to higher level processing. We briefly describe the qualitative considerations behind a proposed quality metric

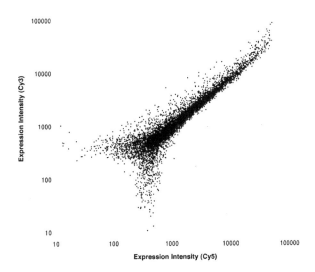

Figure 1.12. Background interference: Scatter plot showing expression intensity of Cy5 vs. Cy3.

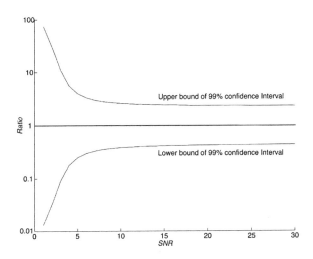

Figure 1.13. Background interference: effect on the confidence interval.

for cDNA expression ratio measurement that combines local and global statistics to describe the quality at each ratio measurement (Chen *et al.*, 2001). The single quality metric enables unified data filtering, and it can be used directly in higher-level data analysis.

For a given cDNA target, the following factors affect ratio measurement quality: (1) Weak fluorescent intensities from both channels result in less stable ratios. This quality problem has been addressed in the last section via the confidence interval if over- or under-expression is the only concern; however, when comparing one experiment to another via a common reference sample, low intensities provide less reliable ratios. (2) A smaller than normal detected target area indicates possible poor quality in clone preparation, printing, or hybridization. (3) A very high local background level may suggest a problematic region in which any intensity measurement may be inaccurate. (4) A high standard deviation of target intensity is usually caused by the contamination of strong fluorescent crud sitting within the target region. Our image processing package extracts all of this information, and for each factor, a quality metric is defined taking a value from 0 (lowest measurement quality) to 1 (highest measurement quality). The final reported quality is the minimum of all individual measurement quality assessments.

4. Conclusions

Various image analysis issues have been addressed: target segmentation, background detection, target detection, and intensity measurement. Since microarray technology is still under development and image quality varies considerably, a robust and precise image analysis algorithm that reduces background interference and extracts precise signal intensity and expression ratios for each gene is critical to the success of further statistical analysis. The overall methodology discussed in this chapter has been developed and enhanced through five years of experience working with cDNA microarray images. It continues to be expanded and revised as new issues arise.

References

Schena, M., Shalon, D., Davis, R. W., and Brown, P. O. Quantitative monitoring of gene expression patterns with a complementary DNA microarray. *Science* **270**, 467-470 (1995).

Schena, M., Shalon, D., Heller, R., Chai, A., Brown, P. O., and Davis, R. W. Parallel human genome analysis: microarray-based expression monitoring of 1000 genes. *Proc Natl Acad Sci U S A* **93**, 10614-10619 (1996).

DeRisi, J., Penland, L., Brown, P. O., Bittner, M. L., Meltzer, P. S., Ray, M., Chen, Y., Su, Y. A., and Trent, J. M. Use of a cDNA microarray to analyze gene expression patterns in human cancer. *Nat Genet* **14**, 457-460 (1996).

DeRisi, J. L., Iyer, V. R., and Brown, P. O. Exploring the metabolic and genetic control of gene expression on a genomic scale. *Science* **278**, 680-686 (1997).

Duggan, D. J., Bittner, M. L., Chen, Y., Meltzer, P. S., and Trent J. M. Expression profiling using cDNA microarrays. *Nat Genet* **21**, 10-14 (1999).

Eisen, M. B., Spellman, P. T., Brown, P. O., and Botstein, D. Cluster analysis and display of genome-wide expression patterns. *Proc. Nat. Acad. Sci. USA* **95**, 14863-14868 (1998).

Khan, J., Simon, R., Bittner, M., Chen, Y., Leighton, S. B., Pohida, T., Smith, P. D., Jiang, Y., Gooden, G. C., Trent, J. M. and Meltzer, P. S. Gene expression profiling of alveolar rhabdomyosarcoma with cDNA microarrays. *Cancer Res* **58**, 5009-5013 (1998).

Golub, T. R., Slonim, D. K., Tamayo, P., Huard, C., Gaasenbeek, M., Mesirov, J. P., Coller, H., Loh, M. L., Downing, J. R., Caligiuri, M. A., Bloomfield, C. D. and Lander, E. S. Molecular classification of cancer: class discovery and class prediction by gene expression monitoring. *Science* **286**, 531-537 (1999).

Ben-Dor, A., Bruhn, L., Friedman, N., Nachman, I., Schummer, M. and Yakhini, Z. Tissue classification with gene expression profiles. *J Comput Biol* **7**, 559-583 (2000).

Bittner, M., Meltzer, P., Khan, J., Chen, Y., Jiang, Y., Seftor, E., Hendrix, M., Radmacher, M., Simon, R., Yakhini, Z., Ben-Dor, A., Dougherty, E., Wang, E., Marincola, F., Gooden, C., Lueders, J., Glatfelter, A., Pollock, P., Gillanders, E., Leja, A., Dietrich, K., Beaudry, C., Berrens, M., Alberts, D., Sondak, V., Hayward, N., and Trent, J. Molecular classification of cutaneous malignant melanoma by gene expression profiling. *Nature* **406**, 536-540 (2000).

Hedenfalk, I., Duggan, D., Chen, Y., Radmacher, M., Bittner, M., Simon. R., Meltzer, P.,Gusterson, B., Esteller, M., Raffeld, Yakhini, Z., Ben-Dor, A., Dougherty, E., Kononen,J., Bubendorf, L., Fehrle, W., Pittaluga, S., Gruvverger, S., Loman, N., Johannsson, O., Olsson, H., Wifond, B., Sauter, G., Kallioniemi, O. P., Borg, A., and Trent, J. Gene expression profiles distinguish hereditary breast cancers. *N Engl J Med* **34**, 539-548 (2001).

Khan, J., Wei, J. S., Ringner, M., Saal, L. H., Ladanyi, M., Westermann, F., Berthold, F., Schwab, M., Antonescu, C. R., Peterson, C. and Meltzer, P. S. Classification and diagnostic prediction of cancers using gene expression profiling and artificial neural networks. *Nat Med* **7**, 673-679 (2001).

Chen, Y., Dougherty, E. R., and Bittner, M. Ratio-based decisions and the quantitative analysis of cDNA microarray images. *Biomedical Optics* **2**, 364-374 (1997).

Schadt, E. E., Li, C., Su, C., and Wong, W. H., Analyzing high-density ologonucleotide gene expression array data, *J. Cell Biochem.*, **80**(2), 192-202 (2000).

Kim, J.H., H.Y. Kim, and Y.S. Lee, A novel method using edge detection for signal extraction from cDNA microarray image analysis. Exp Mol Med, **33**(2): 83-88 (2001).

Bittner, M., Chen, Y. and Dougherty, E. R., (Eds.) *Microarrays: Optical Technologies and Informatics, Proceedings of SPIE*, Vol. 4266, 1-12 (2001).

Newton, M.A., Kendziorshi, C.M., Richmond, C.S., Blattner, F.R., Tsui, K.W., On differential variability of expression ratios: improving statistical inference about gene expression changes from microarray data. J Comput Biol 8(1): 37-52 (2001).

Wang, X., S. Ghosh, and S.W. Guo, Quantitative quality control in microarray image processing and data acquisition. Nucleic Acids Res 29(15): p. E75 (2001).

Bowtell, D., Options available – from start to finish – for obtaining expression data by microarray. *Nat Genet* **21**, 25-32 (1999).

Wang, X.F. and Herman, B., eds., *Fluorescence imaging spectroscopy and microscopy*, John Wiley & Sons, Inc., NY (1996).

Otsu, N. A threshold selection method from gray-level histogram, *IEEE Trans. Syst. Man. Cyber.* SMC-9(1): 62-66 (1979).

Gibbons, J. D. and Chakraborti, S., *Nonparametric Statistical Inference*, 3rd ed. Marcel Dekker, New York (1996).

Feldman, D. and Fox, M. *Probability, the mathematics of uncertainty*, Marcel Dekker, New York (1991).

Sokal, R. R. and Rohlf, F. J. *Biometry: the principles and practice of statistics in biological Research*, 3rd Ed. W. H. Freeman and Co. (1995).

Shanmugalingam, S. On the analysis of the ratio of two correlated normal variables, *The Statistician*, 31(3): 251-258, (1982).

Schneeberger, H. and Fleischer, K. The distribution of the ratio of two variables, *J. Statistical Computation and Simulation*, 47: 227-240, (1993).

Korhonen, P.J. and Narula, S. C. The probability distribution of the ratio of the absolute values of two normal variables. *J. Statistical Computation and Simulation*, 33:173-182, (1989).

Chen, Y., Kamat, V., Dougherty, E.R., Bittner, M.L., Meltzer, P.S., Trent, J.M., Ratio Statistics of Gene Expression Levels and Applications to Microarray Data Analysis. (submitted to *Bioinformatics).*

Chapter 2

STATISTICAL CONSIDERATIONS IN THE ASSESSMENT OF CDNA MICROARRAY DATA OBTAINED USING AMPLIFICATION

Jing Wang[1], Kevin R. Coombes[1], Keith Baggerly[1], Limei Hu[2], Stanley R. Hamilton[2], and Wei Zhang[2]

[1] *Department of Biostatistics*

[2] *Department of Pathology*

The University of Texas M. D. Anderson Cancer Center, Houston, Texas, USA

1. Introduction

Data acquisition is the critical first step in microarray studies to profile gene expression. Successful data acquisition depends on obtaining consistent signals above the level of the background noise on the microarray. Many factors influence the quality and quantity of the fluorescent signals obtained from microarray experiments, including the strength of the fluorescent dyes, the sensitivity of the scanners that detect the signals, and the amount of labeled cDNA targets applied to the microarray. In order to increase signal intensity, substantial efforts have been made to develop stronger dyes and to improve the sensitivity of the scanners. Both laser and CCD based scanners have been developed, but most commercially available scanners have similar performances (Ramdas *et al.*, 2001). Technological improvements in dyes and scanners, by themselves, have provided incomplete solutions to the problem of generating strong, reproducible signals from microarray experiments using small amounts of starting materials.

The single most important factor in microarray data acquisition is the amount of RNA required. In order to obtain consistent hybridizations and adequate signals, conventional protocols require more than 50 μg of total RNA (or 1–2 μg of mRNA) as starting material, from which labeled cDNA targets are generated by incorporating Cy3-dNTP or Cy5-

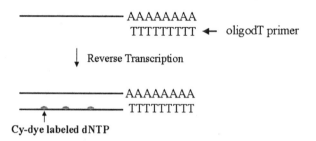

Figure 2.1. Conventional protocol for generating fluorescence labeled cDNA target for microarrays. cDNA is generated by reverse transcription using oligo dT primer. The Cy-dye labelled dNTP is incorporated into the cDNA during reverse transcription.

dNTP during reverse transcription (Figure 2.1). To harvest this amount of RNA, several million cultured cells or several hundred milligrams of tissue are needed. Unfortunately, this requirement cannot realistically be met in many clinical experiments. For clinical specimens obtained by core biopsy or fine needle aspiration, only a small amount of total RNA (around 2 µg) can be obtained. To study gene expression regulation during early embryogenesis, one may want to push the limit to a single cell, which would only yield 20–40 pg of total RNA (Roozemond *et al.*, 1976; Uemura *et al.*, 1980). Developing amplification methods that can reduce the amount of RNA required for successful microarray experiments has become an intensive area of research. However, it is necessary to determine if an amplification protocol produces consistent, reliable results.

We begin this chapter by briefly reviewing some amplification methods that have been proposed. Taking the results of experiments performed with a conventional protocol as "ground truth", we then describe a data analysis method for evaluating how well an amplification protocol works. Finally, we illustrate this analytical method with a set of data obtained in our laboratory.

2.   Amplification methods

Several methods have been developed to acquire strong signals from minute amounts of RNA. The current amplification methods for microarray experiments fall into two main categories: (1) amplification of the starting RNA material before hybridization; and (2) amplification of the fluorescence signals after hybridization for better detection, mainly involving the dendrimer (3DNA) technique (Stears *et al.*, 2000) or an enzyme catalysis approach, such as the TSA^{TM} assay (Alder *et al.*, 2000).

2.1   RNA amplification

A straightforward method for reducing the amount of required starting material is to amplify the RNA, much as DNA is amplified using the polymerase chain reaction (PCR). RNA amplification is based on cDNA synthesis and a template-directed *in vitro* transcription reaction. An RNA polymerase promoter is incorporated into each cDNA molecule by priming cDNA synthesis with a synthetic oligonucleotide containing the phage T7 RNA polymerase promoter sequence. The second strand of cDNA, which will serve as the template for the RNA polymerase, can be generated either by second strand cDNA synthesis or by a combination of a switch mechanism at the 5' end of RNA templates and PCR (Van Gelder *et al.*, 1990; Phillips *et al.*, 1996; Wang *et al.*, 2000). After synthesis of double-stranded cDNA, T7 RNA polymerase is added and anti-sense RNA (aRNA) is transcribed from the cDNA template. The processive synthesis of multiple aRNA molecules from a single cDNA template results in amplification of aRNA. The aRNA is then labeled with Cy-dye-dNTP by reverse transcription using random hexamer primer (Figure 2.2).

In theory, the lowest limit for the total amount of RNA used for representative amplification is at the level of 20–40 pg that is the average amount of RNA from a single cell (Roozemond *et al.*, 1976; Uemura *et al.*, 1980). Gene expression profiling results become unrepresentative when the amount of RNA is below that limit, since some low copy number genes would be lost at the very beginning.

2.2   Fluorescent signal amplification

RNA amplification can increase the amount of labeled cDNA applied to the microarray. However, if the signal can be amplified at the detection step so that one molecule can yield signals equivalent to 10–100 unamplified molecules, then fewer hybridized molecules will be needed. Similar strategies have been used to detect proteins on western blots by chemiluminescence assay. The MICROMAX-TSA® Labeling and Detection System (NEN Life Science Products, Inc.) is based on such a principle. The detailed protocol can be obtained from the manufacturer. Basically, tyramide signal amplification (TSA) achieves signal amplification by catalyzing reporter deposition of labeled tyramide at the site of probe binding (Bobrow *et al.*, 1989). In this system,

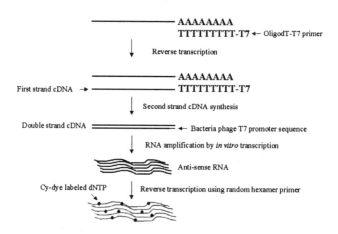

Figure 2.2. Illustration of the RNA amplification approach. Oligo dT primer is modified by flanking bacteria phage T7 sequence at its 5' end, and the sequence is incorporated into the cDNA during reverse transcription. After second strand cDNA synthesis, anti-sense RNA is amplified by *in vitro* transcription using RNA polymerase. Cy-dye labeled dNTP is incorporated into cDNA by reverse transcription using random hexamer primer.

labeled cDNA is prepared from experimental and control samples by incorporating fluorescein-dNTP and biotin-dNTP, respectively, during reverse transcription. The two populations of cDNA are then cohybridized to a microarray. After hybridization, stringent washes are carried out to eliminate excess and non-specifically bound cDNA. Either anti-fluorescein antibody or streptavidin is conjugated with horseradish peroxidase (HRP), an enzyme that catalyses the deposition of Cy-dye labeled tyramide amplification reagent. The reaction results in the deposition of numerous Cy-dye labels immediately adjacent to the immobilized HRP. Using this method, a minimum starting material of 0.5 μg of total RNA yields adequate signal intensities, according to the manufacturer (http://lifesciences.perkinelmer.com).

The dendrimer (3DNA) detection technique, commercialized by Genisphere, is another signal amplification method (Figure 2.3). This technique does not rely on modified nucleotide incorporation in the fluorescent labeling reaction. The dendrimers are complexes of partially double-stranded oligonucleotides, forming stable, spherical structures with a determined number of free ends where fluorescent dyes can be coupled. One dendrimer molecule carries more than 100 fluorescent molecules. A modified oligo dT primer containing dendrimer binding sequence at

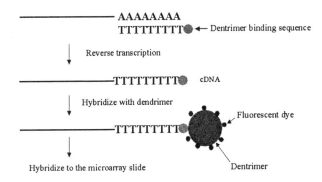

Figure 2.3. Illustration of the dendrimer technique. Oligo dT primer modified with the dendrimer binding sequence is used to generate cDNA by reverse transcription. Then the cDNA is hybridized with the dendrimer through binding the complementary sequence on the dendrimer and the primer. Finally, the complexes are hybridized with the microarray slide. There can be more than 200 Cy dye molecules on each dendrimer molecule and therefore, the fluorescence signal is amplified more than 200 times.

its 5' end is used during reverse transcription to generate cDNA targets. Then the cDNA targets are hybridized with the dendrimer and the DNA spots on the microarray (Stears *et al.*, 2000).

From the published results, it is clear that these amplification methods can lead to amplified signals, thus reducing the requirement for large amounts of biological starting material. However, as with the well-known concern about the faithfulness of PCR amplification, there is concern about the reliability of the data obtained from amplification methods.

3. Data analysis strategy

The goal of an amplification protocol is to enhance the microarray signals obtained from a small amount of RNA. Initial evaluation of the protocol can be achieved by visual inspection of the scanned microarray images. For example, if we use a regular protocol starting with 1 μg of total RNA, we would have an image containing only a few spots representing the most highly expressed genes. When we use the amplification protocol based on the T7 RNA polymerase (RNA amplification), we normally obtain an image with as many bright spots as we would see from the regular protocol with 100 μg total RNA. Such an image clearly indicates successful amplification of the RNA. In a similar fashion, a good image is always the starting point for evaluating whether an indirect labeling method worked.

The danger is that a bright image, by itself, is not an adequate indicator of the faithfulness of the amplification. A similar situation occurs when evaluating gene expression with PCR, which amplifies the signal but often loses quantitative power due to the rapid speed (exponential amplification) at which saturation is reached in the PCR reaction. Even in the absence of formal data analysis, it is not difficult to envision that nonlinear amplification of different RNA molecules can occur for a variety of reasons. First, different double strand cDNA molecules have different sizes, and thus it takes different amounts of time to complete one round of RNA synthesis during the T7 RNA polymerase reaction. This problem is alleviated to some extent by the low processivity associated with T7 RNA polymerase, although variation among cDNA of different lengths is still a factor. Consequently, the amplification procedure can change the relative amounts of transcripts in the cells, favoring short genes and genes whose secondary structure allows an early drop-off of the T7 polymerase. Second, even when amplification is performed during the detection step, nonlinear signal enhancement can still occur. For example, when too many fluorescent dendrimers are deposited at one spot because of a high expression level, fluorescent quenching can occur. This phenomenon may adversely affect the dynamic range of the signal detection.

Consequently, some criteria must be set to assess the outcome of an amplification procedure based on the goal of the intended project. In many of our projects, we want to identify genes that are differentially expressed between tumors and normal tissues or between tumors before and after exposure to drug treatment. Focusing on the ability to detect differentially expressed genes, we have developed criteria in three areas.

1.   Enhancement of signal intensity for low copy number genes.

A good amplification method should increase the signal intensity of low expressing genes; i.e., the number of genes exhibiting measurable signal intensity above background levels using the amplification protocol should be greater than that using the regular protocol. To evaluate this property, we use the following criteria:

a) If a spot can consistently be seen above background using the regular protocol, then it can also be seen using the amplification protocol (measure of consistency).

b) Some spots can be seen above background using amplification that are not seen using the regular protocol (measure of signal enhancement).

To convert these criteria into tools we can use quantitatively, we need a precise definition of what it means to "see" a spot. We use the signal-to-noise (S/N) ratio for this assessment. The S/N ratio for microarray data is computed, in most software packages, by dividing the background-corrected intensity by the standard deviation (SD) of the background pixels. We typically require that S/N > 2.0 in order to say that a gene has measurable signal intensity on an array. In other words, if the difference between signal intensity and background intensity is greater than 2.0 SD of the local background, then the gene gives adequate signal intensity.

2. Reproducibility and reliability of signal intensity.

The reproducibility of microarray results, whether produced using the same protocol or different protocols, is important for evaluating amplification protocols. Replicate experiments performed using the amplification protocol should be as reproducible as experiments performed using the regular protocol. In other words, given an appropriate metric of the reproducibility of microarray experiments, the value of this metric when comparing two amplification experiments should be comparable to the value when comparing two regular experiments. Ideally, we should also see comparable values when comparing an amplification experiment to a regular experiment.

To measure reproducibility, we use the concordance correlation coefficient, which is analogous to the Pearson correlation coefficient but measures how well a set of points matches the identity line (the 45° line). In contrast to other measures such as the Pearson correlation coefficient or the paired t-test, concordance correlation is specifically designed to measure agreement (Lin *et al.*, 1989, 1992). The concordance correlation coefficient (r_c) is computed as

$$r_c = r_p \frac{2s_1 s_2}{s_1^2 + s_2^2 + (\mu_1 - \mu_2)^2},$$

where r_p is the Pearson correlation coefficient. This formula expresses the concordance correlation coefficient as a product of two factors: a measure of precision (r_p, which measures the deviation from the best-fit line) and a measure of accuracy (which determines how far the best-fit line deviates from the identity line).

3. Preservation of patterns of differential gene expression.

One of the major goals of microarray experiments is to find differentially expressed genes. Ideally, the same set of differentially expressed genes should be identified using microarrays produced from both the regular and amplification protocols. In other words, an ideal amplification protocol would satisfy the following criteria:

a) If a spot is identified as differentially expressed using the regular protocol, then it is also identified as differentially expressed using the amplification protocol.

b) If a spot is identified as differentially expressed using the amplification protocol and if the spot can be seen using the regular protocol, then it is also identified as differentially expressed using the regular protocol.

It is, however, overly optimistic to expect to see identical lists of differentially expressed genes generated from the two protocols. We may not even achieve this goal using the same protocol twice, due to the variability associated with microarray experiments (Lockhart *et al.*, 1996; Hess *et al.*, 2001; Baggerly *et al.*, 2001). Nonetheless, we do expect most differentially expressed genes to appear on both lists.

In order to proceed quantitatively, we must be more precise about what it means for "most" genes to behave as expected. The critical step is to estimate the power π of the statistical test used to identify differentially expressed genes. For example, suppose we follow the common practice of computing a t-statistic for each gene and then declaring a gene to be differentially expressed if its t-statistic exceeds some threshold. Based on the number N of genes on the microarray, we can adjust the significance level α assigned to a single t-test to ensure that the number of false positives is small (preferably less than one). We then use α and the number of experiments to set the threshold. By making some assumptions about the intrinsic variability of the measurements and the size of the effect we are hoping to detect, we can then compute the power of this test.

Suppose, to take a specific example, that 100 genes on the microarray exhibit twofold differential expression, and that both the regular protocol and the amplification protocol allow us to perform a test with power $\pi = 0.7$ for detecting these differences. Then each protocol should correctly identify about 70 differentially expressed genes. However, there is no reason for them to identify the same set of 70 genes. In fact, it is likely that they will identify different sets of genes, and that only about 49 genes (since $0.7 \times 0.7 = 0.49$) will be identified by both protocols. In other words, we would only expect 49 of the 70 genes identified by the regular protocol to also be identified by the amplification protocol. The "rate of agreement" in this example is equal to $49/70 = 0.7$. In general, the rate of agreement is expected to equal the power of the test, and so we merely require this level of agreement to conclude that the amplification protocol is consistent with the conventional protocol.

After first weakening our criteria to require only that the fraction of genes identified by both tests should approximately equal the power of the test, we can ask for something slightly stronger. Genes that are identified by only one protocol should at least tend in the correct direction. More precisely, the "sign of the log ratio" should be the same with both protocols whenever a gene is identified as differentially expressed by one of the protocols.

4. An example

In this section, we describe an example involving nine microarrays that illustrates how we evaluated a particular amplification method. Each microarray contains 2304 genes spotted in duplicate, along with positive and negative controls. Among the nine experiments, five used the regular protocol with varying amounts of total RNA (100 μg to 300 μg) and four used an RNA amplification protocol with 1.0 μg total RNA. Each experiment compared two glioma cell lines, U251 (in the Cy5 channel) and LN229 (in the Cy3 channel).

4.1 Data preprocessing

Normalization and threshold. Standard methods to identify differentially expressed genes typically involve taking the ratio between the signals from the two fluorescent dye channels. One problem is that different channels often have different overall intensities. These differences can occur because of different amounts of labeled target hybridized to the array, different rates of Cy3 and Cy5 degradation in storage (Cy5 is less stable than Cy3), or different gains used when scanning images. Therefore, we cannot directly take ratios between the two background-corrected values without going through a normalization process to remove systematic biases from the data.

Normalization can be achieved in various ways. One method is to normalize to a "housekeeping" gene or a mixture of housekeeping genes. One significant drawback to this approach is that housekeeping genes are not always stable in all samples. We prefer a normalization method based on the assumption that the overall expression level of all genes in a high-density array is more stable than any single gene or any small set of genes. Based on this assumption, our normalization approach is to adjust the values on any array so that the median (or another fixed percentile) ratio between channels becomes equal to 1. Because some arrays contain many low expressors, we usually normalize each channel separately by rescaling the background-corrected values to set the 75th percentile of the spots equal to 1000. Under assumptions similar to those described above, this normalization method brings the median ratio between reliably expressed spots close to 1, thus giving results similar to those obtained by normalizing ratios. An advantage of this approach is that it allows us to extract individual channels from different experiments and compare them directly.

After normalization, we replace all negative and all small positive normalized values by a "detection threshold" value. The setting of the detection threshold is data-driven. In our microarray experiments, we have found that a normalized value of 150 corresponds approximately to a spot whose S/N ratio equals 1. Any spot whose background-corrected intensity is below this threshold cannot be reliably distinguished from background noise. Thus, we typically use 150 for the detection threshold.

Smooth t-statistics.   Microarrays contain genes that exhibit a wide spectrum of expression levels. It is, by now, well known that the detection of low signal intensities is much more variable than the detection of high signal values. Taking ratios involving low expressing genes is highly unstable. Therefore, a measure of fold difference based simply on the ratios provides biologists with incomplete (and potentially inaccurate) information. Instead, we apply a statistical approach to determine the gene expression profile on each array. Briefly, we compute "smooth t-statistics" or "studentized log ratios" by rescaling the estimates of the log ratios of gene expression levels between samples to account for the observed variability. In this approach, we estimate both the mean log intensity and the standard deviation of the spots within a channel (using the fact that all genes are spotted in duplicate on our microarrays). We then fit a smooth curve describing the standard deviation as a function of the mean log intensity. After estimating the variability within a channel, the smooth curves that estimate the within-channel standard deviation are pooled to calculate a single curve estimating the standard deviation (σ) across the entire range of intensities on the microarray. We use the pooled estimates to compute a "smooth t-statistic" for each gene by the formula

$$\text{t-score} = [\log_2{(A)} - \log_2{(B)}] / \sigma,$$

where A and B are the geometric means of the expression levels in the two channels. (We use the term "smooth t-statistics" to distinguish these scores from the usual t-statistics that one would obtain by using the raw estimates of the standard deviation on an independent, gene-by-

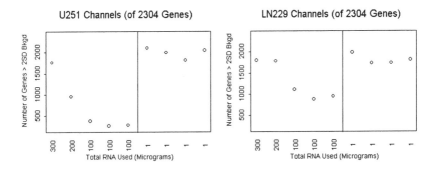

Figure 2.4. Number of genes exhibiting adequate signal (S/N > 2) in experiments comparing the glioma cell lines U251 and LN229. Each point on the graph corresponds to a separate microarray experiment, identified by the amount of total RNA used to prepare the labelled cDNA targets. Arrays produced with 100 to 300 μg total RNA used the regular protocol; arrays produced with 1.0 μg total RNA used the amplification protocol.

gene basis.) A detailed mathematical description of the application of smooth t-statistics to microarray data has been addressed in Baggerly *et al.* (2001), Baldi *et al.* (2001), and in a chapter by Baggerly *et al.* in this book (Chapter 4).

4.2   Data analysis

In this section, we apply the analytical method described in section 3 to the normalized log intensities and smooth t-statistics computed from our microarray experiments.

Enhancement of signal intensity for low copy number genes.
To evaluate whether the amplification protocol preserves signals that are seen using the regular protocol, we evaluated each channel (U251 labeled with Cy5 and LN229 labeled with Cy3) separately. In each channel, we located all the spots that consistently exhibited adequate signal strength (S/N > 2.0) in all experiments conducted using the regular protocol (Figure 2.4). We then computed the percentage of those genes that also have S/N > 2.0 in all microarray experiments conducted using the amplification protocol. High percentage agreements are expected if the amplification protocol succeeded. We found that 95% to 100% of the genes that were consistently found to exhibit adequate signal using the regular protocol were also found to exhibit adequate signal using the amplification protocol.

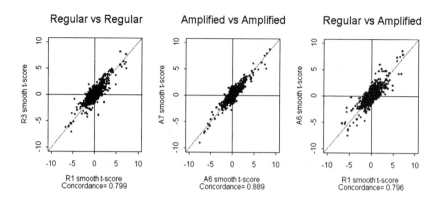

Figure 2.5. Demonstration of concordance correlation of 'smooth t-scores' between (left) two arrays produced using the regular protocol, (middle) two arrays produced using the amplification protocol, and (right) arrays produced using different protocols.

Reproducibility and reliability of signal intensity. We assessed the reproducibility of microarray experiments by computing the concordance correlation coefficient between the smooth t-scores. In our study, the concordance correlation coefficient between experiments using the regular protocol ranged from 0.574 to 0.835 across 5 microarrays, with a median value of 0.739. The concordance correlation coefficient between experiments using the amplification protocol ranged from 0.843 to 0.894 across 4 microarrays, with a median value of 0.873. The concordance correlation coefficient between experiments that used different protocols ranged from 0.562 to 0.796, with a median value of 0.652.

As measured by the concordance correlation coefficient in this study, replicate microarray experiments using the amplification protocol were at least as reproducible as microarray experiments using the regular protocol. In addition, the concordance coefficients when comparing a microarray experiment using the amplification protocol to an experiment using the regular protocol were in the same range as those comparing two experiments that both used the regular protocol. In other words, the results obtained using the amplification protocol were consistent with the results obtained using the regular protocol. Figure 2.5 illustrates the consistency of the smooth t-scores concordance correlation in three typical comparisons of pairs of experiments.

Assessment of differentially expressed genes. To identify differentially expressed genes, we computed a combined set of smooth t-statistics using all 5 microarrays produced using the regular protocol. Similarly, we computed a set of t-statistics using the data from all 4 experiments performed using the amplification protocol. Genes are considered to be differentially expressed if the smooth t-statistic exceeds a threshold value. We chose |t-score| > 4.0 as the cut off for our analysis. Because we measured each gene in duplicate in each channel of each microarray, we can assume that the t-statistics follow a t-distribution with at least 14 degrees of freedom. (There are at least 16 observations of each gene: 4 arrays, 2 channels per array, and 2 observations per channel.) Thus, the cut off value of 4.0 corresponds to a significance level on an individual two-sided t-test of $\alpha = 0.0013$. Since there are 2304 genes on our array, this test should find fewer than 3 false positives from each set of experiments.

In order to compute the power of this test to identify genes that exhibit twofold differential expression, we must estimate the standard deviation. For most genes expressed at moderate to high levels (i.e., with S/N > 2.0), we have observed within-group standard deviations of the base-two log intensities on the order of 0.5 or 0.6. Taking α as above, with at least 8 observations of each gene in each channel, this test has a power between 0.55 and 0.78. As described earlier, we would expect the rate of agreement between the two protocols to equal the power. In our experiments, we identified 26 differentially expressed genes using the amplification protocol and 20 genes using the regular protocol, with 14 genes in common. These observations yield agreement rates of $14/20 = 0.70$ and $14/26 = 0.54$, just as predicted. Moreover, all the genes identified as differentially expressed by one of the protocols "tend in the correct direction", in the sense that the sign of the log ratio is the same (Figure 2.6). In addition, seven of the differentially expressed genes were tested by northern blotting assay and were confirmed (data not shown).

5.   Discussion

Although amplification protocols can produce a multitude of bright spots in the images of microarray experiments, this success does not necessarily mean that the information so acquired is reliable. Systematic biases and nonlinear amplification effects have the potential to distort the profile of gene expression in individual samples. In spite of these risks, there is tremendous pressure to reduce the amount of RNA required to perform microarray experiments. Before deciding to use one of the many amplification protocols that have been proposed, it is important to decide if the protocol can be used to produce consistent, reliable results.

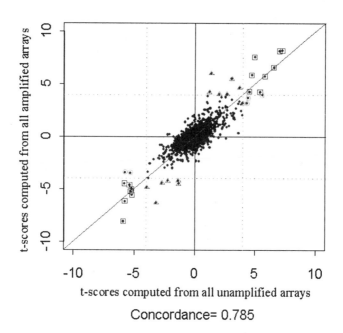

Figure 2.6. Differentially expressed genes identified using two different protocols. Genes identified by both protocols are indicated in squares. Genes identified as differentially expressed on amplified arrays, but not on unamplified (regular) arrays are indicated in triangles. Genes identified as differentially expressed on unamplified arrays but not on amplified arrays are indicated in circles. The vertical and horizontal dots lines correspond to the cutoff values ±4.

In this chapter, we have described a statistical method to evaluate the faithfulness of amplification protocols used to produce microarray data. Our evaluation is based on the idea that microarray experiments conducted using a conventional protocol provide a level of "ground truth" or a "gold standard" to which the amplification results can be compared. From this perspective, we require successful amplification protocols to satisfy three conditions.

First, most spots (more than 95%, say) that are visible on microarrays produced using a conventional protocol should remain visible on arrays produced using the amplification protocol. The utility of this criterion depends on the definition of visibility; we require $S/N > 2.0$. There is no particular reason to prefer the value 2.0; one can set more or less stringent requirements on the signal-to-noise ratio. Alternatively, one can define visibility directly in terms of the background-corrected signal intensity, by finding a level that most effectively separates the intensities of positive and negative controls on the array.

Second, replicated microarrays produced using the amplification protocol should be as reproducible as replicate arrays using the conventional protocol. This reproducibility can be assessed by computing the concordance correlation between normalized or studentized intensities or log ratios. Other measures of agreement can also be used, but we feel that measures of how well the data fits the identity line are more appropriate than statistics that measure more general linear trends.

Third, the set of genes identified as differentially expressed using the amplification protocol should be consistent with the set identified using the conventional protocol. The method described here can be applied whenever it is possible to estimate the power of the statistical test being used to identify differentially expressed genes; the expected level of agreement only depends on the power.

We have illustrated these three criteria by applying them successfully to a set of replicate experiments performed in our laboratory to compare two cell lines using both an amplification protocol and a conventional protocol. The results indicated that the amplification protocol produced reproducible, reliable microarray data that was consistent with the regular protocol.

Fundamental to our approach is the assumption that the conventional protocol provides a gold standard to which the amplification protocol can be compared. An alternative view is that an amplification protocol can be evaluated strictly on its own terms. From this perspective, the critical issue is the reproducibility of the results obtained using the amplification protocol, which only be assessed by performing numerous replicate experiments with the amplification protocol. One advantage to our analysis is that it allows for the possibility of comparing two different amplification methods to determine if one of them more faithfully reproduces results consistent with those obtained using a conventional protocol.

Acknowledgments

This work was partially supported by the Tobacco Settlement Fund appropriated by the Texas Legislature, and by a generous donation from the Kadoorie Foundation to the Cancer Genomics Core Program.

References

Adler, K., Broadbent, J., Garlick, R., Joseph, R., Khimani, A., Mikulskis, A., Rapiejko, P., Killian, J. (2000). MICROMAX^{TM}: A highly sensitive system for differential gene expression on microarray. In Microarray Biochip Technology, Edited by Mark Schena, 221-230.

Baggerly, K.A., Coombes, K.R., Hess, K.R., Stivers, D.N., Abruzzo, L.V., and Zhang,W. (2001). Identifying differentially expressed genes in cDNA microarray experiments. J Comp. Biol., 8, 639-59.

Baldi, P., and Long, A.D. (2001). A Bayesian framework for the analysis of microarray expression data: regularized t-test and statistical inferences of gene changes. Bioinformatics. 7, 509-19.

Bobrow, M.N., Harris, T.D., Shaughnessy, K.J. and Litt, G.J. (1989). Catalyzed reporter deposition, a novel method of signal amplificaiton. Application to immunoassays. J. Immunol. Methods. 125, 279-285.

Hess, K.R., Zhang, W., Baggerly, K.A., Stivers, D.N., Coombes, K.R. (2001). Microarrays: handling the deluge of data and extracting reliable information.Trends Biotechnol. 19, 463-8.

Lin, L. (1989). A Concordance Correlation Coefficient to Evaluate Reproducibility. Biometrics 45, 255-268.

Lin, L. (1992). Assay validation using the concordance correlation coefficient. Biometrics 48, 599-604.

Lockhart, D.J., Dong, H., Byrne, M.C., Follettie, M.T., Gallo, M.V., Chee, M.S., Mittmann, M., Wang, C., Kobayashi, M., Horton, H., Brown, E.L. (1996). Expression monitoring of hybridization to high-density oligonucleotide arrays. Nat. Biotechnol. 14, 1675-80.

Phillips, J. and Eberwine, J. H. (1996). Antisense RNA amplification: a linear amplification method for analyzing the mRNA population from single living cells. Methods 10, 283-8.

Ramdas, L., Wang, J., Hu, L., Cogdell, D., Taylor, E., Zhang, W. (2001). Comparative Evaluation of Laser-based Microarray Scanners. BioTechniques 31, 546-552.

Roozemond, R.C. (1976). Ultramicrochemical determination of nucleic acids in individual cells using the Zeiss UMSP-I microspectrophotometer. Application to isolated rat hepatocytes of different ploidy clases. Histochem J, 625-638.

Stears, R.L., Getts, R.C., and Gullans, S.R. (2000). A novel, sensitive detection system for high-density microarrays using dendrimer technology, Physiol. Genomics, 3, 93-99.

Uemura et al., (1980). Age-related changes in neuronal RNA content in rhesus moneyks (Macaca mulatta). Brain Res Bull 5,117-119.

Van Gelder, R.N., Von Zastrow, M.E., Yool, A., Dement, W.C., Barchas, J.D., Eberwine, J.H. (1990). Amplified RNA synthesized from limited quantities of heterogeneous cDNA. Proc Natl Acad Sci. U S A. 87, 1663-7.

Wang, E., Miller, L.D., Ohnmacht, G.A., Liu, E.T. and Marincola, F.M. (2000). High-fidelity mRNA amplification for gene profiling. Nat. Biotechnol 18, 457-459.

Chapter 3

SOURCES OF VARIATION IN MICROARRAY EXPERIMENTS

Experimental design and the analysis of variance for microarrays

M. Kathleen Kerr[1], Edward H. Leiter[2], Laurent Picard[3], and Gary A. Churchill[2]

[1] *Department of Biostatistics, University of Washington, Seattle, Washington, USA*

[2] *The Jackson Laboratory, Bar Harbor, Maine, USA*

[3] *Corning, Inc., Corning, New York, USA*

1.   Introduction

Multiple factors and interactions of those factors contribute to variation in gene expression microarray data. This fact is indirectly acknowledged by expressing a need for data "normalization." The essence of the problem is to account for different sources of variation in order to "isolate" the effects of the factors of interest. In this paper we use a small experiment to illustrate some principles of experimental design and scientific inference for microarray studies. The focus of this paper is the statistical issues rather than the biological results of the particular data example.

Brown and Botstein (1999) give an introduction to microarray technology. In brief, DNA sequences are spotted and immobilized onto a glass slide or other surface, the array. Each spot contains a single sequence, although the same sequence may be spotted multiple times per array. RNA is removed from cell populations of interest. Purified mRNA is reverse-transcribed into cDNA and a dye-label is incorporated. The standard fluors are Cy3 and Cy5, commonly referred to as "green" and "red." Green- and red-labeled pools of cDNA are mixed together and washed over an array. Single strands of dye-labeled cDNA hybridize to their complementary sequences spotted on the array, and unhybridized cDNA is washed off. The array is scanned, producing a "red" and "green" image for each slide. Analysis of this image (Brown, Goodwin, and Sorger, 2001) produces a "red" and "green" reading for each spot.

The absolute signal intensities for each spot contain little information about gene expression for several reasons. One issue is that different sequences vary in hybridization efficiency. This difficulty is overshadowed, however, by the fact that there is poor control over the amount of DNA contained in each spot. This means it is impossible to tell whether a large fluorescent intensity is the result of high abundance of the transcript in the RNA or simply due to a particularly large or dense spot. However, the *relative* red and green intensity contains information about the relative expression of a gene in the two samples based on the principle that the sample that contained more transcript should produce the higher signal.

2.   The Experiment

Spotted cDNA microarrays were used to study the expression of genes in the liver related to type II diabetes. NZO/HlLt male mice represent a model of maturity-onset type II diabetes. Mice of this strain are large at weaning, and gain weight rapidly thereafter. The rapid development of post-weaning obesity leads to the development of insulin resistance, and eventually type II diabetes in males (Leiter *et al.*, 1998). If the diet fed to weaning mice is supplemented with 0.001% CL316,243, a beta-3 adrenergic receptor agonist, the metabolic defects are in large part normalized such that obesity is blunted and diabetes is averted.

Prototype mouse arrays were printed on CMT-GAPS coated slides using Corning proprietary high-throughput printing technology. Seventy-six genes associated with growth and metabolism were spotted four times on each array and two additional genes were spotted 16 times per array, for a total of seventy-eight unique sequences. In order to screen for genes with dysregulated expression in mice whose diet is unsupplemented with CL316,243, we compared gene expression in livers of treated and control NZO/HlLt male mice. All mice were fed for 4 weeks from weaning on NIH 31 diet (4% fat). The chow of the treated mice was supplemented with 0.001% CL316,243.

3.   Experimental Design

Two arrays were used to compare the treated and control RNA samples. The dye-labeling was switched between the arrays. This design of this experiment is variously referred to as a "dye swap," "flip fluor,"

	Dye 1/"Green"	Dye 2/"Red"
Array 1	Treatment	Control
Array 2	Control	Treatment

Table 3.1. Experimental Design

or Latin Square design (Kerr, Martin, and Churchill, 2000). Table 3.1 shows the experimental layout.

It turns out that the Latin Square design has a structure that is conducive to efficient estimation of various effects in the data. Following Kerr, Martin, and Churchill (2000) and Kerr and Churchill (2001a and 2001b), we identify four fundamental experimental factors: arrays (A), dyes (D), the treated and control RNA varieties (V), and genes (G). With four factors there are $2^4 = 16$ possible factorial effects. However, not all of these effects are identifiable. Table 3.2 gives the confounding structure of the Latin Square design (Cochran and Cox, 1992). Each experimental effect is confounded with one other effect, its *alias*. Non-aliased effects are orthogonal.

Table 3.2. Confounding structure of the Latin Square design. The symbol $\sim$ identifies aliased effects. A =Array, D =Dye, V =Variety of mRNA, G =Gene.

mean	$\sim$	ADV	G	$\sim$	ADVG
A	$\sim$	DV	AG	$\sim$	DVG
D	$\sim$	AV	DG	$\sim$	AVG
V	$\sim$	AD	VG	$\sim$	ADG

The effects of interest are expression levels of genes specifically attributable to varieties, i.e. variety×gene (VG) interactions. In order to measure VG effects it is necessary to assume that ADG effects are 0. A three-way interaction of arrays, dyes, and genes, would mean that every measurement depends on the channel and array where it is taken, and that this dependence is different for every gene. In effect, this says there is no consistency across arrays and that the technology is ineffective. An advantage of the Latin Square design is that VG effects are orthogonal to all other effects. This means that other factorial effects will not bias estimates of VG effects, and accounting for other sources of variation does not reduce the precision of VG estimates.

The orderly structure of the Latin Square design is a consequence of the balance among the factors. In detail:

1 Varieties and dyes are balanced. Each variety is labeled with both dyes and each dye-labeled sample is used equally often in the experiment.

2 Varieties and arrays are balanced. Each variety appears on each array.

3 Arrays and dyes are balanced. Each array has a red channel and a green channel.

4 Genes and arrays are balanced. The same set of genes is spotted on each array.

5 Genes and dyes are balanced. For every gene there are an equal number of red signals and green signals.

6 Genes and varieties are balanced. Associated with every variety are fluorescent intensities for the same set of genes.

This level of balance is not shared by other strategies for microarray experiments, including some commonly used designs (Kerr and Churchill, 2001a). It is reasonable to assume that the same set of genes is spotted on every array in an experiment because of the automated printing process. It follows that genes are balanced with respect to the other factors and properties (4)-(6) hold. Also, by the nature of the technology, each array has a red channel and a green channel, so property (3) holds. Property (1) may or may not hold — this is determined by the design. If each sample is labeled with both the red and green dyes and the design uses each equally often in the experiment, then varieties and dyes will be balanced. Kerr and Churchill (2001a) calls such designs *even* designs.

In general microarray experiments, property (2) will not hold. This is because each array is strictly limited to two varieties since there are two fluorescent dyes. When there are more than two varieties, every variety cannot be assayed on every array, so arrays and varieties will not be balanced. In this case variety effects and array effects are not orthogonal, but partially confounded. Consequently, variety×gene effects (the effects of interest) and array×gene effects are also partially confounded. As discussed in the next section, array×gene effects correspond to variation due to spots and are expected to be an important source of variation in microarray data. Therefore this partial confounding means that the experimental design will be a key determinant in how precisely the effects of interest, VG, are estimated (Kerr and Churchill, 2001a).

4. Data Analysis

We began the analysis of the experiment with simple plots to get an overview of the data. Following the recommendation of Yang *et al.* (2000), we plotted the difference of the two readings for each spot against the mean with the data on the log scale, i.e. the log-ratio against the mean log intensity. Ideally, this plot should appear as a horizontal band of points representing genes that are not differentially expressed, with some points above or below the band representing genes that are differentially expressed. Instead, we see curvature in these plots (Figure 3.1(a) and (b)). Consequently, we applied the "shift-log" data transformation (Kerr *et al.*, 2001), which assumes there is a differences in the dyes that is additive on the raw data scale. The transformation requires estimating a single parameter, the "shift," for each array. Fix an array i and let R be the vector of red signal intensities and G be the vector of green signal intensities. We estimate a constant shift s_i for array $i = 1, 2$ to minimize the absolute deviation from the median of $\log(G + s_i) - \log(R - s_i)$. The estimated shifts are $s_1 = 1728.8$ and $s_2 = 1801.6$. In other words, for array 1 the transformed data is $\log(G + 1728.8), \log(R - 1728.8)$ and similarly for array 2. Figure 3.1(c) and (d) show the ratio versus intensity plots for the shift-log data. Comparing (a) with (c) and (b) with (d), we see the shift-log transform straightens these plots.

Like symbols were used in Figure 3.1 to plot the four (or sixteen) points corresponding to the multiple spots of each gene. Consider the four squares above the mass of the data in the plots for array 1. Notice that these squares are below the mass of points in array 2. It is logical that the points are in opposite positions in the two plots because the dyes were switched. In contrast, consider the diamonds in the upper-left corner of the plots for array 1. Under close inspection we see these points stay in approximately the same position for array 2. This gene, *Igf2*, which encodes the insulin-like growth factor-2 (IGF-2), consistently produced higher signal in the green fluor, regardless of which sample was labeled green. Note that if only one array had been used in this experiment instead of two, there would have been no way to see that the differential signal was an artifact of the dyes and not due to differential expression between the varieties. This is a clear demonstration of confounding, since on a single array dye and variety are completely confounded. If only one array had been used, one would have been tempted to conclude that *Igf2* is differentially expressed based on this dye interaction.

Here we could identify this gene-specific dye effect by inspecting simple scatterplots because we spotted only 78 genes. However, it is more

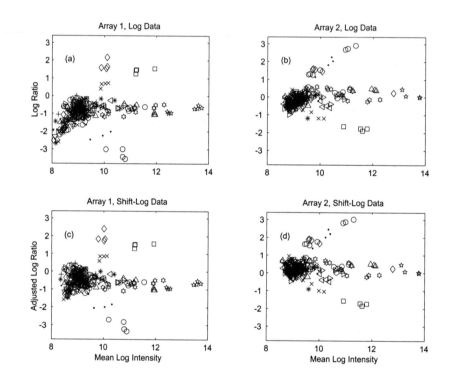

Figure 3.1. Plots of the (adjusted) log ratio against the mean intensity of every spot for the log data and the shift-log data. Like symbols represent the four or sixteen spots for the same sequence.

typical for thousands of genes to be spotted. With such a large amount of data it is impractical to expect to identify such anomalies simply by inspecting scatterplots. Analysis of variance (ANOVA) methods (Kerr, Martin, and Churchill, 2000) provide a tool to automatically detect and quantify such artifacts in the data. Let y_{ijkgr} be the shift-log transformed data value from array $i = 1, 2$, dye $j = 1, 2$, variety $k = 1, 2$, and spot $r = 1, \dots, s_g$ of gene $g = 1, \dots, 78$, where $s_g = 4$ or 16 depending on how many times gene g was spotted. Consider the ANOVA model

$$y_{ijkgr} = \mu + A_i + D_j + V_k + G_g + (AG)_{igr} + (VG)_{kg} + (DG)_{jg} + \epsilon_{ijkgr},$$

where μ is the overall mean, A_i is the overall array effect, D_j is the overall dye effect, V_k is the overall variety effect, and G_g is the overall gene effect across the other factors. The $(AG)_{igr}$ terms capture spot effects, which arise because there is variation in the amount of DNA deposited in each spot during the printing process. The $(DG)_{jg}$ terms capture gene-specific dye effects such as those observed for *Igf2*. The $(VG)_{kg}$ effects represent levels of signal intensity for genes that can specifically be attributed to the treatment and control varieties. These are the effects of interest. The error ϵ_{ijkgr} is assumed to be independent have mean 0. We note that all sixteen possible experimental effects of array, dye, variety, and gene are directly or indirectly accounted for in this model because of the confounding described in Table 3.2.

Table 3.3 gives the analysis of variance for the transformed data. Because genes were spotted multiple times on the arrays, degrees of freedom remain to estimate error even though all factorial effects are represented. The dye×gene effect has the smallest mean square, but it is 11 times larger than the residual mean square. Figure 3.2(a) shows a histogram of the estimated dye×gene effects for the 78 genes. There is one outlier in the distribution, corresponding to *Igf2*. This demonstrates how ANOVA automatically detects effects in the data.

The experimental objective is to identify genes that are differentially expressed between the treated and control samples. In statistical language, we are interested in genes with significant variety×gene interactions. In order to decide which genes show convincing evidence of differential expression, we need to put error-bars on the estimates of differential expression. The residuals from this analysis are heavier-tailed than the normal distribution, so standard confidence intervals based on the normal distribution are not appropriate. Instead, we use a bootstrapping procedure (Efron and Tibshirani, 1986) (Kerr, Martin, and Churchill, 2000). Figure 3.2(b) is a histogram of the estimates of $(VG)_{2g} - (VG)_{1g}$ for the 78 genes. The vertical lines on the histogram show the threshold for differential expression determined by the bootstrapping procedure for

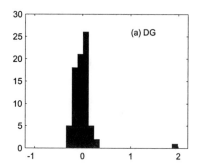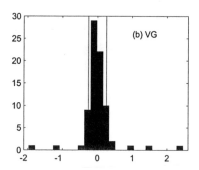

Figure 3.2. Histograms of estimated effects for differences in (a) dye-by-gene interactions and (b) variety-by-gene interactions. Vertical lines in (b) mark the limits for a gene to be found differentially expressed according to bootstrap 99.9% confidence bounds. Thus the proportion of the histogram outside these bounds indicates the proportion of genes for which there was evidence of differential expression at the 99.9% confidence level.

Table 3.3. Analysis of Variance. Notation: SS=sum of squares; df=degrees of freedom; MS=mean square.

Source	SS	df	MS
Array	0.11	1	0.11
Dye	3.35	1	3.35
Variety	45.69	1	45.69
Gene	917.53	77	11.92
Spot	145.19	289	0.46
Variety*Gene	66.46	77	0.86
Dye*Gene	21.30	77	0.28
Residual	19.72	820	0.0241
Adjusted Total	1219.35	1343	

4-fold spot replication. Ten-thousand bootstrap simulations were used. Figure 3.3 plots the 78 estimates of $(VG)_{2g} - (VG)_{1g}$ in ascending order with 99.9% bootstrap confidence bounds. The two genes spotted with 16-fold replication instead of 4-fold replication are easily identified since their confidence intervals are half as wide. There is strong evidence that five genes, which are labeled in the plot, are differentially expressed. The gene whose expression shows the highest increase in expression in response to drug treatment is *Mt2*, encoding the metallothionein-2 protein involved in detoxification and metabolism of heavy metals. The gene whose expression shows the greatest decrease in *Ppara*, encoding the peroxisome proliferator associated receptor alpha, a transcription

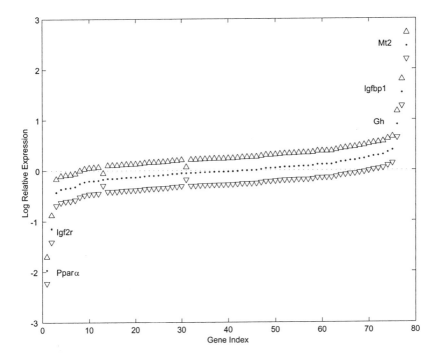

Figure 3.3. Estimated log relative expression with 99.9% bootstrap confidence bounds.

factor regulating hepatic genes associated with fatty acid oxidation in the liver. In addition to the five labeled genes, a handful of other genes have confidence bounds that do not contain 0. These genes are candidates for further study, but the evidence of differential expression is much less compelling.

5. Discussion

The analysis of variance provides a statistical foundation for analyzing microarray data. Standard methods of computing simple ratios of signal intensity do not provide a means of determining which ratios are significant. It is common for an arbitrary cut-off of two- or three-fold to be used. In contrast, in the ANOVA framework one uses the data to both estimate relative expression and also to learn the magnitude of the noise in the data. This allows one to put error-bars on estimates of relative expression and gauge the strength of the evidence for differential expression.

The importance of replication in microarray experiment has been stressed by many authors (Kerr and Churchill, 2001b) (Lee *et al.*, 2000). In this experiment, replication was achieved by using two arrays and spotting genes multiple times per array. This provided degrees of freedom to measure all factorial effects with positive residual degrees of freedom. Multiple spotting may not always be possible when there are thousands of genes to be studied because there may not be space. In this case we advocate replication by adding more arrays to the experimental design.

In this paper we have used "replication" to mean taking additional measurements on a sample of interest. This allows one to identify different sources of variation and to reduces the effect measurement error on estimates. To clarify terminology, we have proposed calling this kind of replication *subsampling* (Kerr *et al.*, 2001). Subsampling is different from replication in the classical meaning of the word in statistics, which refers to sampling individuals from a population of interest. "Replication" addresses biological variability, whereas "subsampling" only addresses measurement error. Since the experimental design described in this paper included subsampling but not replication, inferences are limited to the two particular RNA samples that were assayed. "True," classical replication is required to extend inferences to a greater population.

In addition to replication, another fundamental aspect of good design is balance. In the Latin Square design there is total balance among arrays, dyes, varieties, and genes. This means the effects of interest, VG, are orthogonal to all other factorial effects except one (ADG), which is reasonably assumed to be zero. In particular, VG effects are orthogonal to gene-specific dye effects DG. In this experiment we uncovered a striking example of DG effects. We have seen DG effects in other data, and thus advocate *even* designs — designs where RNA varieties and dyes are balanced (Kerr and Churchill, 2001a). This balance ensures that DG and VG are orthogonal so that accounting for the former does not alter the latter. By minding the principles of good design, balance and replication, experimenters can make the most of their microarray studies.

Acknowledgments

The authors thank Valerie Lemée and Bruno Caubet for assistance printing and testing these microarrays, Peter Reifsnyder for running the experiment, and Chunfa (Charles) Jie for critical reading of a previous version of this manuscript. This work was supported by a Burroughs-Wellcome post-doctoral fellowship from the Program in Mathematics and Molecular Biology to MKK, NIH grants DK56853 and RR08911 to EHL, and NIH grants CA88327, DK58750, and CA34196 to GAC.

References

Brown, P.O., and Botstein, D. (1999) Exploring the New World of the Genome with DNA Microarrays. Nature Genetics, 21, 33-37.

Brown, C.S., Goodwin, P.C., and Sorger, P.K. (2001) Image Metrics in the Statistical Analysis of DNA Microarray Data. Proceedings of the National Academy of Sciences of the USA, 98, 8944-8949.

Cochran, W.G., and Cox, G.M. (1992) Experimental Design, 2nd edition, John Wiley & Sons.

Efron, B., and Tibshirani, R. (1986) Bootstrap Methods for Standard Errors, Confidence Intervals, and Other Measures of Statistical Accuracy. Statistical Science, 1, 54-77.

Kerr, M.K., Afshari, C.A., Bennett, L., Bushel, P., Martinez, J., Walker, N., and Churchill, G.A. Statistical Analysis of a Gene Expression Microarray Experiment with Replication. Statistica Sinica, in press. http://www.jax.org/research/churchill

Kerr, M.K. and Churchill, G.A. (2001a) Experimental Design for Gene Expression Microarrays. Biostatistics, 2, 183-201.

Kerr, M.K. and Churchill, G.A. (2001b) Statistical Design and the Analysis of Gene Expression Microarray Data. Genetical Research, 77, 123-128.

Kerr, M.K., Martin, M., and Churchill G.A. (2000) Analysis of Variance for Gene Expression Microarray Data. Journal of Computational Biology, 7, 819-837.

Lee, M.-L.T., Kuo, F.C., Whitmore, G.A., and Sklar, J. (2000) Importance of Replication in Microarray Gene Expression Studies: Statistical Methods and Evidence from Repetitive cDNA Hybridizations. Proceedings of the National Academy of Sciences of the USA, 97, 9834-9839.

Leiter, E. H., Reifsnyder, P. C., Flurkey, K., Partke, H.-J., Junger, E., and Herberg, L. (1998) NIDDM Genes in Mice: Deleterious Synergism by Both Parental Genomes Contributes to Diabetogenic Thresholds. Diabetes, 47, 1287-1295.

Yang, Y.H., Dudoit, S., Luu, P., and Speed, T.P. (2000) Normalization for cDNA Microarray Data. Technical report 589, Department of Statistics, University of California, Berkeley, 2000. http://www.stat.berkeley.edu/users/terry/zarray

Chapter 4

STUDENTIZING MICROARRAY DATA

Keith A. Baggerly[1],Kevin R. Coombes[1], Kenneth R. Hess[1],
David N. Stivers[1], Lynne V. Abruzzo[2], and Wei Zhang[3]
[1]*Section of Bioinformatics, Department of Biostatistics*
[2]*Department of Hematopathology*
[3]*Cancer Genomics Laboratory, Department of Pathology*
University of Texas M. D. Anderson Cancer Center, Houston, Texas, USA

1. Introduction

Microarrays let us measure relative expression levels of thousands of genes simultaneously. While there are a variety of microarray platforms (e.g., radioactively labeled cDNA on nylon membranes, Affymetrix gene chips) we will focus on spotted cDNA arrays. These involve cDNA samples labeled with fluorescent dyes, hybridized to probes spotted by a robotic arrayer onto a glass substrate. Two samples are hybridized to each array, with the two samples labeled with different dyes. The most commonly used dyes are Cy5 (red) and Cy3 (green). Scanning the array produces a greyscale intensity image for each sample; we will refer to these as the two "channels" of the array. While these two samples may both be of interest, it is often the case that there will be one sample of primary interest and the other sample will be from a "control" or "reference" substance (typically a mixture of cell lines).

The question addressed by microarrays is whether the expression level of a given gene in the sample of interest is different from the level in the control sample. This difference is measured as a ratio or log ratio. Data from multi-array experiments is often summarized in matrix form, with rows corresponding to genes, columns corresponding to arrays, and entries corresponding log ratios. Clustering and "gene hunting" are then performed on this matrix.

One problem with such a summary of the data is that it treats all measurements symmetrically, and thus fails to take into account the fact that microarray data is of widely varying quality. Within an array,

one gene (or our measurement of it) may be inherently more variable
than another. One array may be visibly "dirtier" than another for a
variety of reasons (the quality of the RNA used comes immediately to
mind). This problem can be addressed if we can measure this changing
variability, and couple a matrix of standard deviations with the matrix
of measurements. Rescaling the fold differences by their associated standard deviations "studentizes" the data, and provides a more robust way
of assessing which differences are in fact significant.

In the remainder of this chapter, we will describe how the scale of
the standard deviation varies as a smooth function of the underlying
intensity of the spots, and how we can estimate this function on a single
array by using replicate spottings of the same gene at different positions
on the array. These replicates can further be used for quality control
checks, and can provide visual evidence of the "effectiveness" of a reference channel. A detailed example is given; theoretical details of the
form of the error model are available elsewhere (Baggerly *et al*, 2001).

2.   Fold Differences and Error Models

The first few papers on spotted cDNA microarrays (e.g., Schena *et al*,
1995, Schena *et al*, 1996, DeRisi *et al*, 1996) flagged genes as interesting
if the detected fold difference in expression exceeded a fixed arbitrary
threshold. Later, (Chen *et al*, 1997) this threshold was made less arbitrary by using replicated housekeeping genes to assess the standard
deviation of replicates on an array. Both of these methods implicitly
assume that the coefficient of variation for the gene expression values is
constant. More recent work (Baggerly *et al*, 2001, Hughes *et al*, 2000,
Newton *et al*, 2001, Rocke and Durbin, 2001) has shown that this is not
the case; the standard deviation associated with a log ratio decreases
with increasing intensity. An idealized version of the situation is shown
in Figure 4.1. We can measure fold differences associated with bright
spots quite precisely, while the same is not true for faint spots. Faint
spots can undergo large fold changes that are almost entirely due to
random variation. Consequently, the fold change associated with significance varies with intensity.

This trend in variation with intensity can be made visually apparent
on a single array through the use of replicate spots.

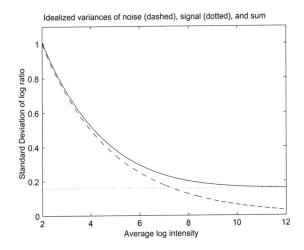

Figure 4.1. Idealized standard deviation of log ratios as a function of spot intensity. There are different factors contributing to the error. A multiplicative component is associated with a lognormal or binomial model for target counts; this contributes a constant amount of variance to the log ratio (dotted line). An additive component is associated with the estimation of background; on the log ratio scale this contributes a large amount to the variance of weak signals and a negligible amount to the variance of strong signals (dashed line). The observed variance is a sum of these. This phrasing (additive plus multiplicative error) is given in Rocke and Durbin, 2001. Note that the vertical scale was chosen arbitrarily here, and should not be expected to match that observed in an actual experiment.

3.   A case study

3.1   Array layout and preprocessing

On our arrays, every spot has been printed in duplicate; one chip design involves a 4 by 12 grid of patches, where each patch is an 8 by 8 grid of spots, and where the top 4 rows of each patch are replicated in the bottom 4 rows. All told, this yields 1152 genes spotted in duplicate, together with 96 positive controls (32 each of GAPDH, beta actin, and tubulin), 96 negative controls (all yeast DNA), and 576 blanks (buffer solution).

At each spot, intensity is measured as the sum of all pixel intensities within a fixed radius circle; local background is assessed as the median pixel intensity in the four diamond-shaped interstices between a spot and its neighbors. This background value is subtracted from each pixel in the circle, giving a subtracted volume, or "sVol". The raw image files have pixel intensities that range from 0 to $2^{16} - 1$; these are scaled in processing so that the range of sVol values is from 0 to about 2^{12}.

Values of sVol less than 1 are set to 1 (our threshold) and the data are log-transformed (base 2).

Replicate agreement can be assessed by plotting the log sVol value for one replicate against another, and looking for concentration about the 45 degree identity line. A better display is achieved by taking this plot and rotating it 45 degrees clockwise, as advocated by Dudoit *et al*, 2000, so that we are plotting the difference between replicates on the y-axis and the average of the replicates on the x-axis. This corresponds to plotting the standard deviation of the replicates as a function of average intensity, letting us see how the variability changes. We have chosen a fairly faint image to use for our examples below to illustrate the points more clearly.

3.2   Single channel images

We tend to produce plots first for each channel separately. We begin here with the Cy5 channel, and check replicate agreement for the genes, Figure 4.2, the positive and negative controls, Figure 4.3, the blanks, Figure 4.4, and all of them superimposed, Figure 4.5. Plotting the genes illustrates the "spread effect" as we move from high to low intensities (right to left); the V-shape at the left end of the plots is due to the threshold at 1.

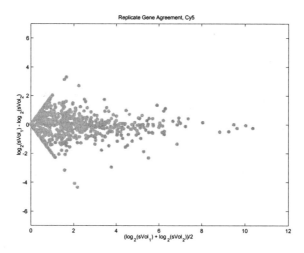

Figure 4.2. Agreement (y-axis) between replicate spottings of the same gene as a function of intensity (x-axis). Each spot on the plot corresponds to a pair of spots on the array, with the identity of the spots the same. Note that agreement is poorer when intensity is low.

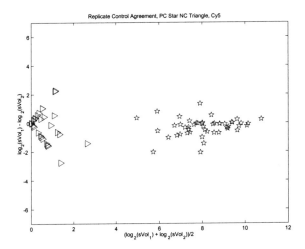

Figure 4.3. Just as we can check agreement between replicate spottings of a gene, we can look at the agreement between positive controls (PCs) and negative controls (NCs) paired following the same geographic pattern (top half to bottom half). There should be clear separation between the two, and the positive controls give us some feel for the variation in intensity possible on an array. Variation in intensity is corrected for by using ratios.

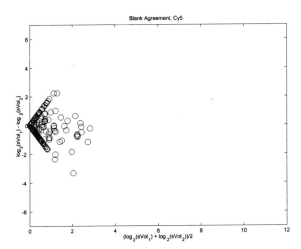

Figure 4.4. We can also check agreement at the paired blank values. These are at the same level as the negative controls in this example, but that is not always the case. Paying attention only to those genes that are present at levels higher than either blanks or negative controls is a conservative rule.

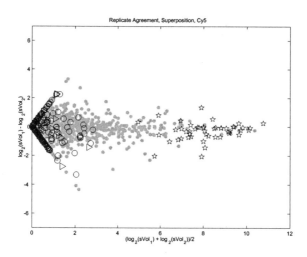

Figure 4.5.   Superimposing the results shows where we can focus on potentially usable results.

Plotting the controls gives us some idea of the stability of the array; the ranges of the negative controls and the blanks give us some idea of the range of intensity values that are believable. A gene must be higher in intensity than either blanks or negative controls before we pay attention to it.

Next, we check the green channel (which tends to be a bit brighter) and then the replicate ratios. The fan shape is more apparent in the green channel, Figure 4.6.

3.3   Replicate Ratios

The fan shape is even more evident on the replicate ratio plot, Figure 4.7, than it was on either of the individual channel plots.

Note that the variance at the high end decreases markedly as we move from the single channel replicates to the ratio replicates. This is visual evidence of the fact that ratios are more stable than intensities at the high end. This implies that there are trends in hybridization efficiency across the array, but that these trends are paralleled in the two channels so that taking ratios cancels out the trends. Looking for a decrease in variance at the high end is one way of checking the "efficiency" of a reference channel. The story at the left end of the plot is different; there the variability is being driven by nonsystematic background "white noise" which does not cancel so well across channels. Zooming in on the high end of the ratio plot, Figure 4.8 shows that there are some noticeable

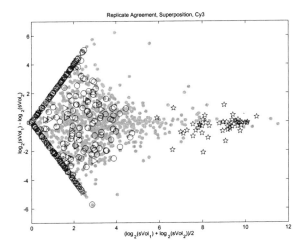

Figure 4.6. Superimposing the results shows where we can focus on potentially usable results. The flare is more apparent in the green channel for this array.

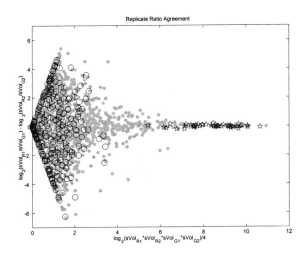

Figure 4.7. The change in variance with intensity is more apparent on the ratio plot than on either of the individual channel plots. The most marked change is that the variance at the high end has decreased on the ratio plot. The variance here is associated with multiplicative errors (a slight preference for one replicate spot over another) and these errors are correlated between channels and hence cancel. Noise at the low end is more random, and cancels to a much lesser degree.

outliers. Note that outliers on replicate agreement plots, either single channel or ratio, are not indicative of features of biological interest!

Rather, these correspond to data artifacts on the array such as dust specks or fibers.

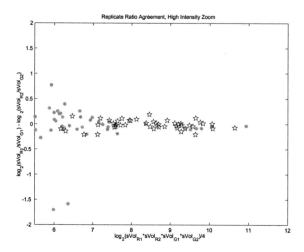

Figure 4.8. Zooming in on the high intensity replicate ratios. Agreement here is quite good, and this lets us estimate the variance associated with a log ratio measurement. Assessment must use robust methods so as to take into account the presence of occasional artifacts such as the two in the lower left - these correspond to replicate spots where the ratio is quite different even though the gene is the same.

Checking the two outliers from the bottom left of Figure 4.8 confirms their artifactual nature; one is due to a fiber, the other is due to a dust speck. These are shown in detail in Figure 4.9.

3.4   Variance Fitting and Studentization

Thus far, we have exploited the plots primarily for quality control. However, the variance estimable from the replicate ratios can be used to define the precision with which a ratio measurement can be made, and this in turn will allow us to scale the differences between channels, which are "average ratios". We estimate the standard deviation by fitting a smooth curve to the spread of the replicate ratios, capturing the precision with which a ratio is measured as a function of the average channel intensity. This is done using a robust loess fit to the absolute values of the differences between replicate log ratios. Loess (Cleveland, 1979) uses only a windowed fraction of the data, so that the standard deviation is "local"; here the fraction is 40%. This fit is applied to the genes only, not to the positive controls, negative controls, or blanks. Further, so as not to be caught by the "V" at the left end, we impose a monotonicity condition: variance cannot decrease as the signal gets

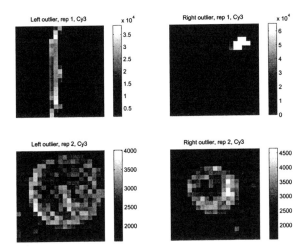

Figure 4.9. Visually checking the outliers from the replicate ratio plot (just looking at the green channel here) confirms that these are artifacts. Note the pixel intensity scales are different in the different subimages.

weaker. The results are shown in Figure 4.10. Note that the outliers are clearly visible. We then shift to plotting the average difference between

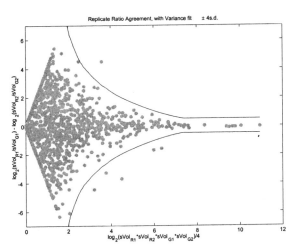

Figure 4.10. Loess fit to the standard deviation of the replicate ratios. We fit the mean absolute replicate difference, which corresponds to roughly 0.8 s.d. with normal data. The bands shown correspond to ±4 s.d.

channels. There is a vertical displacement corresponding to a normalization effect, indicated by the horizontal line. We assess the normalization offset using the median displacement for expressed genes; expressed here corresponds to an average signal value greater than 3. The results are shown in Figure 4.11. To assess the degree to which we believe a gene to be differentially expressed, we take the difference between the observed channel difference and the normalization line and divide it by our estimate of the local standard deviation. This produces locally studentized values. While the extreme case in the lower right (transfected p53)

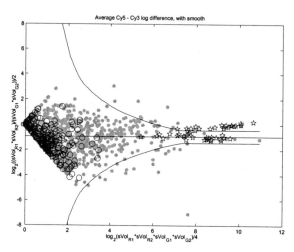

Figure 4.11. Average differences in log intensities for the Cy5 and Cy3 channels, with local standard deviation bounds. Spots outside the bands correspond to differentially expressed genes. There is a clear extreme case at the bottom, corresponding to p53 which had been transfected into one of the two samples used in this experiment. Note that the positive controls appear to have some differences.

would have been caught even without adjustment, many false positives would have been identified using a fixed fold difference. It is interesting to note that this image illustrates a potential conflict with the positive controls Zooming in to the high-intensity region and labeling the different types of positive controls with different symbols shows that GAPDH and tubulin are essentially on the normalization line, whereas beta actin was expressed at different levels in the two samples. This is shown in Figure 4.12.

4.   Discussion

Studentizing microarray data allows us to incorporate the inherently different scales of variation into our assessment of differential expression. This makes our results more stable. Further, for comparing different arrays, the studentized values incorporate overall levels of "noisiness" of

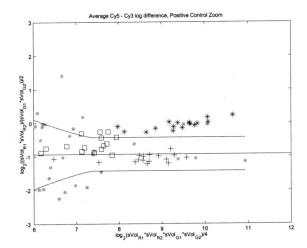

Figure 4.12. Channel difference zoom on positive controls. Crosses are GAPDH, stars are beta actin, and squares are Tubulin. GAPDH is expressed at a different relative level than beta actin, but replicates within type are fairly consistent.

the arrays. This scaling of variation with intensity arises from the combination of different types of measurement error, which can be assessed using replicate spots. Replicating entire arrays can further allow for the partitioning of between and within array variation. Efficiently identifying and removing sources of variation that are not of interest allows us to focus more clearly on the information that is the biological focus of our experiments.

Acknowledgments

Research at the MDACC Cancer Genomics Core Laboratory is partially supported by Tobacco Settlement Funds as appropriated by the Texas State Legislature and by a generous donation from the Michael and Betty Kadoorie Foundation.

References

Baggerly, K.A., Coombes, K.R., Hess, K.R., Stivers, D.N., Abruzzo, L.V., and Zhang, W. (2001). "Identifying Differentially Expressed Genes in cDNA Microarray Experiments", *J. Comp. Biol.*, *to appear.*

Chen, Y., Dougherty, E.R., and Bittner, M.L. (1997). "Ratio-based decisions and the quantitative analysis of cDNA microarray images", *J. Biomed. Optics*, 2:364–374.

Cleveland, W.S. (1979). "Robust locally weighted regression and smoothing scatterplots", *J. Amer. Statist. Assoc.*, 74:829–836.

DeRisi, J.L., Penland, L., Brown, P.O., Bittner, M.L., Meltzer, P.S., Ray, M., Chen, Y., Su, Y.A., and Trent, J.M. (1996). "Use of a cDNA microarray to analyze gene expression patterns in human cancer", *Nature Genet.*, 14:457–460.

Dudoit, S., Yang, Y.H., Callow, M.J., and Speed, T.P. (2000), "Statistical methods for identifying differentially expressed genes in replicated cDNA microarray experiments", *Statistica Sinica, to appear.*

Hughes, T.R., Marton, M.J., Jones, A.R., Roberts, C.J., Stoughton, R., Armour, C.D., Bennett, H.A., Coffey, E., Dai, H., He, Y.D., Kidd, M.J., King, A.M., Meyer, M.R., Slade, D., Lum, P.Y., Stepaniants, S.B., Shoemaker, D.D., Gachotte, D., Chakraburtty, K., Simon, J., Bard, M., and Friend, S.H. (2000). "Functional discovery via a compendium of gene expression profiles", *Cell*, 102:109–126.

Kerr, M.K., Martin, M., and Churchill, G.A. (2000). Analysis of variance for gene expression microarray data. *J. Comp. Biol.*, 7:819–837.

Lee, M.-L., Kuo, F.C., Whitmore, G.A., and Sklar, J. (2000). "Importance of replication in microarray gene expression studies: Statistical methods and evidence from repetitive cDNA hybridizations", *Proc. Nat. Acad. Sci. USA*, 97:9834–9839.

Newton, M.A., Kendziorski, C.M., Richmond, C.S., Blattner, F.R., and Tsui, K.W. (2001). "On differential variability of expression ratios: Improving statistical inference about gene expression changes from microarray data", *J. Comp. Biol.*, 8:37–52.

Rocke, D.M., and Durbin, B. (2001). "A model for measurement error for gene expression arrays", Technical Report, UC Davis.

Schena, M., Shalon, D., Davis, R.W., and Brown, P.O. (1995). "Quantitative monitoring of gene expression patterns with a complementary DNA microarray", *Science*, 270:467–470.

Schena, M., Shalon, D., Heller, R., Chai, A., Brown, P.O., and Davis, R.W. (1996). "Parallel human genome analysis: Microarray-based expression monitoring of 1000 genes", *Proc. Nat. Acad. Sci. USA*, 93:10614–10619.

Chapter 5

EXPLORATORY CLUSTERING OF GENE EXPRESSION PROFILES OF MUTATED YEAST STRAINS

Merja Oja[1], Janne Nikkilä[1], Petri Törönen[2], Garry Wong[2], Eero Castrén[2], and Samuel Kaski[1]

[1] *Neural Networks Research Centre, Helsinki University of Technology, Helsinki, Finland*

[2] *A. I. Virtanen Institute, University of Kuopio, Kuopio, Finland*

1.   Introduction

Genome projects are continuously revealing novel genes and the function of many of the genes discovered previously is unknown. The clarification of the function of these genes will be a great challenge and will require coordinated work by many different disciplines. Because many genes share a relatively high homology in different organisms, simple model organisms have been exploited in the analysis of gene function. For example, projects are underway to mutate every single gene in baker's yeast, *Saccharomyces cerevisiae* (Winzeler et al., 1999). One of the problems that has hampered functional analysis is that simple organisms display only a limited number of observable phenotypes and many mutations do not produce any phenotype at all.

Gene expression analysis has been recently proposed as an alternative means to analyze phenotypes of mutated yeast strains. Hughes *et al.* (2000) used the large mutation data set in an exploratory fashion to suggest inferences about gene function, for example about which pathway an uncharacterized gene belongs to. The global transcriptional response, i.e. the expression level of all the genes, of a cell after a mutation can be regarded as a detailed molecular phenotype. Tentative inferences on the similarity of function can be made based on similarity of the gene expression profiles, and the hypotheses can then be tested experimentally.

Hughes *et al.* (2000) clustered the mutations according to similarities in their expression profiles, to make similarities hidden within the large data set explicit. Hierarchical agglomerative clustering, a popular method for grouping gene expression data, was used. The hierarchical clustering methods visualize a tree of subclusters of increasing detail. The problem with the trees is that they are sensitive to small variations in the data that may affect the agglomeration process, and for large data sets the trees become huge. It is then hard to extract the essential cluster structures from them.

We have been using Self-Organizing Map (SOM)-based clustering and information visualization for the same purpose (Törönen et al., 1999; Kaski et al., 2001a). The SOM (Kohonen, 1982; Kohonen, 1995) projects the genes or mutations onto a two-dimensional display which is ordered according to the similarity of the expression profiles. Perhaps the most immediate advantage over alternative methods is the easily understandable visualizability of the result: The SOM display provides an overview of the similarity relationships within the large genome-wide data set. Clusters are visualized with gray shades on the display, and for example mutation clusters can be characterized in terms of which kinds of genes the mutations affect.

This paper has two interrelated goals. We show how to use Self-Organizing Map-based data exploration tools to analyze a genome-wide expression data set and visualize its essential properties. By directly comparing the results with those obtained with hierarchical clustering we hope to convince experimentalists on the utility of the tools in other gene expression analyses as well. More specifically, we show that the same things that have been found using hierarchical clustering can be found easily with the SOM. The second goal is to demonstrate methods for exploring the gene expression data.

2.   The Data

We used the data set analyzed earlier by Hughes *et al.* (2000) . The set consists of expression measurements for all yeast (*Saccharomyces cerevisiae*) genes in 300 different treatments, in which one gene was either removed or its function was blocked. The expression of each gene in the mutated yeast is compared to the expression of a wild type yeast. The expression ratio and noise estimate for the ratio is measured for each gene-mutation pair. Thus the amount of data is 300×6297 ratios, and for each ratio there is an estimate for the variance of the measurement noise.

Hughes *et al.* analyzed first the probability that a given expression measurement deviates from pure background noise, in order to concentrate on expressed genes. They call the process significance-cutting. They constructed a gene-specific error model to calculate the P-value that a given measurement (for a gene-mutation pair) deviates from noise. Genes for which both the P-value and the actual measured expression ratio did not exceed the specified thresholds were left out of the analysis, and likewise for the mutations.

In our work the significance threshold was chosen to be $P < 0.01$, and two different thresholds are used for the expression ratios: When choosing the metric a looser test was used to leave enough data to make inferences: the ratio had to exceed 2. When clustering the data the threshold was stricter, 3.2, in accordance with the study of Hughes *et al.* The size of the resulting larger set was 179×1482, and the size of the smaller set was 127×578.

Each yeast strain was assigned a MIPS[1] functional class according to the class of the gene that had been mutated. The MIPS classification is hierarchical and allows a gene to belong to several functional classes. The original classification (249 classes) was reduced to 70 by picking the top level of homogeneous main classes, and second level of more heterogeneous classes. The functional classes which we considered heterogeneous were metabolism, cell growth, transcription, cellular transport, cellular biogenesis and cellular organization. For example, in the cellular organization class the second level distinguished the genes based on which cellular components they affect, and for the metabolism class the second level subclassifies the genes to the metabolism of different biomolecules like fatty acids or amino acids.

3. Choosing the Metric

Clustering methods group together similar data items, here gene expression profiles, while maximizing dissimilarity between clusters. Hence, the results depend heavily on the metric. The most common distance or similarity measures for real-valued data, such as gene expression profiles, are based on either the Euclidean distance or the correlation coefficient. In this paper we will confine ourselves to variants of these two measures. To our knowledge, there are no comparisons of different metrics with gene expression data in the literature.

The suitability of different measures depends both on the accuracy of the data and on the goal of the analysis. The questions that need

---

[1] http://www.mips.biochem.mpg.de/proj/yeast

Absolute magnitudes \ Zero level	Reliable	Unreliable
Interesting	Euclidean metric	(Euclidean with mean subtracted)
Not interesting	Inner product	Correlation

Figure 5.1. Distance or similarity measures for gene expression profiles. Suitability of the measures depends on whether the zero level of the expressions is reliable or not, and whether the absolute or relative magnitudes are important. Inner product is taken of normalized vectors.

be answered include: (i) Is the zero level of the measurements reliable enough, i.e., is there possibly a (large) shift in the baseline? (ii) Which is interesting, the absolute magnitude of expression ratios or the relative values? Some suitable measures for each case are listed in Figure 5.1.

If the baseline is not very reliable then the measures should be invariant to it. In the correlation coefficient the average of the measurements is effectively subtracted before the analysis, and the same can in principle be done for the Euclidean measure as well.

The Euclidean distance measures both the absolute and relative differences in gene expression. If one is only interested in the latter, then the (Euclidean length of the) expression profiles can be normalized. It is easy to show that, for normalized profiles, similarity orderings according to increasing Euclidean distance and decreasing *inner products* $x^T y$ are equivalent. Here x and y are expression profiles considered as real-valued vectors. The correlation coefficient, on the other hand, is equivalent to the inner product of vectors from which the average has been subtracted.

In summary, correlation is invariant to both the zero level and the scale of gene expressions. Inner product (of normalized vectors) is invariant to scale. Euclidean distances takes into account differences in both.[2]

Note that the absolute value of correlation is sometimes used. This is very sensible within one time series, assuming that there is some common underlying factor that inhibits the activity of one gene and enhances the other. In this paper the advantages are not as salient, however, since the multitude of factors may cause different directions for each

---

[2] Correlation can be viewed as a method of reducing both additive and multiplicative noise, whereas inner product reduces only multiplicative noise. Note, however, that all noise reduction models may also remove essential information.

gene/treatment combination. Hence we decided to avoid potential interpretation difficulties in the later stages of analysis, and did not use the absolute value.

Plain Euclidean or correlation-based distances weight the expression of each gene at each treatment equally. A standard way of preprocessing variables in the absence of prior information is to scale their variance to unity. Hughes *et al.* (2000) have additionally measured the noise of each gene-mutation pair individually, and the expression values can be downscaled by the standard deviation of the noise.

In this section we measure empirically which distance measures are most appropriate for gene expression data. As a criterion we use the ability of a simple non-parametric classifier to distinguish between functional classes of the genes.

3.1   Methods

The k-nearest neighbor (KNN) classifier stores its teaching data set, and classifies a new expression profile $\mathbf{x}$ according to majority voting between the classes of the K closest samples in the teaching set. That is, if there are more samples from class C than from the other classes, then the prediction will be that $\mathbf{x}$ belongs to class C. We used K=5.

The different metrics were compared by cross-validation: The data set was divided into 27 subsets, and the classification result was computed 27 times for each metric. Each subset was used for evaluating the accuracy of the classifier in turn, while the teaching set consisted of the remaining 26 subsets.

Since the genes may belong to several functional classes we had to modify the basic KNN classifier slightly. The new expression profile was considered to be randomly assigned to one of its classes, and the classifier chooses the predicted class randomly from a distribution it has estimated. The percentage of correct classification was then estimated by the probability that the two classes coincide. The classifier estimates the distribution by searching for the K closest samples and calculating the proportion of the classes in them. A gene $\mathbf{x}$ belonging to $N_C(\mathbf{x})$ functional classes contributes for each of the classes with a proportion of $1/N_C(\mathbf{x})$. The classification error is estimated by the average of $(1 - \hat{p}(C(\mathbf{x})))$ over $\mathbf{x}$ and its correct classes within the test set. Here $C(\mathbf{x})$ is the correct class of $\mathbf{x}$, and $\hat{p}(C)$ is the proportion of samples of class C within the neighborhood of K samples.

		Preprocessing		
Metric	None	Normalized noise	Normalized variance	Normalized noise & variance
Euclidean	0.9172	0.9138	0.9152	0.9186
Inner product	0.8898	0.8878	0.8899	0.8899
Correlation	0.8880	0.8866	0.8904	0.8905

Table 5.1. Comparison of different metrics for classifying mutations. The figures are estimated classification errors. For the metrics see Figure 5.1. Normalized noise: Each expression value is scaled by the inverse of the estimated standard deviation of the noise. Normalized variance: The variance of the expression of each gene is normalized to unity. Note that the number of classes is very large compared to the size of the data set; classification error of the naive classifier that predicts according to prior class probabilities is 0.9566.

3.2   Results

The classification errors for the three metrics, with and without the different preprocessing methods are shown in Table 5.1. The best results were obtained by the correlation of expression profiles, which were normalized with respect to the measurement noise.

A paired t-test was used to measure significance of the difference between the best result, and the best results of the other two metrics. Only the difference from the Euclidean metric was significant ($P < 0.01$). The effects of preprocessing were additionally tested for the correlation measure, and they were not significant.

Although the correlation measure was slightly better than the inner product distance, we chose the latter one for the further analysis in the next section to keep the results comparable to the results of Hughes *et al.* (2000) .[3]

4.   Self-Organizing Map-Based Exploratory Clustering

4.1   Self-Organizing Maps

The Self-Organizing Map (SOM; Kohonen, 1982; Kohonen, 1995) is a discrete grid of map units. Each map unit represents certain kinds of data, in the present paper different mutations of yeast, the genes of which are expressed in similar ways. The input data is represented in an ordered fashion on the map: Map units close-by on the grid represent more similar expression profiles and units farther away progressively

---

[3]Note that our terminology is different; they call the inner product measure correlation.

more different profiles. The map is a similarity diagram that presents an overview of the mutual similarity of the large amount of high-dimensional expression profiles.

The mapping is defined by associating an n-dimensional *model vector* $\mathbf{m}_i$ with each map unit i, and by mapping each expression profile $\mathbf{x} \in R^n$ to the map unit having the closest model vector. Here we use the inner product metric: Each expression measurement is first normalized by the estimate of the noise standard deviation, after which the length of the vectors is normalized to unity. The inner product is then used as the similarity measure, and the closest model vector is the vector c fulfilling the condition

$$\mathbf{x}^T \mathbf{m}_c \geq \mathbf{x}^T \mathbf{m}_i \qquad (5.1)$$

for all i. The mapping becomes ordered and represents the data after the values for the model vectors are computed in an iterative process. In "on-line" type of computation at step t one expression profile $\mathbf{x}(t)$ is selected at random, the closest model vector c is found by (5.1), and the model vectors are adapted according to

$$\mathbf{m}_i(t+1) = \mathbf{m}_i(t) + h_{ci}(t)\mathbf{x}(t) , \qquad (5.2)$$

and normalized to unit length. Here $h_{ci}(t)$ is the *neighborhood function*, a decreasing function of the distance of the units c and i on the map grid. We have used a Gaussian function that decreases in width and height during the iterative computation. For more details, variants, and different methods of computing SOMs see (Kohonen, 1995; Kaski et al., 1998).

The SOM can be used both as a nonlinear projection method for dimensionality reduction, and as a clustering method. According to recent evidence (Venna and Kaski, 2001) the similarity diagrams formed by the SOM are more trustworthy than multidimensional scaling-based alternatives, in the sense that if two data points are close-by on the display they are more likely to be close-by in the input space as well.

Compared with most clustering methods the SOM has the advantage that instead of extracting a set of clusters it visualizes the cluster structures. The SOM display is an overview of the similarity relationships and cluster structures of the data set. Such visualizations will be used below to study clusters in gene expression patterns.

Several measurements are missing from the mutation data set, as is common for other real-word data sets as well. For SOM there exists a simple method for computing with missing values: In each step in the iteration, simply neglect the values missing from the current input vector. This approach works well at least if only a small percentage of the values is missing (Samad and Harp, 1992). For the inner product metric we have carried out all computations within the subspace defined by the available components.

4.2   Overview of the Cluster Structure of the Data

The self-organizing map used in this work was the inner-product version of the SOM consisting of 13 times 9 map units arranged in a regular hexagonal grid. The map was computed in two phases. In the first ordering phase of 120,000 iterations the standard deviation of the Gaussian-shaped neighborhood function decreased linearly from 5 to 2 while its height decreased from 0.2 to 0.02. In the second fine-tuning phase of 1,200,000 iterations the standard deviation decreased to 1 and the height to zero.

The SOM overview of the yeast gene mutation data is presented in Figure 5.2. Dense cluster-like areas are indicated in light shades and the sparser areas separating them in data space are indicated in dark shades. The lightness is proportional to the distances between the model vectors of neighboring map units in the data space. Because the density of the model vectors in the data space reflects the density of the data, the distances reflect the density structure of the data. The distance matrix is called the U-matrix (Ultsch, 1993).

The clusters formed by hierarchical clustering of gene expression patterns in mutated yeast strains by Hughes *et al.* (2000) have been manually drawn on the U-matrix in Figure 5.2. Overall, clusters derived from hierarchical clustering nicely fall into dense areas of the SOM-based display, indicating that these two clustering methods in general produce very comparable results.

Closer analysis of the SOM-based U-matrix reveals features which highlight the usefulness of the ability of SOM to visualize the spatial relationships between neighboring clusters. First, mutations in genes related to mitochondrial respiration are clustered in the upper middle region of the SOM display, perfectly agreeing with the cluster of mutations affecting mitochondrial function derived by Hughes *et al.* (2000). At the lower right corner of this cluster is a subcluster, which includes cup5 and vma8, genes which affect mitochondrial function indirectly by affecting iron homeostasis. Whereas these genes were clustered in a separate cluster from the other mitochondrial mutants by hierarchical clustering (Hughes *et al.*, 2000), SOM places them into a separate group, yet in the close vicinity of the main cluster of mutations affecting mitochondrial respiration.

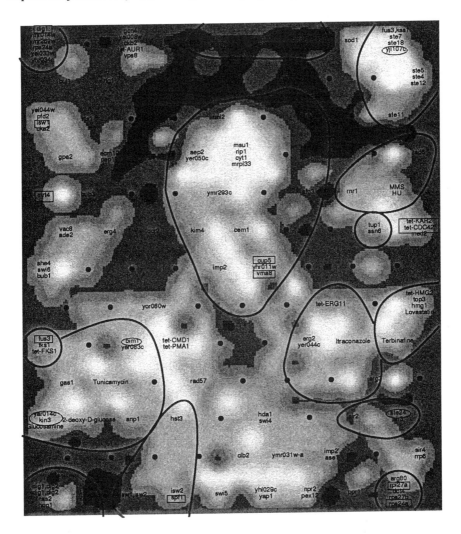

Figure 5.2. Overview of the cluster structure of the data: Light shades denote dense areas in the data space and dark shades sparse areas, respectively. Dots denote the map units without any data vectors. Clusters found by Hughes et al. (2000) with hierarchical clustering have been encircled with the thick lines. Yeast strains that did not belong to the original clusters have been encircled with the thin lines, and boxed strains have moved away from their original clusters.

To the right of the mitochondrial respiration cluster is an area which in SOM analysis appears as a single cluster, but was divided into two distinct albeit spatially related clusters by hierarchical clustering (Hughes et al., 2000). All the mutations in this cluster are related to DNA damage or block of DNA replication (Zhou and Elledge, 1992), which indicates that clustering into a single cluster is functionally justified.

Finally, SOM places sir2, sir3 and sir4 mutants in neighboring neurons within a single cluster in the lower right end of the SOM display. These three genes directly interact to form a protein complex involved in telomeric silencing. Whereas sir2 and sir3 were clustered also by hierarchical clustering, sir4 was not clustered together with them.

In conclusion, in contrast to methods which simply extract a set of clusters, SOM-based display offers more information in the form of hints about the neighbor relations and topography of the data space.

4.3   Interpretation of the Clusters

The overview display (Fig. 5.2) reveals clusters and visualizes their similarity relationships: Clusters close-by on the map are similar. Ultimately the clusters need be analyzed based on biological knowledge and further experimentation, but there exist also SOM-based exploration tools that may help in understanding properties of the clusters.

Perhaps the most simple and intuitive method for characterizing the clusters is to compute the average expression profiles in each interesting map area, and then to visualize their differences in different map areas. It is of particular interest to measure what changes when moving along the map to neighboring map units; note that the map is a non-linear description of the data space and hence the coordinates are non-linear.

We have previously (Kaski et al., 2002) characterized differences between neighboring map units in terms of original data variables, here expression of different genes, using *local factors*. The Self-Organizing Map can be regarded as an approximative *principal surface*, a non-linear generalization of principal components, and measuring changes along the map thus reveals the local principal variation in the data.

In Figure 5.3 we have characterized two areas of the map. About the cluster area in the top right corner of the map we already have some biological knowledge based on the functional similarity of the mutated genes. In contrast, the cluster area in the lower end of the map is a previously unknown cluster structure found with the SOM-based overview in Figure 5.2. We characterize the clusters by analyzing which genes have the most different expression values in two neighboring clusters, and how those genes behave in other mutations. In this manner the similar, or dissimilar, behavior of the genes in the previously identified cluster structures may clarify the functionality of the mutated genes in the still unknown clusters.

The results of the analysis of the original variables, the expressions of different genes, in the top right corner of the map are in accordance with the previous biological identification of these cluster structures. In Figure 5.3, panel **A** reveals that the components having the largest positive differences, when moving into the upper cluster from the lower cluster, characterize specifically the lower cluster. Indeed, the expression of these genes, many of which are related to metabolism of lipids and carbohydrates, is much lower in this cluster than anywhere else in the data space. Conversely, genes which are downregulated (panel **B**) when moving from the lower to the upper cluster and generally expressed at a lower level in the upper cluster than in other parts of the data space. Many of these downregulated genes are related to mating and pheromone response, which is consistent with the clustering of mating-related mutations in the upper right corner.

Figure 5.2 reveals also a new, possibly interesting cluster in the middle of the lower end of the map. This cluster was not analyzed in Hughes *et al.* (2000). The mutated genes in this cluster do not form a clear functional group, but many of them are related to transcriptional control, including transcription factors, histone deacetylases and cell-cycle regulated genes. In Figure 5.3 the panels **C** and **D** reveal the components that are up or downregulated when moving from the mid-left to the mid-right cluster at the lower part of the map. Upregulated genes are diffusely localized not only to the right cluster, but also to a large area above it, further indicating that the right cluster does not form a functionally uniform group. Downregulated genes highlight the left cluster, where the genes are expressed on average at a higher level than in the other parts of the data space. Many of the regulated genes encode for heat shock proteins indicating the activation of heat shock reaction in response to mutations clustered in the lower left region.

5. Conclusions

We have shown that cluster displays constructed of self-organizing maps correspond well with clusterings formed earlier by hierarchical clustering (Hughes et al., 2000). The main difference between the methods seems to be in the ways they visualize the data: trees vs. 2-dimensional maps.

Map displays enable focusing of the further analysis and provide a framework for intuitive exploration of the data. We demonstrated such exploration by investigating known and still unknown cluster structures.

We compared the performance of different distance measures, Euclidean metric, inner product, and correlation coefficient, in functional

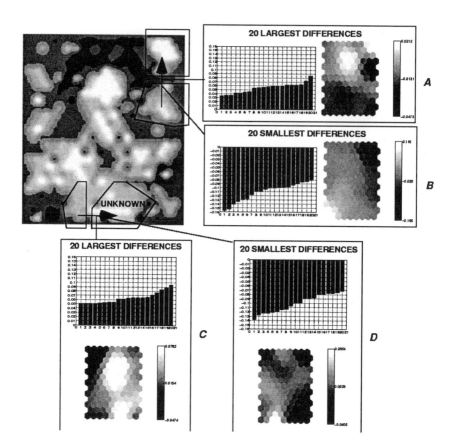

Figure 5.3. Characterization of a sample cluster (panels **A** and **B**) and a previously unknown cluster (panels **C** and **D**). The bar-plots represent changes in average gene expressions when moving according to the arrow from one cluster to another. The 20 largest positive differences and 20 largest negative differences between the average expression vectors of the clusters are shown. The grayshaded map grids in the same panels show the average expression of the same 20 components plotted on the whole map. Light shade denotes high average value and dark shade low average value.

classification of the expression profiles. The correlation coefficient was the best, although only the difference from Euclidean metric was significant. This suggests that the zero level in the signals may not be reliable, and that only the relative values of the expression ratios are important. Scaling the expressions by the inverse of a noise estimate improved the results, but not significantly.

The metrics we studied in this paper were global, i.e., the same everywhere in the data space. We will later apply metrics that learn to concentrate on the most important kinds of differences (Kaski et al., 2001b; Sinkkonen and Kaski, 2001). The importance is derived from auxiliary data.

Acknowledgments

This work was supported by the Academy of Finland, in part by the grants 50061 and 1164349.

References

Hughes, T. R., Marton, M. J., Jones, A. R., Roberts, C. J., Stoughton, R., Armour, C. D., Bennett, H. A., Coffrey, E., Dai, H., He, Y. D., Kidd, M. J., King, A. M., Meyer, M. R., Slade, D., Lum, P. Y., Stepaniants, S. B., Shoemaker, D. D., Gachotte, D., Chakraburtty, K., Simon, J., Bard, M., and Friend, S. H. (2000). Functional discovery via a compendium of expression profiles. *Cell*, 102:109–126.

Kaski, S., Kangas, J., and Kohonen, T. (1998). Bibliography of self-organizing map (SOM) papers: 1981–1997. *Neural Computing Surveys*, 1(3&4):1–176.

Kaski, S., Nikkilä, J., and Kohonen, T. (2002). Methods for exploratory cluster analysis. In P. S. Szczepaniak, J. Segovia, J. Kacprzyk, and L. A. Zadeh, editors, *Intelligent Exploration of the Web*. Springer, Berlin. Accepted for publication.

Kaski, S., Nikkilä, J., Törönen, P., Castrén, E., and Wong, G. (2001a). Analysis and visualization of gene expression data using self-organizing maps. In *Proceedings of NSIP-01, IEEE-EURASIP Workshop on Nonlinear Signal and Image Processing*.

Kaski, S., Sinkkonen, J., and Peltonen, J. (2001b). Bankruptcy analysis with self-organizing maps in learning metrics. *IEEE Transactions on Neural Networks*, 12:936–947.

Kohonen, T. (1982). Self-organized formation of topologically correct feature maps. *Biological Cybernetics*, 43:59–69.

Kohonen, T. (1995). *Self-Organizing Maps*. Springer, Berlin. (Third, extended edition 2001).

Samad, T. and Harp, S. A. (1992). Self-organization with partial data. *Network: Computation in Neural Systems*, 3(2):205–212.

Sinkkonen, J. and Kaski, S. (2001). Clustering based on conditional distributions in an auxiliary space. *Neural Computation*. In press.

Törönen, P., Kolehmainen, M., Wong, G., and Castrén, E. (1999). Analysis of gene expression data using self-organizing maps. *FEBS Letters*, 451:142–146.

Ultsch, A. (1993). Knowledge extraction from self-organizing neural networks. In Opitz, O., Lausen, B., and Klar, R., editors, *Information and Classification*, pages 301 – 306. Springer Verlag.

Venna, J. and Kaski, S. (2001). Neighborhood preservation in nonlinear projection methods: An experimental study. In Dorffner, G., Bischof, H., and Hornik, K., editors, *Artificial Neural Networks—ICANN 2001*, pages 485–491. Springer, Berlin.

Winzeler, E. A., Shoemaker, D. D., Astromoff, A., Liang, H., Anderson, K., Andre, B., Bangham, R., Benito, R., Boeke, J. D., Bussey, H., et al. (1999). Functional characterization of the S. Cerevisiae genome by gene deletion and parallel analysis. *Science*, 285:901–906.

Zhou, Z., and Elledge, S.J. (1992). Isolation of *crt* mutants constitutive for transcription of the DNA damage inducible gene *RNR3* in *Saccharomyces Cerevisiae*. *Genetics*, 131:851–866.

Chapter 6

SELECTING INFORMATIVE GENES FOR CANCER CLASSIFICATION USING GENE EXPRESSION DATA

Tatsuya Akutsu[1] and Satoru Miyano[2,1]
[1] *Bioinformatics Center, Institute for Chemical Research, Kyoto University, Japan*
[2] *Human Genome Center, Institute of Medical Science, University of Tokyo, Japan*

1.   Introduction

Accurate classification of tumor types is very important in order to apply adequate therapies to patients. Therefore, many studies have been done for cancer classification. Most of such studies were based primarily on morphological appearance of the tumor. Recently, a new approach based on global gene expression analysis using DNA microarrays (Brown, 1997) has been proposed (Cole, 1999; Golub, 1999; Kononen, 1998; Tsunoda, 2000). Although effectiveness of such an approach was already demonstrated, the proposed information processing methods were rather heuristic (Golub, 1999; Tsunoda, 2000). Therefore, further studies should be done for making more accurate classification.

Golub *et al.* divided cancer classification into two problems (Golub, 1999): *class discovery* and *class prediction*. Class discovery is to define previously unrecognized tumor subtypes, whereas class prediction is to assign particular tumor samples to already-defined classes. Although class discovery is more challenging, we consider class prediction in this chapter because class prediction seems to be more basic and thus information processing methods for class prediction should be established earlier. In the previous methods (Golub, 1999; Tsunoda, 2000), predictions were made by means of the *weighted voting*. All genes were not used for weighted votes, but several tens of genes relevant to class distinction were selected and used. Use of selected genes seems better because of several reasons. For example, computational cost for determining parameters and cost for measurement of gene expression levels are much lower if the selected genes are used. Golub *et al.* called these selected genes *in-*

(A)

	gene 1	gene 2	gene 3	gene 4
sample 1	high	high	high	high
sample 2	high	high	high	high
sample 3	low	low	low	low
sample 4	high	high	high	high
sample 5	high	high	high	high
sample 6	low	low	low	low
sample 7	low	low	low	low
sample 8	high	high	high	high
sample 9	low	low	low	low

Class 1: sample 1, sample 2, sample 3, sample 4, sample 5
Class 2: sample 6, sample 7, sample 8, sample 9

(B)

	gene 5	gene 6	gene 7	gene 8
	high	high	high	high
	low	high	high	high
	high	low	high	high
	high	high	low	high
	high	high	high	low
	high	low	low	low
	low	high	low	low
	low	low	high	low
	low	low	low	high

Figure 6.1. Selection of informative genes. Set of genes {gene 5, gene 6, gene 7, gene 8} is better than {gene 1, gene 2, gene 3, gene 4} because wrong predictions might be made for sample 3 and sample 8 if the latter set is used.

formative genes. However, selection methods of informative genes were rather heuristic. Indeed, Golub *et al.* wrote that *the choice to use 50 informative genes in the predictor was somewhat arbitrary.* Tsunoda *et al.* selected informative genes based on dependency between every pair of genes and clustering results. Moreover, various other techniques in AI, machine learning and statistics were recently applied to selection of informative genes (see http://bioinformatics.duke.edu/CAMDA/). However, most of these methods are still heuristic. Therefore, this chapter focuses on the selection of informative genes.

In order to demonstrate the importance of selection of informative genes, consider an extreme example shown in Figure 6.1. In this figure, "high" means that the expression level of a gene is high and "low" means that the expression level is low. If informative genes were to be selected based on correlation between the classification result and the expressions of each gene only, every gene would have the same significance and thus (A) and (B) would be equivalent. However, if class prediction were made by the majority voting using set (A), sample 3 and sample 7 would be classified into wrong classes. On the other hand, all samples would be classified into correct classes if class prediction were made using set (B). Therefore, informative genes should be determined by considering dependencies among multiple genes. Although use of pairwise dependencies is helpful (Tsunoda, 2000), there may be a difficult case: the correlation between gene 1 and gene 2 and the correlation between gene 1 and gene 3 are low whereas the correlation between gene 2 and gene 3 is high.

In this chapter, we treat this selection problem as an inference problem of *threshold functions* for Boolean variables. We treat class prediction as a problem of deciding whether or not a given sample belongs to the target class. Note that class prediction with multiple classes can be treated by making class prediction for each class independently. We do not use real values because it is difficult to give an appropriate mathematical definition using real values and it is widely recognized that gene expression data obtained by DNA microarrays contain large noises. Instead, each value is simplified to either 1 (high expression level) or 0 (low expression level), where the method of simplification is described in Section 2.

In particular, we consider an inference problem of r-*of-k threshold functions* (Littlestone, 1988; Littlestone, 1994). This function outputs 1 if at least r variables among k variables are 1. For the example of Fig. 1, we can make correct predictions using set (B) by letting $r = 3$ or $r = 2$, where $k = 4$ in this case. The inference problem is to select k genes so that the r-of-k function makes correct predictions for all samples when Boolean values of n genes (usually $n \gg k$) for m samples are given as an input (i.e., a training data set).

Since threshold functions are useful and important for classification, many studies have been done in the field of machine learning. Among them, the WINNOW algorithm (Littlestone, 1988) and its variants (Littlestone, 1994) are famous and widely used. For example, a variant of WINNOW was applied to protein-coding region prediction (Furukawa, 1996) and inference of genetic networks (Noda, 1998). We applied WINNOW to selection of informative genes. However, as shown in Section 4, the results were not satisfactory. Therefore, we developed a new algorithm. We compared the algorithm with WINNOW and a very simple algorithm, using gene expression data (Golub, 1999) obtained from human acute leukemia patients. The results show that the greedy algorithm is as good as the other two algorithms for the test data set and is much better for the training data set.

2. Selection of Informative Genes

As mentioned in the Introduction, we do not use real values. Instead, each value is rounded to either 1 (high expression level) or 0 (low expression level) by a simple comparison with a threshold. The threshold values are also determined by a very simple method. The following is the method.

Assume that expression data for n genes are available. Let $\{g_1, \ldots, g_n\}$ denote the set of genes. Let $\{s_1, \ldots, s_m\}$ denote the set of samples from patients. Assume that it is known whether each sample s_j belongs to the target cancer class. We let $class(s_j) = 1$ if s_j belongs to the class, otherwise we let $class(s_j) = 0$. Let M be the number of samples that belong to the target class (i.e., $M = \#\{s_j | class(s_j) = 1\}$). Let $e_{i,j}$ denote the observed expression level of gene g_i for sample s_j. For each gene i, $e_{i,1}, e_{i,2}, \ldots, e_{i,m}$ are sorted in the ascending order. Let $e_{i,p}$ and $e_{i,q}$ denote the M-th largest value and the $(M+1)$-th largest value among $e_{i,1}, e_{i,2}, \ldots, e_{i,m}$, respectively. We define the threshold value $\hat{e}_i$ for gene g_i by $\hat{e}_i = (e_{i,p} + e_{i,q})/2$. Then we round $e_{i,j}$ to a Boolean value $x_{i,j}$ by

$$x_{i,j} = \begin{cases} 1, & \text{if } e_{i,j} > \hat{e}_i, \\ 0, & \text{otherwise.} \end{cases}$$

Note that, in an ideal case (i.e., $e_{i,j} > e_{i,j'}$ hold for all j, j' such that $class(s_j) = 1$ and $class(s_{j'}) = 0$), the following property holds: $x_{i,j} = 1$ if and only if $class(s_j) = 1$. The symmetric case in which the expression levels for samples belonging to the target class are lower than those for the other samples can be treated in an analogous way.

Next, we formalize the selection problem using threshold functions for Boolean variables. In particular, we use *r-of-k threshold functions* (Littlestone, 1988; Littlestone, 1994). An r-of-k threshold function $f(z_1, \ldots, z_n)$ is defined by selecting a set of k significant variables. The value of f is 1 whenever at least r of these k variables are 1. If the k selected variables are $z_{i_1}, \ldots, z_{i_k}$, then f is 1 exactly when $z_{i_1} + \ldots + z_{i_k} \geq r$. For example, consider a case of $n = 5$, $k = 3$, $r = 2$ and $i_1 = 1, i_2 = 2, i_3 = 5$. Then, $f(1,1,1,1,1) = 1$, $f(0,0,0,0,0) = 0$, $f(1,0,1,1,1) = 1$, $f(1,0,1,1,0) = 0$, and $f(1,1,0,0,0) = 1$.

An r-of-k threshold function can be considered as a special case of *weighted voting functions* or *linearly separable functions*. These functions are usually defined by

$$f(z_1, \ldots, z_n) = 1 \text{ if and only if } \sum_{i=1}^{n} \mu_i z_i \geq 1,$$

where μ_i and z_i can take real values. If we let $\mu_i = 1/r$ for each of $z_{i_1}, \ldots, z_{i_k}$ and let $\mu_i = 0$ for the other variables, the linearly separable function becomes an r-of-k threshold function. Recall that the weighted voting was successfully applied to classification of cancer genes (Golub, 1999). Thus, it is reasonable to define the selection problem of informative genes using r-or-k functions when Boolean values are used.

We define the selection problem of informative genes as follows. Assume that expression data and k are given as an input. Then, the problem is to determine a set of k genes $\{g_{i_1}, \ldots, g_{i_k}\}$ which maximizes r under the condition that

$$x_{i_1,j} + \ldots + x_{i_k,j} \geq r, \qquad \text{if } class(s_j) = 1,$$
$$(1 - x_{i_1,j}) + \ldots + (1 - x_{i_k,j}) \geq r, \quad \text{otherwise .}$$

The latter case means that at least r variables must be 0 if the corresponding sample does not belong to the target class. It should be noted the above problem can be transformed into the inference problem of r-of-k functions by inverting $x_{i,j}$ for all i and for all j such that $class(s_j) = 0$.

It is expected that predictions can be done more robustly if r is larger because the prediction result for each s_j by the majority voting does not change even if values of $r - \lceil k/2 \rceil - 1$ input variables are changed.

3. Algorithms for the Selection Problem

In this section, we describe algorithms for the selection of informative genes. We review the WINNOW algorithm first and then we show a new algorithm.

Before reviewing the WINNOW algorithm, we define a very simple algorithm (denoted by SIMPLE) because it is also compared in Section 4. For each gene g_i, SIMPLE computes the number $err(g_i)$ defined by

$$err(g_i) \;=\; \#\{s_j | x_{i,j} \neq class(s_j)\}.$$

Then, SIMPLE sorts g_i in the increasing order of $err(g_i)$ and selects first k genes in the sorted list. That is, genes with k smallest $err(g_i)$ values are selected by SIMPLE.

3.1 The WINNOW Algorithm

The WINNOW algorithm is simple too. The algorithm maintains non-negative real-valued weights $w_1, \ldots, w_n$. At the beginning, each w_i is set to 1. For each sample from the training data set, it makes a prediction by the following rule:

$$pred(s_j) = 1, \quad \text{if } \sum_{i=1}^n w_i x_{i,j} > \theta,$$
$$pred(s_j) = 0, \quad \text{otherwise,}$$

where *this prediction is made only for learning parameters*. The weights are changed only if the prediction result is not correct (i.e., $pred(s_j) \neq class(s_j)$). The amount by which the weights are changed depends on a fixed parameter α, where $\alpha = 1 + \frac{1}{2r}$ and $\theta = n$ are suggested for inferring r-of-k threshold functions (Littlestone, 1988).

WINNOW algorithm {
    let $w_i \leftarrow 1$ for all $i = 1, \ldots, n$;
    for $j = 1$ to m
    {

        if $(class(s_j) = 0$ and $pred(s_j) = 1)$
            {let $w_i \leftarrow w_i/\alpha$ for all i such that $x_{i,j} = 1$};
        else if $(class(s_j) = 1$ and $pred(s_j) = 0)$
            {let $w_i \leftarrow \alpha w_i$ for all i such that $x_{i,j} = 1$}.

    }

}

WINNOW has a very nice property: the number of times that wrong predictions are made is bounded. It is proven (Littlestone, 1988) that the number is at most $8r^2 + 5k + 14kr \ln n$ when we let $\alpha = 1 + \frac{1}{2r}$ and $\theta = n$.

Since WINNOW does not select variables but outputs the weights of variables only, we must select k informative genes using the weights w_i. In order to select k informative genes, we employ a very simple method. We simply select k genes with k highest weights.

3.2   A Simple Greedy Algorithm

Although WINNOW has a nice mathematical property, the number of mistakes is not low for a small number of samples. It is difficult to get many samples because of costs (microarrays are expensive) and other reasons. Indeed, as shown in Section 4, the performance of WINNOW was not so good even for the training data set. Therefore, we developed an algorithm whose performance was good for a small number of samples. This algorithm is a kind of *greedy algorithm*, where greedy algorithms are used for various problems as heuristic algorithms. It should be noted that inference of an *r-of-k* function consistent with training data is known to be computationally hard (NP-hard) (Akutsu, 2000).

Therefore, development of heuristic algorithms is a reasonable choice. This new algorithm is denoted by GREEDY in this chapter.

GREEDY maintains non-negative real-valued weights $w_1, \ldots, w_m$. It should be noted that, in this case, the weights are not assigned to genes, but are assigned to samples. We say that gene g_i *covers sample* s_j if $class(s_j) = x_{i,j}$. First, genes which do not cover most samples are removed and are not considered as candidates for informative genes. Precisely, any gene g_i such that $\#\{s_j | class(s_j) \neq x_{i,j}\} > \theta_0$ is removed, where θ_0 is a threshold and we are currently using $\theta_0 = 7 \sim 9$. Next, GREEDY selects informative genes iteratively (i.e., one gene is selected per iteration). For h-th iteration, gene g_{i_h} which maximizes the score is selected. The score is defined by

$$\sum_{j=1}^{m} \left(\beta^{\,\delta(class(s_j), x_{i_h,j}) \times (h - w_j)} \right),$$

where $\delta(x, y) = 1$ if $x = y$, otherwise $\delta(x, y) = 0$. β is a constant defined based on the experience and we are currently using $\beta = 1.5$. Weight w_j is increased by 1 if the selected gene g_i contributes to the classification of sample s_j. Precisely, w_j is updated by $w_j \leftarrow w_j + 1$ if $x_{i_h,j} = class(s_j)$. Thus, w_j represents the number of genes (among $g_{i_1}, \dots, g_{i_h}$) which cover s_j. GREEDY tries to cover each sample as many times as possible. Although there was no mathematical proof, GREEDY showed good performance as mentioned in Section 4. The following is the description of the GREEDY algorithm.

> **GREEDY algorithm** {
>   remove all g_i such that $err(g_j) > \theta_0$;
>   let $w_i \leftarrow 0$ for all $i = 1, \dots, m$;
>   for $h = 1$ to k
>   {
>   select g_{i_h} which maximizes the score
> $$\sum_{j=1}^{m} \left(\beta^{\delta(class(s_j), x_{i_h,j}) \times (h - w_j)} \right),$$
>   where $g_{i_h} \notin \{g_{i_1}, \dots, g_{i_{h-1}}\}$;
>   let $w_j \leftarrow w_j + 1$ for all j such that $x_{i_h,j} = class(s_j)$.
>   }
> }

It is easy to see that GREEDY runs in $O(kmn)$ time. It is fast enough since the total size of the training data set is $O(mn)$ and k is not so large (usually less than 100).

3.3   Prediction by the Majority Voting

For each of the three algorithms mentioned above, predictions for samples are made by the *majority voting*. Let s_j be a sample for which class prediction is to be made. Let $g_{i_1}, g_{i_2}, \dots, g_{i_k}$ be the informative genes selected by either one of SIMPLE, WINNOW and GREEDY. We define V_{win} (resp. V_{lose}) to be the number of genes g_{i_h} for which $x_{i_h,j} = 1$ (resp. $x_{i_h,j} = 0$). We define *prediction strength* (PS) by $\frac{V_{win} - V_{lose}}{V_{win} + V_{lose}}$ (Golub, 1999). If PS is less than some threshold value Θ_{PS}, we do not make prediction (i.e., s_j is assigned as *uncertain*). If PS$\geq \Theta_{PS}$ and

$V_{win} > V_{lose}$, we predict that s_j belongs to the target class. Otherwise, we predict that s_j does not belong to the target class.

4.   Computational Results

In this section, we show the results of computational experiments on SIMPLE, WINNOW and GREEDY. For implementation and comparison, we used a PC with a 700 MHz AMD Athlon processor. In each case, the inference could be done within ten seconds.

We used the data set obtained by Golub *et al.* (Golub, 1999). Golub *et al.* chose *acute leukemias* as a test case. Acute leukemias are basically classified into two classes: those arising from lymphoid precursors (acute lymphoblastic leukemia, ALL) and those arising from myeloid precursors (acute myeloid leukemia, AML). They used two data sets: one for training and the other for test. The training data set (TR) consisted of 38 bone marrow samples (27 ALL, 11 AML) obtained from acute leukemia patients. The test data set (TS) consisted of 24 bone marrow and 10 peripheral blood samples. Note that the test data set included a much broader range of samples than the training data set because it included samples from peripheral blood, from childhood AML patients. For each sample, expression levels for 6817 genes were measured by microarrays produced by Affymetrix.

4.1   Comparison of Prediction Methods

Before comparing the selection methods, we compared prediction methods using the set of 50 informative genes shown in (Golub, 1999). Comparison of prediction methods should be done because the prediction accuracy may heavily depend on the prediction methods rather than the informative genes. We implemented the following methods:

Simple Majority Voting: As mentioned in Section **??**, V_{win} and V_{lose} are computed and compared.

Support Vector Machines (SVM):   Support vector machines (Cortes, 1995) are considered as one of the most powerful methods for estimating parameters for the weighted voting. In this method (using a simple kernel function), the weighted sum $a_0 + a_1 \cdot e_{1,j} + a_2 \cdot e_{2,j} + \cdots + a_{50} \cdot e_{50,j}$ is computed for each sample s_j, and s_j is assigned to the target class if the sum is positive. $a_0, a_1, \ldots, a_{50}$ are determined from the training data set TR by using quadratic programming.

We compared the prediction results for the test data set TS with those by Golub *et al.*. In order to apply SVM to the prediction, input

Table 6.1. Comparison of Prediction Methods

	Correct	Incorrect	Uncertain
Golub *et al.* ($\Theta_{PS} = 0.30$)	29	0	5
Golub *et al.* ($\Theta_{PS} = 0.0$)	32	2	0
Majority Voting ($\Theta_{PS} = 0.30$)	25	0	9
Majority Voting ($\Theta_{PS} = 0.0$)	31	3	0
SVM	32	2	0

expression levels were normalized. We counted the number of samples for which correct predictions were made, the number of samples for which incorrect predictions were made, and the number of samples for which "uncertain" was assigned (i.e., PS was less than Θ_{PS}). Since we did not know how to define PS for SVM, we did not consider PS for SVM. Table 6.1 shows the results.

From the table, it is seen that SVM and the voting method by Golub *et al.* were better than the simple majority voting. However, the differences were not so large. Even the simple majority voting method made three wrong predictions for the case of $\Theta_{PS} = 0.0$. It should be noted that the definition of PS for the majority voting is different from that by Golub *et al.* and thus the comparison result for the cases of $\Theta_{PS} = 0.30$ does not have a significant meaning.

It is also seen that SVM is as good as the method by Golub *et al.* This suggests that the prediction method by Golub *et al.* is not necessarily the best although further computational experiments must be done for detailed comparison of these two methods.

It should also be noted that all algorithms (in cases of $\Theta_{PS} = 0.0$) made wrong predictions for the same sample (sample No. 66). All algorithms assigned this sample as ALL (with low PS) although this sample belonged to AML. Thus, this sample may be a very special case of AML.

4.2 Comparison of Selection Methods

In this subsection, we show the results of comparison on the selection methods. We compared three algorithms: SIMPLE, WINNOW and GREEDY. Since the majority voting with Boolean variables was not so bad, we used the majority voting as a prediction method.

As for data sets, we used the following three pairs.

(A) The training data set (TR) and the test data set (TS) by Golub *et al* were used. TR consisted of 38 samples, among which 27 samples (resp. 11 samples) were classified into ALL (resp. AML). TS consisted of 34 samples, among which 20 samples (resp. 14 samples) were classified into ALL (resp. AML).

(B) TR was used as the test data set and TS was used as the training data set.

(C) It is known that ALL samples are further classified into T-cell ALL and B-cell ALL. Golub *et al.* made successful prediction on the distinction between T-cell ALL and B-cell ALL. 27 ALL samples in TR were used as the training data set, among which 19 samples (resp. 8 samples) were classified into B-cell ALL (resp. T-cell ALL). 20 ALL samples in TS were used as the test data set, among which 19 samples (resp. 1 sample) were classified into B-cell ALL (resp. T-cell ALL).

Since the rounded Boolean values were biased when the number of the training samples belonging to the target class was different from the number of the other training samples, we made the numbers to be equal by duplicating samples not belonging to the target class. Thus, 27+27 training samples were used in case (A), 20+20 training samples were used in case (B), and 19+19 training samples were used in case (C).

For each of SIMPLE, WINNOW and GREEDY, we examined four cases: $k = 20$, $k = 30$, $k = 40$ and $k = 50$. Recall that k is the number of informative genes to be selected. We used $r = 0.8 \cdot k$ as a parameter in WINNOW, where we examined several values for r and obtained similar results for the other r. It should be note that WINNOW does not necessarily select k genes satisfying the condition of the r-of-k threshold function.

First we measured the qualities of the sets of informative genes by means of r. In Table 6.2 the maximum r computed from each set of informative genes is shown. Recall that we defined the selection problem as a maximization problem on r. From this table, it is seen that GREEDY is much better than SIMPLE and WINNOW. It is also seen that there is no significant difference between SIMPLE and WINNOW. Therefore, it is confirmed that GREEDY is the best among three algorithms under the definition of Section 2. However, it is unclear whether or not the selected informative genes are useful for predictions. So, we made computational experiments on predictions.

Using the informative genes computed by each algorithm, we made predictions on both the samples in the training data set and the samples

Table 6.2. Qualities of Sets of Informative Genes Measured by r

		$k = 20$	$k = 30$	$k = 40$	$k = 50$
	SIMPLE	15	19	24	29
(A)	WINNOW	14	20	24	29
	GREEDY	16	24	33	40
	SIMPLE	13	17	22	29
(B)	WINNOW	12	18	22	27
	GREEDY	17	25	34	43
	SIMPLE	10	15	18	21
(C)	WINNOW	9	15	20	27
	GREEDY	15	23	31	38

in the test data set, by means of the majority voting. As in (Golub et al.), the samples with PS(< 0.30) were assigned as uncertain (i.e., $\Theta_{PS} = 0.30$).

Results of the predictions are shown in Table 6.3. In this table, the number of samples that were assigned as uncertain is shown for each case. The number of samples which were classified into the wrong class is also shown (after the symbol '+'). For example, 5+1 means that 5 samples were assigned as uncertain and 1 sample was classified into the wrong class.

From this table, it is seen that GREEDY always made correct predictions for samples in the training data set. Therefore, it is confirmed again that GREEDY is the best for the training data set. However, for the test data set, there was no significant difference among SIMPLE, WINNOW and GREEDY. Therefore, we can not discuss about the performances of the algorithms for the test data set.

5. Discussions

We have studied the selection problem of informative genes for molecular classification of cancer. We defined this problem by using threshold functions on Boolean variables. We also proposed a new algorithm (GREEDY) for the selection problem. The effectiveness of GREEDY for the training data set was confirmed using real expression data from leukemia patients. Therefore, GREEDY might be useful when the training data set covers a wide range of samples.

For the test data set, there was no significant difference among SIMPLE, WINNOW and GREEDY. It is surprising that SIMPLE was not

Table 6.3. Comparison of Selection Algorithms

		TRAINING				TEST			
		$k = 20$	30	40	50	$k = 20$	30	40	50
	SIMPLE	0	1	1	1	6+1	7	5+1	7
(A)	WINNOW	0	0	1	1	5+1	7	6+1	7
	GREEDY	0	0	0	0	7+1	9	6+1	7
	SIMPLE	2	2	2	2	5	6	7	9
(B)	WINNOW	2	2	2	2	5	7	7	11
	GREEDY	0	0	0	0	5	7	7	8
	SIMPLE	0	1	1	1	1	1	1	1
(C)	WINNOW	1	1	1	1	1	1	1	1
	GREEDY	0	0	0	0	1	1	1	1

worse than WINNOW and GREEDY for the test data set because correlations between genes were not taken into account by SIMPLE. Of course, if errors (items such that $x_{i,j} \neq class(s_j)$) are distributed uniformly, SIMPLE will work well because all genes will be correlated similarly. This can be listed as a reason why there was no significant difference. Another possible reason is that the test data set included a broader range of samples than the training data set (Golub, 1999) and thus the informative genes selected from the training data set were not so useful for the test data set. Of course, we examined the case (case (B)) in which the training data set and the test data set were exchanged. But, in this case, the number of samples in the training data set might not be enough because this data set included a broad range of samples. Overfitting can also be listed as a possible reason. Since we do not have other appropriate data sets now, further comparison using additional data sets is left as future work.

In this chapter, we defined the selection problem using Boolean variables. However, the definition and algorithms using real-valued variables should also be studied because the prediction results for the test data set using real-valued variables were slightly better than those using Boolean variables.

Acknowledgments

This work was partially supported by the Grant-in-Aid for Scientific Research on Priority Areas, "Genome Science", "Genome Information Science", and Grant-in-Aid for Scientific Research (B)(1)(No. 12480080) of the Ministry of Education, Culture, Sports, Science and Technology (MEXT), Japan.

References

Akutsu, T., Miyano, S. and Kuhara, S. (2000) Algorithms for identifying Boolean networks and related biological networks based on matrix multiplication and fingerprint function. *Journal of Computational Biology*, 7, 331-343.

DeRisi, J. L., Lyer, V. R. and Brown, P. O. (1997) Exploring the metabolic and genetic control of gene expression on a genomic scale. *Science*, 278, 680–686.

Cole, K. A., Krizman, D. B. and Emmert-Buck, M. R. (1999) The genetics of cancer - a 3D model. *Nature Genetics*, 21 (Supplement), 38-41.

Cortes, C. and Vapnik, V. (1995) Support-vector networks. *Machine Learning*, 20, 273-297.

Furukawa, N., Matsumoto, S., Shinohara, A., Shoudai, T. and Miyano, S. (1996) HAKKE: A Multi-Strategy Prediction System for Sequences. Genome Informatics 7, 98-107.

Golub, T. R., Slonim, D. K., Tamayo, P., Huard, C., Gaasenbeek, M., Mesirov, J. P., Coller, H., Loh, M. L., Downing, J. R., Caligiuri, M. A., Bloomfield, C. D. and Lander, E. S. (1999) Molecular classification of cancer: class discovery and class prediction by gene expression monitoring. *Science*, 286, 531–537.

Kononen, J., Bubendorf, L., Kallioniemi, A., Barlund, M., Schrami, P., Leighton, S., Torhorst, J., Mihatsch, M. J., Sauter, G. and Kallioniemi, O-P. (1998) Tissue microarrays for high-throughput molecular profiling of tumor specimens. *Nature Medicine*, 4, 844-847.

Littlestone, N. (1988) Learning quickly when irrelevant attributes abound: a new linear-threshold algorithm. *Machine Learning*, 2, 285-318.

Littlestone, N. and Warmuth, M. K. (1994) The weighted majority algorithm. *Information and Computation*, 108, 212-261.

Noda, K., Shinohara, A., Takeda, M., Matsumoto, S., Miyano, S. and Kuhara, S. (1998) Finding Genetic Network from Experiments by Weighted Network Model. *Genome Informatics 9*, 141-150.

Tsunoda, T., Hojo, Y., Kihara, C., Shiraishi, N., Kitahara, O., Ono, K., Tanaka, T., Takagi, T. and Nakamura, Y. (2000) Diagnosis System of Drug Sensitivity of Cancer Using cDNA Microarray and Multivariate Statistical Analysis. *Currents in Computational Molecular Biology* (Universal Academy Press), 16-17.

Chapter 7

DESIGN ISSUES AND COMPARISON OF METHODS FOR MICROARRAY-BASED CLASSIFICATION

Edward R. Dougherty and Sanju N. Attoor
Department of Electrical Engineering
Texas A&M University, College Station, TX, USA

1.   Introduction

A key goal for microarray-based analysis of gene expressions is to perform classification via different expression patterns (for instance, cancer classification) (Golub *et al.*, 1999; Ben-Dor *et al.*, 2000; Bittner *et al.*, 2000; Hedenfalk *et al.*, 2001; Khan *et al.*, 2002; Kim *et al.*, 2002). This requires designing a classifier that takes a vector of gene expression levels as input, and outputs a class label. Classification can be between different kinds of cancer, different stages of tumor development, or a host of such differences. Classifiers are designed from a sample of expression vectors. This requires assessing expression levels from RNA obtained from the different tissues with microarrays, determining genes whose expression levels can be used as classifier variables, and then applying some rule to design the classifier from the sample microarray data. Three critical issues arise: (1) Given a set of variables, how does one design a classifier from the sample data that provides good classification over the general population? (2) How does one estimate the error of a designed classifier when data are limited? (3) Given a large set of potential variables, such as the large number of expression level determinations provided by microarrays, how does one select a set of variables as the input vector to the classifier? The problem of error estimation impacts variable selection in a devilish way. An error estimator may be unbiased but have a large variance, and therefore often be low. This can produce a large number of variable sets and classifiers with low error estimates. For the near future, small samples will remain a critical issue for microarray-based classification (Dougherty, 2001).

2.   Classification Rules

Classification involves a *classifier* ψ, a *feature vector* $\mathbf{X} = (X_1, X_2, ..., X_d)$ composed of random variables, and a binary random variable Y to be predicted by $\psi(\mathbf{X})$. The values, 0 or 1, of Y are treated as class labels. The error, $\varepsilon[\psi]$, of ψ is the probability, $P(\psi(\mathbf{X}) \neq Y)$, that the classification is erroneous. It equals the expected (mean) absolute difference, $E[|Y - \psi(\mathbf{X})|]$, between the label and the classification. $X_1, X_2, ..., X_d$ can be discrete or real-valued. In the latter case, the domain of ψ is d-dimensional Euclidean space $\Re^d$. An optimal classifier, $\psi_\bullet$, is one having minimal error, $\varepsilon_\bullet$, among all binary functions on $\Re^d$. $\psi_\bullet$ and $\varepsilon_\bullet$ are called the *Bayes classifier* and *Bayes error*, respectively. Classification accuracy, and thus the error, depends on the probability distribution of the feature-label pair $(\mathbf{X}, Y)$ – how well the labels are distributed among the variables being used to discriminate them, and how the variables are distributed in $\Re^d$.

The Bayes classifier is defined in a natural way: for any specific vector $\mathbf{x}$, $\psi_\bullet(\mathbf{x}) = 1$ if the expected value of Y given $\mathbf{x}$, $E[Y|\mathbf{x}]$, exceeds $1/2$, and $\psi_\bullet(\mathbf{x}) = 0$ otherwise. Formulated in terms of probabilities, $\psi_\bullet(\mathbf{x}) = 1$ if the conditional probability of $Y = 1$ given $\mathbf{x}$ exceeds the conditional probability of $Y = 0$ given $\mathbf{x}$, and $\psi_\bullet(\mathbf{x}) = 0$ otherwise; that is, $\psi_\bullet(\mathbf{x}) = 1$ if and only if $P(Y = 1|\mathbf{x}) > P(Y = 0|\mathbf{x})$. The label 1 is predicted upon observation of $\mathbf{x}$ if the probability that $\mathbf{x}$ lies in class 1 exceeds the probability that $\mathbf{x}$ lies in class 0. Since the sum of the probabilities is 1, $P(Y = 1|\mathbf{x}) > P(Y = 0|\mathbf{x})$ if and only if $P(Y = 1|\mathbf{x}) > 1/2$. The problem is that we do not know these conditional probabilities, and therefore must design a classifier from sample data.

Supervised classifier design uses a *sample* $S_n = \{(\mathbf{X}^1, Y^1), (\mathbf{X}^2, Y^2), ..., (\mathbf{X}^n, Y^n)\}$ of feature-label pairs and a *classification rule* to construct a classifier ψ_n whose error is hopefully close to the Bayes error. The Bayes error $\varepsilon_\bullet$ is estimated by the error ε_n of ψ_n. Because $\varepsilon_\bullet$ is minimal, $\varepsilon_n \geq \varepsilon_\bullet$, and there is a *design cost* (estimation error), $\Delta_n = \varepsilon_n - \varepsilon_\bullet$. Since it depends on the sample, ε_n is a random variable, as is Δ_n. We are concerned with the expected value of Δ_n, $E[\Delta_n] = E[\varepsilon_n] - \varepsilon_\bullet$. Hopefully, $E[\Delta_n]$ gets closer to 0 as the sample size grows. This will depend on the classification rule and the distribution of the feature-label pair $(\mathbf{X}, Y)$.

A classification rule is said to be *consistent* for the distribution of $(\mathbf{X}, Y)$ if $E[\Delta_n] \to 0$ as $n \to \infty$, where the expectation is relative to the distribution of the sample. The expected design error goes to zero as the sample size goes to infinity. This is equivalent to $P(\Delta_n > \tau) \to 0$ as $n \to \infty$ for any $\tau > 0$, which says that the probability of the design error exceeding τ goes to 0. As stated, consistency depends upon the relation between the classification rule and the joint feature-label distribution. If $E[\Delta_n] \to 0$ for any distribution, then the classification rule is said to be *universally consistent*. Since we often lack an estimate of the distribution, universal consistency is desirable.

3. Some Specific Classification Rules

Since the Bayes classifier is defined by $\psi_\bullet(\mathbf{x}) = 1$ if and only if $P(Y = 1|\mathbf{x}) > 1/2$, an obvious way to proceed is to obtain an estimate $P_n(Y = 1|\mathbf{x})$ of $P(Y = 1|\mathbf{x})$ from the sample S_n. The *plug-in rule* designs a classifier by $\psi_n(\mathbf{x}) = 1$ if and only if $P_n(Y = 1|\mathbf{x}) > 1/2$. If the data is discrete, then there is a finite number of vectors and $P_n(Y = 1|\mathbf{x})$ can be defined to be the number of times the pair $(\mathbf{x}, 1)$ is observed in the sample divided by the number of times $\mathbf{x}$ is observed. Unfortunately, if $\mathbf{x}$ is observed very few times, then $P_n(Y = 1|\mathbf{x})$ is not a good estimate. Even worse, if $\mathbf{x}$ is never observed, then $\psi_n(\mathbf{x})$ must be defined by some convention. The rule is consistent, but depending on the number of variables, may require a large sample to have $E[\Delta_n]$ close to 0, or equivalently, $E[\varepsilon_n]$ close to the Bayes error. Consistency alone is of little consequence for small samples.

For continuous data, $P(y|\mathbf{x})$ is a conditional probability density estimated by $P_n(y|\mathbf{x})$. Many classification rules partition $\Re^d$ into a disjoint union of cells. $P_n(Y = 1|\mathbf{x})$ is the number of 1-labeled sample points in the cell containing $\mathbf{x}$ divided by the total number of points in the cell. A *histogram rule* is defined by the plug-in rule: $\psi_n(\mathbf{x})$ is 0 or 1 according to which is the majority label in the cell. If $\mathbf{x}$ is a point in cell C, then $\psi_n(\mathbf{x}) = 1$ if there are more 1-labeled sample points in C and $\psi_n(\mathbf{x}) = 0$ if there are more 0-labeled sample points in C. The cells may change with n and may depend on the sample points. They do not depend on the labels. To obtain consistency for a distribution, two conditions are sufficient when stated with the appropriate mathematical rigor: (1) the partition should be fine enough to take into account local structure of the distribution, and (2) there should be enough labels in each cell so that the majority decision reflects the decision based on the true conditional probabilities.

The *cubic-histogram rule* partitions $\Re^d$ into same-size cubes. On each cube, the designed classifier is defined to be 0 or 1, according to which is the majority among the labels of the points in the cube. If the cube edge length approaches 0 and n times the common volume approaches infinity as $n \to \infty$, then rule is universally consistent (Devroye *et al.*, 1996; Gordon and Olshen, 1978). For discrete data, the cubic histogram rule reduces to the plug-in rule if the cubes are sufficiently small.

The cubic-histogram rule partitions the space without reference to the actual data. One can instead partition the space based on the data, either with our without reference to the labels. Tree classifiers are a common way of performing data-dependent partitioning. Since any tree can be transformed into a binary tree, we need only consider binary classification trees. A tree is constructed recursively based on some criteria. If S represents the set of all data, then it is partitioned according to some rule into $S = S_1 \cup S_2$. There are then four possibilities:

- S_1 is partitioned into $S_1 = S_{11} \cup S_{12}$ and S_2 is partitioned into $S_2 = S_{21} \cup S_{22}$.

- S_1 is partitioned into $S_1 = S_{11} \cup S_{12}$ and partitioning of S_2 is terminated.

- S_2 is partitioned into $S_2 = S_{21} \cup S_{22}$ and partitioning of S_1 is terminated.

- Partitioning of both S_1 and S_2 is terminated.

In the last case, the partition is complete; in any of the others, it proceeds recursively until all branches end in termination, at which point the leaves on the tree represent the partition of the space. On each cell (subset) in the final partition, the designed classifier is defined to be 0 or 1, according to which is the majority among the labels of the points in the cell.

As an example of a classification tree, we consider the median tree. Based on one of the coordinates, say the first, split the observation vectors into two groups based on the median value of the selected coordinate in the sample. If the first coordinate is used, then the x_1-axis is split at the median, $\tilde{x}_1$, thereby partitioning the data into $S = S_1 \cup S_2$: for $\mathbf{x} = (x_1, x_2, \dots, x_d) \in S$, $\mathbf{x} \in S_1$ if $x_1 \le \tilde{x}_1$, and $\mathbf{x} \in S_2$ if $x_1 > \tilde{x}_1$. The second level of the tree is constructed by partitioning S_1 and S_2 according to the median values of the second coordinate among those points in S_1 and S_2, respectively, where the point selected for the first level is not considered. The partitioning is continued to some chosen level k. If the level goes beyond the dimension d of the space, then one recycles through coordinates again. Note that the median tree does not depend on the labels of the sample data. Median-tree classifiers are consistent if $k \to \infty$ in such a way that $n/k2^k \to \infty$, so long as $\mathbf{X}$ possesses a

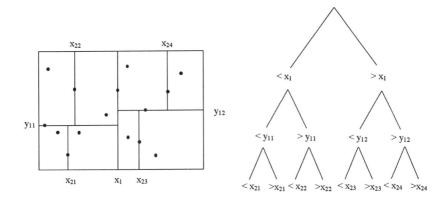

Figure 7.1. Median tree: (a) data partition; (b) tree.

density (Devroye *et al.*, 1996). Figure 7.1 illustrates a median-based tree classifier, both the data partitioning and the binary tree.

The *nearest-neighbor* (NN) *rule* forms a partition by defining $\psi_n(\mathbf{x})$ as the label of the sample point closest to $\mathbf{x}$. The result is a data-dependent partition having a cell for each sample point. That cell consists of all points in the space closest to the sample point. The partition is known as a *Voronoi partition* and the cells are called *Voronoi cells*. The nearest-neighbor rule is simple, but not consistent. An extension of this rule is the *k-nearest-neighbor* (*k*NN) *rule* (see also Chapter 14). For k odd, the k points closest to $\mathbf{x}$ are selected and $\psi_n(\mathbf{x})$ is defined to be 0 or 1 according to which is the majority among the labels of the chosen points. A slight adjustment is made if k is even. The kNN is universally consistent if $k \to \infty$ in such a way that $k/n \to 0$ as $n \to \infty$ (Stone, 1977).

Instead of taking the majority label among a pre-decided number of nearest neighbors as with the NN rule, the *moving-window rule* pre-sets a distance and takes the majority label among all sample points within that distance of $\mathbf{x}$.

The moving-window rule can be "smoothed" by giving more weight to sample points closer to $\mathbf{x}$. A histogram rule based on the majority between 0 and 1 labels is equivalent to the median of the labels, where in case of a tie a 0 is given. More generally, a weighted median of binary values is computed by adding up the weights associated with the 0- and 1-labeled points separately, and defining the output to be the larger sum. A kernel rule is constructed by defining a weighting kernel based on the distance of a sample point from $\mathbf{x}$, in conjunction with a smoothing (scaling) factor. The *Gaussian*, and *Epanechnikov* kernels are given by

$$K_h(\mathbf{x},\mathbf{x}_k) = \exp\left[-\left\|\frac{\mathbf{x}-\mathbf{x}_k}{h}\right\|^2\right] \tag{7.1}$$

$$K_h(\mathbf{x},\mathbf{x}_k) = \begin{cases} \left(1-\|(\mathbf{x}-\mathbf{x}_k)/h\|^2\right), & \text{if } \|(\mathbf{x}-\mathbf{x}_k)\| \le h \\ 0, & \text{if } \|(\mathbf{x}-\mathbf{x}_k)\| > h \end{cases} \tag{7.2}$$

respectively, where $\|\cdot\|$ denotes Euclidean distance, $\mathbf{x}$ is the point at which the classifier is being defined, and $\mathbf{x}_k$ is a sample point. Since the Gaussian kernel is never 0, all sample points get some weight. The Epanechnikov kernel is 0 for sample points more than h from $\mathbf{x}$, so that, like the moving-window rule, only sample points within a certain radius contribute to the definition of $\psi_n(\mathbf{x})$. The moving-window rule is a special case of a kernel rule with the weights being 1 within a specified radius. The three kernel rules mentioned are universally consistent if $h \to \infty$ in such a way that $nh^d \to \infty$ as $n \to \infty$ (Devroye and Kryzak, 1989).

4.   Constrained Classifiers

To reduce design error, one can restrict the functions from which an optimal classifier must be chosen to a class C. This leads to trying to find an optimal *constrained* classifier, $\psi_C \in C$, having error ε_C. Constraining the classifier can reduce the expected design error, but at the cost of increasing the error of the best possible classifier. Since optimization in C is over a subclass of classifiers, the error, ε_C, of ψ_C will typically exceed the Bayes error, unless the Bayes classifier happens to be in C. This *cost of constraint (approximation)* is $\Delta_C = \varepsilon_C - \varepsilon_\bullet$. A classification rule yields a classifier $\psi_{n,C} \in C$ with error $\varepsilon_{n,C}$, and $\varepsilon_{n,C} \ge \varepsilon_C \ge \varepsilon_\bullet$. Design error for constrained classification is $\Delta_{n,C} = \varepsilon_{n,C} - \varepsilon_C$. For small samples, this can be substantially less than Δ_n, depending on C and the rule. The error of the designed constrained classifier is decomposed as $\varepsilon_{n,C} = \varepsilon_\bullet + \Delta_C + \Delta_{n,C}$. The expected error of the designed classifier from C can be decomposed as

$$E[\varepsilon_{n,C}] = \varepsilon_\bullet + \Delta_C + E[\Delta_{n,C}] \tag{7.3}$$

The constraint is beneficial if and only if $E[\varepsilon_{n,C}] < E[\varepsilon_n]$, which means $\Delta_C < E[\Delta_n] - E[\Delta_{n,C}]$. If the cost of constraint is less than the decrease in expected design cost, then the expected error of $\psi_{n,C}$ is less than that

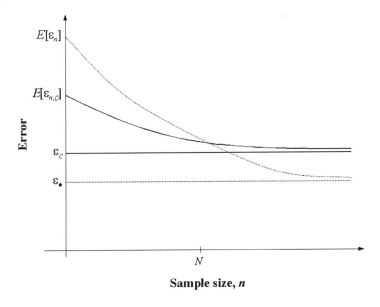

Figure 7.2. Relation between sample size and constraint.

of ψ_n. The dilemma: strong constraint reduces $E[\Delta_{n,C}]$ at the cost of increasing ε_C.

The matter can be graphically illustrated. For the discrete-data plug-in rule and the cubic histogram rule with fixed cube size, $E[\Delta_n]$ is non-increasing, meaning that $E[\Delta_{n+1}] \leq E[\Delta_n]$. This means that the expected design error never increases as sample sizes increase, and it holds for any feature-label distribution. Such classification rules are called *smart*. They fit our intuition about increasing sample sizes. The nearest-neighbor rule is not smart because there exist distributions for which $E[\Delta_{n+1}] \leq E[\Delta_n]$ does not hold for all n. Now consider a consistent rule, constraint, and distribution for which $E[\Delta_{n+1}] \leq E[\Delta_n]$ and $E[\Delta_{n+1,C}] \leq E[\Delta_{n,C}]$. Figure 7.2 illustrates the design problem. The axes correspond to sample size and error. The horizontal solid and dashed lines represent $\varepsilon_\bullet$ and ε_C, respectively; the decreasing solid and dashed lines represent $E[\varepsilon_n]$ and $E[\varepsilon_{n,C}]$, respectively. If n is sufficiently large, then $E[\varepsilon_n] < E[\varepsilon_{n,C}]$; however, if n is sufficiently small, then $E[\varepsilon_n] > E[\varepsilon_{n,C}]$. The point N at which the decreasing lines cross is the cut-off: for $n > N$, the constraint is detrimental; for $n < N$, it is beneficial. When $n < N$, the advantage of the constraint is the difference between the decreasing solid and dashed lines.

5.   Perceptrons and Neural Networks

A classical way of constructing classifiers is to use parametric representation. The classifier is postulated to have a functional form $\psi(x_1, x_2, \ldots, x_d; a_0, a_1, \ldots, a_r)$, where the parameters $a_0, a_1, \ldots, a_r$ are to be determined by some estimation procedure based on the sample data. For parametric representation, we assume the labels to be -1 and 1. The most basic functional form involves a linear combination of the coordinates of the observations. A binary function is obtained by thresholding. A *perceptron* has the form

$$\psi(x) = \mathrm{T}\left[a_0 + \sum_{i=1}^{d} a_i x_i\right] \tag{7.4}$$

where $\mathbf{x} = (x_1, x_2, \ldots, x_d)$ and T thresholds at 0 and yields -1 or 1. A perceptron divides the space into two half-spaces determined by the hyperplane defined by the parameters $a_0, a_1, \ldots, a_d$. The hyperplane is determined by the equation formed from setting the linear combination equal to 0. Using the dot product, $\mathbf{a} \cdot \mathbf{x}$, which is equal to the sum in the preceding equation absent the constant term a_0, the hyperplane is defined by $\mathbf{a} \cdot \mathbf{x} = -a_0$. An optimal perceptron minimizes the error $\varepsilon[\psi] = P(\psi(\mathbf{X}) \neq Y)$. A specific design procedure finds parameters that hopefully define a perceptron whose error is close to the optimal perceptron. Often, analysis of the design procedure depends on whether the sample data is separable by a hyperplane. The sample is *linearly separable* if there exists a hyperplane such that points with label -1 lie on one side of the hyperplane and the points with label 1 lie on the other side.

The classical *Rosenblatt perceptron algorithm* uses an iterative procedure to find a separating hyperplane in the case of a linearly separable sample (Rosenblatt, 1962). Let $\{(\mathbf{x}^1, y^1), (\mathbf{x}^2, y^2), \ldots, (\mathbf{x}^n, y^n)\}$ be the sample-data set, $\mathbf{a}$ be a parameter vector defining the perceptron, and $\{(\mathbf{x}_t, y_t)\}$ be a sequence of observation-label pairs formed by cycling repeatedly through the sample data. Proceed iteratively by forming a sequence $\mathbf{a}(0), \mathbf{a}(1), \mathbf{a}(2), \ldots$. Let $\mathbf{a}(0) = (0, 0, \ldots, 0)$. Given $\mathbf{a}(t)$, define $\mathbf{a}(t+1)$ by

$$\mathbf{a}(t+1) = \begin{cases} \mathbf{a}(t), & \text{if } \psi_t(\mathbf{x}_{t+1}) = y_{t+1} \\ \mathbf{a}(t) + y_{t+1}\mathbf{x}_{t+1}, & \text{if } \psi_t(\mathbf{x}_{t+1}) \neq y_{t+1} \end{cases} \tag{7.5}$$

where ψ_t is the perceptron formed using the parameter vector $\mathbf{a}(t)$. The parameter vector changes if and only if the label for time $t+1$ is not correctly given by applying the perceptron for time t to the observation vector for time $t+1$. In finite time, the procedure produces a hyperplane that correctly separates the data. The time can be bounded in terms of the maximum norm of the observation vectors and the degree to which the -1- and 1-labeled vectors are separated. For data sets that are not linearly separable, another method has to be employed to estimate the optimal perceptron. Gradient-based methods lead to consistent design under very general conditions.

The Wiener filter provides non-iterative computationally efficient perceptron design. It has been incorporated into a massively parallel method for finding promising very small gene sets for microarray-based classification that has been applied to breast cancer (Kim *et al.*, 2002). Given the joint feature-label distribution, the optimal mean-square-error linear estimator of Y based on $\mathbf{X}$ is determined by a weight vector $\mathbf{a} = (a_0, a_1, \ldots, a_d)^t$, '$t$' denoting transpose. The autocorrelation matrix for $\mathbf{X}$ and the cross-correlation vector for $\mathbf{X}$ and Y are given by

$$\mathbf{R}_X = \begin{pmatrix} E[X_1X_1] & E[X_1X_2] & \cdots & E[X_1X_d] \\ E[X_2X_1] & E[X_2X_2] & \cdots & E[X_2X_d] \\ \vdots & \vdots & \ddots & \vdots \\ E[X_dX_1] & E[X_dX_2] & \cdots & E[X_dX_d] \end{pmatrix} \tag{7.6}$$

$$E[\mathbf{X}Y] = (E[X_1Y], E[X_2Y], ..., E[X_dY])^t \tag{7.7}$$

respectively. If $\mathbf{R}_X$ is nonsingular, then the optimal weight vector is given by $\mathbf{a} = \mathbf{R}_X^{-1} E[\mathbf{X}Y]$. If $\mathbf{R}_X$ is singular, then $\mathbf{R}_X^{-1}$ is replaced by the pseudoinverse of $\mathbf{R}_X$. An approximation of the optimal perceptron is given by $\mathrm{T}[\mathbf{a}^t \mathbf{X}_0]$, where $\mathbf{X}_0 = (1, \mathbf{X}^t)^t$. The sample-based classification rule for the weight vector is determined by estimating $\mathbf{R}_X$ and $E[\mathbf{X}Y]$ from the sample data.

The *support vector machine* (*SVM*) provides another method for designing perceptrons (Vapnik *et al.*, 1997; Vapnik, 1998). It has been used with microarray data to identify gene sets possessing common functionality (Brown *et al.*, 2000). Figure 7.3 shows a linearly separable data set and three hyperplanes (lines). The outer lines pass through points in the sample data, and the third, called the *maximal-margin hyperplane* (*MMH*) is equidistant between the outer lines. It has the property that distance from it to the nearest -1-labeled vector is equal to the distance from it to the nearest 1-labeled vector. The vectors closest to it are called *support vectors*. The distance from the MMH to any support vector is called the *margin*. The matter is formalized by recognizing that differently labeled sets are separable by the hyperplane $\mathbf{u} \cdot \mathbf{x} = c$,

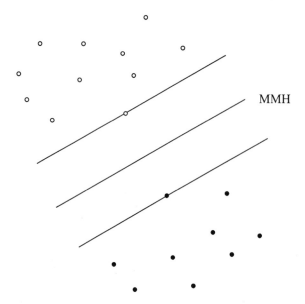

Figure 7.3. Support-vector hyperplane.

where $\mathbf{u}$ is a unit vector and c is a constant, if $\mathbf{u} \cdot \mathbf{x}^k > c$ for $y_k = 1$ and $\mathbf{u} \cdot \mathbf{x}^k < c$ for $y_k = -1$. For any unit vector $\mathbf{u}$, define

$$c_1(\mathbf{u}) = \min_{\{\mathbf{x}^k : y^k = 1\}} \mathbf{u} \cdot \mathbf{x}^k \qquad (7.7)$$

$$c_0(\mathbf{u}) = \min_{\{\mathbf{x}^k : y^k = -1\}} \mathbf{u} \cdot \mathbf{x}^k$$

Let $\mathbf{u}_0$ be the unit vector that maximizes $\rho(\mathbf{u}) = 1/2[c_1(\mathbf{u}) - c_0(\mathbf{u})]$ over all unit vectors $\mathbf{u}$. The MMH is given by the vector $\mathbf{u}_0$ and the constant $c_0 = \rho(\mathbf{u}_0)$. This hyperplane is unique.

A method is needed to find the MMH. It can be found by solving the following quadratic optimization problem: among the set of all vectors $\mathbf{v}$ for which there exists a constant b such that

$$\mathbf{v} \cdot \mathbf{x}^k + b \geq 1, \text{ if } y_k = 1 \qquad (7.8)$$
$$\mathbf{v} \cdot \mathbf{x}^k + b \leq -1, \text{ if } y_k = -1$$

find the vector $\mathbf{v}_0$ of minimum norm, $\|\mathbf{v}_0\|$. Then the vector defining the MMH and the margin are given by $\mathbf{u}_0 = \mathbf{v}_0 / \|\mathbf{v}_0\|$ and $\rho(\mathbf{u}_0) = \|\mathbf{v}_0\|^{-1}$, respectively.

The SVM methodology can be extended to the situation in which the sample is not linearly separable. We leave a detailed explanation to the literature. It can also be extended to other functions besides perceptrons. We will also not pursue that issue here. Instead, we turn to the complex parametric classifiers defined by neural networks.

A (*feed-forward*) *two-layer neural network* has the form

$$\psi(x) = \mathrm{T}\left[c_0 + \sum_{i=1}^{k} c_i \sigma[\psi_i(\mathbf{x})]\right] \tag{7.10}$$

where σ is a sigmoid function, and

$$\psi_i(\mathbf{x}) = \sum_{j=0}^{d} a_{ij} x_j \tag{7.11}$$

where x_0 is the constant 1. A sigmoid function is nondecreasing with limits -1 and $+1$ at $-\infty$ and ∞, respectively. Each operator in the sum of Eq. 7.11 is called a *neuron*. These form the *hidden layer*. We consider neural networks with the threshold sigmoid, $\sigma(x) = -1$ if $x \leq 0$ and $\sigma(x) = 1$ if $x > 0$. Other sigmoid functions include the logistic sigmoid, $\sigma(x) = (1 - e^{-x})/(1 + e^{-x})$, and the arctan sigmoid, $\sigma(x) = (2/\pi) \arctan x$.

Increasing the complexity of the neural network by placing more functions in the hidden layer provides increasing approximation to the Bayes classifier. This approximation can be obtained to any desired degree (Cybenko, 1989; Funahashi, 1989). Greater accuracy of the best classifier for a network structure results in greater design cost. Not only does the increase in network complexity result in the need for larger data sets, it also makes estimation of the weights more problematic. Typically, the method of steepest decent on the error surface (as a function of the weights) is used (Bishop, 1995). In the case of linear filters, the mean-square error is a quadratic and the resulting error surface (as a function of the weights) has a single local minimum that is also the global minimum; however, the error surface for a neural network can have many local minima, so that a gradient method such as steepest descent can get stuck at a local minimum, making the matter much more problematic. Finally, a neural network is consistent if $k \to \infty$ such that $(k \log n)/n \to 0$ as $n \to \infty$ (Farago and Lugosi, 1993).

6.   Error Estimation

The error of a designed classifier needs to be estimated. If there is an abundance of data, then it can be split into *training* and *test* data. A classifier is designed on the training data. Its estimated error is the proportion of errors it makes on the test data. The estimate is unbiased and its variance tends to zero as the amount of training data goes to infinity. If the test-data error estimate is $\bar{\varepsilon}_n$ and there are m sample pairs in the test data, then

$$E[|\bar{\varepsilon}_n - \varepsilon_n|^2] \le \frac{1}{4m} \qquad (7.12)$$

It is necessary to use 25 test sample pairs to get the corresponding standard-deviation bound down to 0.1. The problem is that, for small samples, one would like to use all the data for design.

One small-sample approach is to use all the sample data to design a classifier ψ_n, and estimate ε_n by applying ψ_n to the same data. The *resubstitution estimator*, $\underline{\varepsilon}_n$, is the fraction of errors made by ψ_n. It is typically quite low-biased. Here we specifically consider histogram rules for fixed (non-data-dependent) partitions. For these, $\underline{\varepsilon}_n$ is biased low, meaning $E[\underline{\varepsilon}_n] \le E[\varepsilon_n]$. For small samples, the bias can be severe. It improves for large samples. The variance of the resubstitution estimator satisfies the bound, $\text{Var}[\underline{\varepsilon}_n] \le 1/n$. If the number of cells in the partition is bounded by K, then an upper bound for the mean-square error of $\underline{\varepsilon}_n$ as an estimator of ε_n is given by

$$E[|\underline{\varepsilon}_n - \varepsilon_n|^2] \le \frac{K}{n} \qquad (7.13)$$

(see (Devroye *et al.*, 1996) for the theoretical details of the bounds in this section). In the case of discrete binary features, $K = 2^d$ and the bound is exponential in terms of the number of variables. Figure 7.4 shows a generic situation for the inequality $E[\underline{\varepsilon}_n] \le E[\varepsilon_\bullet] \le E[\varepsilon_n]$ for increasing sample size.

Another small-sample approach is *cross-validation*. Classifiers are designed from parts of the sample, each is tested on the remaining data, and ε_n is estimated by averaging the errors. For *leave-one-out estimation*, n classifiers are designed from sample subsets formed by leaving out one sample pair, each is applied to the left-out pair, and the estimator $\hat{\varepsilon}_n$ is $1/n$ times the number of errors made by the n classifiers. Since the classifiers are designed on sample sizes of $n - 1$, $\hat{\varepsilon}_n$ actually estimates the error ε_{n-1}. It is an unbiased estimator of ε_{n-1}, meaning that $E[\hat{\varepsilon}_n] = E[\varepsilon_{n-1}]$. Unbiasedness is important, but of critical concern is the variance of the estimator for small n.

For a sample of size n, $\hat{\varepsilon}_n$ estimates ε_n based on the same sample. Performance depends on the classification rule. For histogram rules with fixed partitions,

$$E[|\hat{\varepsilon}_n - \varepsilon_n|^2] \le \frac{1 + 6/e}{n} + \frac{6}{\sqrt{\pi(n-1)}} \qquad (7.14)$$

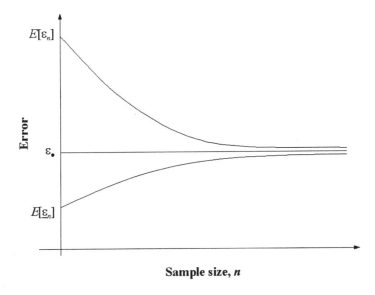

Figure 7.4. Relation between expected design error and resubstitution error.

From Eqs. 7.13 and 7.14, $E[|\underline{\varepsilon}_n - \varepsilon_n|^2]$ is of the order $n^{-1/2}$ for leave-one-out estimation as opposed to only n^{-1} for resubstitution. Unbiasedness comes with the cost of increased variance. There is a certain tightness to the bound of Eq. 7.14. For any partition, there exists a distribution for which

$$E[|\hat{\varepsilon}_n - \varepsilon_n|^2] \geq \frac{1}{e^{1/12}\sqrt{2\pi n}}. \tag{7.15}$$

Performance can be very bad for small n.

To appreciate the difficulties inherent in the leave-one-out bounds, we will simplify them in a way that makes them more favorable to precise estimation. The performance of $\hat{\varepsilon}_n$ guaranteed by Eq. 7.14 becomes better if we lower the bound. A lower bound than the one in Eq. 7.14 is $(1.8)/\sqrt{n-1}$. The corresponding standard-deviation bounds for $n = 50$ and 100 exceed 0.5 and 0.435, respectively. These are essentially useless. The minimum worst-case-performance bound of Eq. 7.15 would be better if it were lower. A lower bound than the one given is $(0.35)/\sqrt{n}$. The corresponding standard-deviation bounds for $n = 50$ and 100, exceed 0.22 and 0.18, respectively.

7.   Feature Selection

Given a large set of potential features, such as the set of all genes on a microarray, it is necessary to find a small subset with which to classify. There are various methods of choosing feature sets, each having advantages and disadvantages. The typical intent is to choose a set of variables that provides good classification. The basic idea is to choose nonredundant variables.

A critical problem arises with small samples. Given a large set of variables, every subset is a potential feature set. For v variables, there are $2^v - 1$ possible feature vectors. Even for choosing from among 200 variables and allowing at most 20 variables, the number of possible vectors is astronomical. One cannot apply a classification rule to all of these; nonetheless, even if the classes are moderately separated, one may find many thousands of vectors for which $\hat{\varepsilon}_n \approx 0$. It would be wrong to conclude that the Bayes errors of all the corresponding classifiers are small.

Adjoining variables stepwise to the feature vector decreases the Bayes error but can increase design error. For fixed sample size n and different numbers of variables d, Figure 7.5 shows a generic situation for the Bayes error $\varepsilon_\bullet(d)$ and the expected error $E[\varepsilon_n(d)]$ of the designed filter as functions of d. $\varepsilon_\bullet(d)$ decreases; $E[\varepsilon_n(d)]$ decreases and then increases. Were $E[\varepsilon_n(d)]$ known, then we could conclude that $\varepsilon_\bullet(d)$ is no worse than $E[\varepsilon_n(d)]$; however, we have only an estimate of $\varepsilon_n(d)$, which for small samples can be well below (or above) $\varepsilon_\bullet(d)$. Thus, the estimate curve $\hat{\varepsilon}_n(d)$ might drop far below the Bayes-error curve $\varepsilon_\bullet(d)$, even being 0 over a fairly long interval.

Regarding the general issue of the number of variables, the expected design error is written in terms of n and C in Eq. 7.3, but C depends on d. A celebrated theorem of pattern recognition provides bounds for $E[\Delta_{n,C}]$ (Vapnik and Chervonenkis, 1974; Vapnik and Chervonenkis, 1971). The *empirical-error* *rule* chooses the classifier in C that makes the least number of errors on the sample data. For this rule, $E[\Delta_{n,C}]$ satisfies the bound

$$E[\Delta_{n,C}] \leq 4\sqrt{\frac{V_C \log n + 4}{2n}} \qquad (7.16)$$

where V_C is the *VC* (*Vapnik-Chervonenkis*) *dimension* of C. Details of the VC dimension are outside the scope of this chapter. Nonetheless, it is clear from Eq. 7.16 that n must greatly exceed V_C for the bound to be small. The VC dimension of a perceptron is $d + 1$. For a neural network with an even number k of neurons, the VC dimension has the lower bound $V_C \geq dk$. If k is odd, then $V_C \geq d(k - 1)$. To appreciate

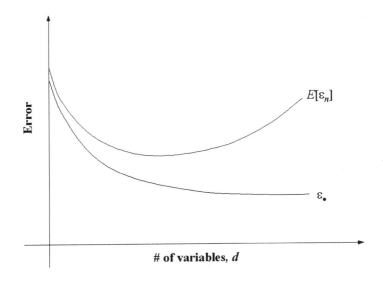

Figure 7.5. Effect of increasing numbers of variables.

the implications, suppose $d = k = 10$. Setting $V_C = 100$ and $n = 5000$ in Eq. 7.16 yields a bound exceeding 1, which says nothing. Admittedly, the bound of Eq. 7.16 is worst-case because there are no distributional assumptions. The situation may not be nearly so bad. Still, one must proceed with care, especially in the absence of distributional knowledge. Adding variables and neurons is often counterproductive unless there is a large sample available. Otherwise, one could end up with a very bad classifier whose error estimate is very small!

8. Illustration of Classification Techniques on Microarray Data

This section illustrates classification and error estimation using cDNA-microarray gene-expression data that have previously been used to classify childhood small, round blue cell tumors (SRBCTs) into four cancer types: neuroblastoma (NB), rhabdomyosarcoma (RMS), non-Hodgkin lymphoma (NHL), and the Ewing tumor family (EWS) (Khan *et al.*, 2002). We will break down the data in the same manner as the original study. There are 6,567 genes on each microarray, and omitting genes that do not satisfy a threshold level of expression reduces the number to 2,308. There is a total of 83 microarrays, 63 for training

Model	Test Error	Leave-one-out	Resubstitution
Perceptron	0.05	0	0
Neural Network	0.1	0.0469	0.0156
Nearest Neighbor	0.1	0.0313	0
5-Nearest Neighbor	0	0.0156	0.0156
Linear SVM	0.05	0.0313	0
Nonlinear SVM	0.15	0.0156	0
Gaussian Kernel	0.1	0.0313	0
Moving Window	0.15	0.0625	0.0313
Epanechnikov Kernel	0.05	0.0156	0

Table 7.1. Classifier errors

and 20 for testing. Principle component analysis has been performed using the training samples and the 10 dominant components have been selected as features. To this point, that data have been handled exactly as in the original study. Henceforth, we will illustrate classification rules discussed in the present chapter.

The leftmost column of Table 7.1 lists the classifier models applied to the data: (1) perceptron using the thresholded Wiener filter, (2) neural network with 3 hidden nodes and the logistic sigmoid activation function for both layers, (3) nearest-neighbor, (4) k-nearest neighbor with $k = 5$, (5) linear support vector machine, (6) polynomial nonlinear support vector machine, (7) moving window with a Gaussian kernel with $h = 0.5$, (8) moving window, and (9) moving window with the Epanechnikov kernel. Four-way classification has been achieved by first grouping NB (class 1) with RMS (class 2) and RMS (class 3) with NHL (class 4), and building a classifier for the two groups, after which, classifiers are built to separate the classes within the groups. As reported in the table, three error measures have been computed: test-data error, leave-one-out error, and resubstitution error. We note that the leave-one-out error has been computed using the 10 principle components found from the full training set. This has been done for two reasons: (1) to keep throughout the same components used in the original study; and (2) to ease the computational burden. In fact, since feature reduction is part of the training, to be fully rigorous, principle component analysis should be redone on each of the 63 sample subsets used for computing the leave-one-out error.

The relationship between small samples and classifier complexity is illustrated by the results. The perceptron (using a very simple training method) outperforms the more complex neural network for all error measurements. The differing conclusions that can be drawn according to which error estimate is used can also be seen. The linear support vector machine has a lower test error than the nonlinear support vector machine, but the situation is reversed for leave-one-out error. A similar situation occurs with respect to the nearest-neighbor and 5-nearest-neighbor classifiers. Moreover, with respect to the latter, the test error is less than the resubstitution error. Owing to the excellent results achieved by the perceptrons (Wiener and support vector machine), it is clear that the classes (in principle-component representation) are readily separated by a very small number of hyperplanes.

9. Conclusion

Except in situations where the amount of data is large in comparison to the number of variables, classifier design and error estimation involve subtle issues. This is especially so in applications such as cancer classification where there is no prior knowledge concerning the vector-label distributions involved. It is clearly prudent to try to achieve classification using small numbers of genes and rules of low complexity (low VC dimension), and to use cross-validation when it is not possible to obtain large independent samples for testing. Even when one uses a cross-validation method such as leave-one-out estimation, one is still confronted by the high variance of the estimator. In many applications, large samples are impossible owing to either cost or availability. Therefore, it is unlikely that a statistical approach alone will provide satisfactory results. Rather, one can use the results of classification analysis to discover gene sets that potentially provide good discrimination, and then focus attention on these. In the same vein, one can utilize the common engineering approach of integrating data with human knowledge to arrive at satisfactory systems.

References

Ben-Dor, A., Bruhn, L., Friedman, N., Nachman, I., Schummer, M. and Yakhini, Z. (2000) Tissue classification with gene expression profiles. *Computational Biology*, **7**, 559-583.

Bishop, C. M., (1995) *Neural Networks for Pattern Recognition*, Oxford University Press, Oxford.

Bittner, M., Meltzer, P., Khan, J., Chen, Y., Jiang, Y., Seftor, E., Hendrix, M., Radmacher, M., Simon, R., Yakhini, Z., Ben-Dor, A., Dougherty, E., Wang, E., Marincola, F., Gooden, C., Lueders, J., Glatfelter, A., Pollock, P., Gillanders, E., Leja, A., Dietrich, K., Beaudry, C., Berrens, M., Alberts, D., Sondak, V., Hayward, N., and Trent, J. (2000) Molecular classification of cutaneous malignant melanoma by gene expression profiling. *Nature*, **406**, 536-540.

Brown, M. P. S., Grundy, W. N., Lin, D., Cristianini, N., Sugnet, C. W., Furey, T. S., Ares, Jr., M., and D. Haussler. (2000) Knowledge-Based Analysis of Microarray Gene Expression Data by Using Support Vector Machines. *Proc. National Academy Science*, **97** (1), 262-267.

Cybenko, G. (1989) Approximation by Superposition of Sigmoidal Functions. *Mathematics Control, Signals, Systems*, **2**, 303-314.

Devroye, L., Gyorfi, L., and G. Lugosi. (1996) *A Probabilistic Theory of Pattern Recognition*. Springer-Verlag, New York.

Devroye, L., and Kryzak, A. (1989) An Equivalence Theorem for L_1 Convergence of the Kernel Regression Estimate, *Statistical Planning and Inference*, **23**, 71-82.

Dougherty, E. R. (2001) Small Sample Issues for Microarray-Based Classification. *Comparative and Functional Genomics*, **2**, 28-34.

Farago, A., and Lugosi, G. (1993) Strong Universal Consistency of Neural Network Classifiers. *IEEE Trans. on Information Theory*, **39**, 1146-1151.

Funahashi, K. (1989) On the Approximate Realization of Continuous Mappings by Neural Networks. *Neural Networks*, **2**, 183-192.

Golub, T. R., Slonim, D. K., Tamayo, P., Huard, C., Gaasenbeek, M., Mesirov, J. P., Coller, H., Loh, M. L., Downing, J. R., Caligiuri, M. A., Bloomfield, C. D. and Lander, E. S. (1999) Molecular classification of cancer: class discovery and class prediction by gene expression monitoring. *Science*, **286**, 531-537.

Gordon, L. and R. Olshen (1978) Asymptotically Efficient Solutions to the Classification Problem, *Annals of Statistics*, **6**, 525-533.

Hedenfalk, I., Duggan, D., Chen, Y., Radmacher, M., Bittner, M., Simon. R., Meltzer, P.,Gusterson, B., Esteller, M., Raffeld, Yakhini, Z., Ben-Dor, A., Dougherty, E., Kononen,J., Bubendorf, L., Fehrle, W., Pittaluga, S., Gruvverger, S., Loman, N., Johannsson, O., Olsson, H., Wifond, B., Sauter, G., Kallioniemi, O. P., Borg, A., and Trent, J. (2001) Gene expression profiles distinguish hereditary breast cancers. *New England J. Medicine*, **34**, 539-548.

Khan, J., Wei, J. S., Ringner, M., Saal, L. H., Ladanyi, M., Westermann, F., Berthold, F., Schwab, M., Antonescu, C. R., Peterson, C. and Meltzer, P. S. (2002) Classification and diagnostic prediction of cancers using gene expression profiling and artificial neural networks. *Nature Medicine*, **7**, 673-679.

Kim, S., Dougherty, E. R., Barrera, J., Chen, Y., Bittner, M., and J. M. Trent (2002) Strong Feature Sets From Small Samples. *Journal of Computational Biology*, **9**(1).

Rosenblatt, F. (1962) *Principles of Neurodynamics: Perceptrons and the Theory of Brain Mechanisms*, Spartan, Washington DC.

Stone, C. (1977) Consistent Nonparametric Regression. *Annals of Statistics*, **5**, 595-645.

Vapnik, V. N., Golowich, S. E., and A. Smola (1997) Support Vector Method for Function Approximation, Regression, and Signal Processing. in *Advances in Neural Information Processing Systems*, **9**, MIT Press, Cambridge.

Vapnik, V. N. (1998) *Statistical Learning Theory*, John Wiley, New York.

Vapnik, V., and A. Chervonenkis (1974) *Theory of Pattern Recognition*, Nauka, Moscow.

Vapnik, V., and A. Chervonenkis (1971) On the Uniform Convergence of Relative Frequencies of Events to Their Probabilities *Theory of Probability and its Applications*, **16**, 264-280.

Chapter 8

ANALYZING PROTEIN SEQUENCES USING SIGNAL ANALYSIS TECHNIQUES

Karen M. Bloch[1] and Gonzalo R. Arce[2]

[1] *DuPont Company, Wilmington, Delaware, USA*
[2] *Department of Electrical & Computer Engineering, University of Delaware, Newark, Delaware, USA*

1. Introduction

Genomes carry all information of life from one generation to the next for every organism on earth. Each genome, which is a collection of DNA molecules, can be represented as a series of strings comprised of four letter symbols. Less than 10 years ago, determining the sequence of these letters to read a single gene was a slow tedious process. But today, through the use of new strategies, genome sequencing is a billion-dollar worldwide effort in both industry and academia.

At the end of 1998, researchers had completely read the genome of only one multicellular organism, a worm known as *C. elegans.* Now, sequences exist for the fruit fly, the human and for the weed important to plant geneticists known as *Arabidopsis.* Drafts of the genomes of the mouse, rat, zebrafish, and pufferfish are soon to be completed. Researchers have also been working on simpler organisms. Several dozen microbial genomes are now available, including those that cause cholera and meningitis. Most of these data are accessible free of charge, encouraging the exploration of this data. However, it is not the genes, but the proteins they code for that actually do all the work. The search for protein function has lead to the era of proteomics, the identification and characterization of each protein and its structure, and of every protein-protein interaction (Pennisi, 2000).

Proteins are the molecules that accomplish most of the functions of living cells. All proteins are constructed from linear sequences of smaller molecules called amino acids. There are twenty naturally occurring amino acids and they can be represented in a protein sequence as a string of alphabetic symbols. Protein molecules fold to form specific three dimensional shapes which specify their particular chemical function (Hunter, 1993).

Analysis of protein sequences can provide insights into function and can also lead to knowledge regarding biologically active sites of the protein. While analysis of protein sequences is often performed directly on the symbolic representation of the amino acid sequence, patterns in the sequence are often too weak to be detected as patterns of symbols. Alternative sequence analysis techniques can be performed by assigning numerical values to the amino acids in a protein. The numerical values are derived from the physicochemical properties of the amino acid and are relevant to biological activity. It has been shown that the EIIP, as one such measure, correlates with certain biological properties (Veljkovic, 1985). Once a numerical mapping for a protein sequence is achieved, the sequence can be treated as a signal.

From a mathematical point of view, a signal can be described in a variety of ways. For example, a signal can be represented as a function of time which shows how the signal magnitude changes over time. Alternatively, a signal can be written as a function of frequency by performing a Fourier transform. This tells how quickly a signal's magnitude changes (Qian, 1996). For many real world applications, it is useful to characterize a signal in the time and frequency domains simultaneously. Such signal analysis methods can provide fingerprints which indicate the existence of some event of importance. In the case of protein sequences represented as numerical signals, such an event might be the existence of a binding site.

This chapter illustrates the use of frequency and time-frequency signal analysis techniques with two classes of proteins, fibroblast growth factors and homeodomain proteins. Fibroblast growth factors constitute a family of proteins that affect the growth, migration, differentiation, and survival of certain cells. Homeodomain proteins contain a single 60-amino acid DNA binding domain. It is the numerical representation of these amino acid sequences, along with various frequency and time-frequency analysis methods which we describe herein.

2.   Frequency Analysis of Proteins

The Resonant Recognition Model (RRM) (Cosic, 1994) is a physico-mathematical model that analyses the interaction of a protein and its target using signal processing methods. One application of this model involves prediction of a protein's biological function. In this technique, a Fourier transform is applied to a numerical representation of a protein sequence and a peak frequency is determined for a protein's particular function. The Discrete Fourier Transform (DFT) is defined as follows:

$$X(n) = \sum_{m=0}^{N-1} x(m)e^{-j(2/N)nm} \qquad n = 1, 2, \ldots, N/2 \qquad (8.1)$$

where $x(m)$ is the mth member of a numerical series, N is the total number of points in the series, and $X(n)$ are coefficients of the DFT. The coefficients describe the amplitude, phase, and frequency of sinusoids which make up the original signal. In the RRM, the genetic sequences are treated as discrete signals, and it is assumed that the points are equidistant with the distance $d = 1$. In this case, the maximal frequency in the spectrum is $F = 1/2d = 0.5$.

The aim of this method is to determine a single parameter that correlates with a biological function expressed by a set of genetic sequences. To determine such a parameter, it is necessary to find common characteristics of sequences with the same biological function. The cross-spectral function determines common frequency components of two signals. For a discrete series, the cross-spectral function is defined as:

$$S(n) = X(n)Y^*(n), \qquad n = 1, 2, \ldots, N/2 \qquad (8.2)$$

where $X(n)$ are the DFT coefficients of the series $x(n)$ and $Y^*(n)$ are the complex conjugate DFT coefficients of the series $y(n)$. Peak frequencies in the cross-spectral function define common frequency components for analyzed sequences. The common frequency components for a group of protein sequences can be defined as follows:

$$|S(n)| = |X_1(n)||X_2(n)| \ldots |X_M(n)|, \qquad n = 1, 2, \ldots, N/2. \qquad (8.3)$$

where M is the number of sequences.

This methodology can be illustrated via an example. We have chosen to study fibroblast growth factors (FGF) which constitute a family of proteins that affect the growth, differentiation, and survival of certain cells. The symbolic representations of two FGF amino acid sequences are shown below:

```
> Basic bovine FGF
PALPEDGGSGAFPPGHFKDPKRLYCKNGGFFLRIHPDGRVDGVREKSDPH
IKLQLQAEERGVVSIKGVCANRYLAMKEDGRLLASKCVTDECFFFERLES
NNYNTYRSRKYSSWYVALKRTGQYKLGPKTGPGQKAILFLPMSAKS
```

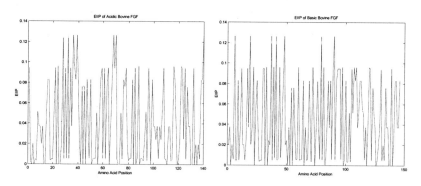

Figure 8.1. Numerical EIIP Representations of FGF Proteins

```
> Acid bovine FGF
FNLPLGNYKKPKLLYCSNGGYFLRILPDGTVDGTKDRSDQHIQLQLCAES
IGEVYIKSTETGQFLAMDTDGLLYGSQTPNEECLFLERLEENHYNTYISK
KHAEKHWFVGLKKNGRSKLGPRTHFGQKAILFLPLPVSSD
```

Symbolic representations, such as these, can be translated into numerical sequences using the EIIP index (Tomii, 1996). It has been shown that the EIIP correlates with certain biological properties (Veljkovic, 1985). The graphical representation of the corresponding numerical sequences for the FGF proteins obtained by replacing every amino acid with its EIIP value can be see in Figure 8.1. A DFT is performed on each numerical sequence. The resulting spectra are shown in Figure 8.2. The cross-spectral function of the 2 FGF spectra generates the consensus spectrum shown in Figure 8.3. For the spectra, the x-axis represents the RRM frequencies and the y-axis represents the normalized intensities. The prominent peak denotes the common frequency component for this family of proteins.

The presence of a peak frequency in a consensus spectrum implies that all the analyzed sequences have one frequency component in common. This frequency is related to the biological function provided the following conditions are met:

- One peak only exists for a group of protein sequences sharing the same biological function.

- No significant peak exists for biologically unrelated protein sequences.

- Peak frequencies are different for different biological functions.

What is lacking in this technique is the ability to reliably identify the individual amino acids that contribute to the resonant recognition peak

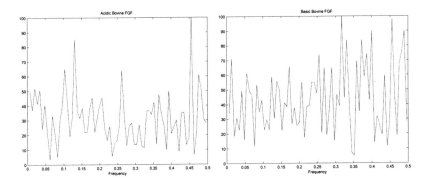

Figure 8.2. FFT Representations of FGF Proteins

frequency. Like signals in general, signals associated with biological sequences are often non-stationary, meaning that the frequency components change along the primary protein sequence chain. These issues can be addressed with time-frequency analysis techniques.

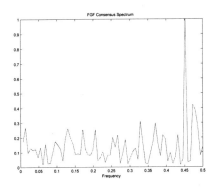

Figure 8.3. Consensus Spectrum of Bovine FGF Proteins

3. Time-Frequency Analysis

Frequency analysis alone cannot handle the transitory nature of non-stationary signals. A time-frequency (or space-frequency) representation of a signal provides information about how the spectral content of the signal evolves with time (space), and therefore provides a tool to analyze non-stationary signals.

3.1 Non-Stationary Signals

Let us consider as a first example the sum of two synthetic signals with constant amplitude, and a linear frequency modulation. This type of signal is called a chirp, and as its frequency content is varying in time, it is a non-stationary signal. An example of such a signal can be seen in Figure 8.4. It consists of the sum of 2 linear chirps varying from [0-0.5] and [0.2-0.5]. >From this time-domain representation, it is difficult to determine what kind of modulation is contained in this signal. If we now consider the energy spectrum of this signal by squaring the modulus of its Fourier transform, as illustrated in Figure 8.5, we still cannot determine anything about the evolution in time of the frequency components. In order to more accurately describe such signals, it is better to directly represent their frequency content while still keeping the time (or spatial) parameter. This is the aim of the time-frequency analysis methods discussed in the next sections.

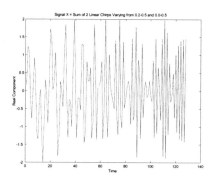

Figure 8.4. Sum of 2 Chirps   *Figure 8.5.* FFT of the signal

3.2   Wavelet Transform

A method for protein analysis proposed by Fang and Cosic (Fang, 1999) uses a continuous wavelet transform to analyze the EIIP representations of protein sequences. The continuous wavelet transform (CWT) is one example of a time-frequency or space-frequency representations. Because the CWT provides the same time/space resolution for each scale, the CWT can be chosen to localize individual events such as active site identification. The amino acids that comprise the active site(s) are identified as the set of local extrema of the coefficients in the wavelet transform domain. The energy concentrated local extrema are the locations of sharp variation points of the EIIP and are proposed as the most critical locations for a protein's biological function (Fang, 1999). Experiments have shown that the potential cell attachment sites of FGF's are between residues 46-48 and 88-90. Figure 8.6 is a CWT

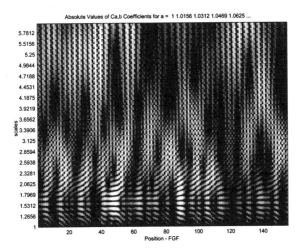

Figure 8.6.   Continuous Wavelet Spectrogram of Basic Bovine FGF Protein (see color plates)

spectrogram of basic bovine FGF protein. It can be observed that there are two bright regions at the higher frequencies from scale 1.266 to scale 2.062 which correspond to the amino acids at the active sites.

The use of wavelet transforms shows promise for identifying amino acids at potential biologically active sites, but does not reveal the characteristic frequency component of the Resonant Recognition Model. An additional weakness exists in that the spectrogram of the CWT can often be difficult to interpret. The weakness of this approach can be addressed by the use of a different time-frequency transform.

3.3   Wigner-Ville Distribution

Quadratic time-frequency representations are powerful tools for the analysis of non-stationary signals. The Wigner-Ville distribution (WVD),

$$W_x(t, f) = \int_\tau x(t + \tau/2)x^*(t - \tau/2)e^{-j2\pi f\tau}d\tau \qquad (8.4)$$

for example satisfies a number of desirable mathematical properties and possesses optimal resolution in time-frequency space (Arce, 2000). Application of the Wigner-Ville distribution allows more subtle signal features to be detected, such as those having short length and high frequency variation. However, the use of the Wigner-Ville distribution has been limited because of the presence of cross or interference terms. The

Wigner-Ville distribution of the sum of two signals $x(t) + y(t)$

$$W_{x+y}(t, f) = W_x(t, f) + 2\Re(W_{xy}(t, f)) + W_y(t, f) \qquad (8.5)$$

has a cross term $2\Re(W_{xy}(t, f))$ in addition to the two auto components. Because the cross term usually oscillates and its magnitude is twice as large as that of the auto components, it can obscure useful time dependent spectral patterns. Figure 8.7 shows the Wigner-Ville transform[1] of the sum of the two chirps presented in Figure 8.4. One can observe the individual components of this signal, but there is also a great deal of cross-term interference evident in the figure. Reduction of the cross term interference without destroying the useful properties of the WVD is very important to time-frequency analysis (Qian, 1996).

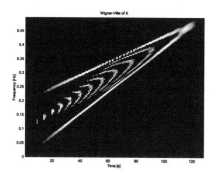
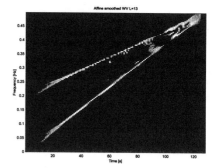

Figure 8.7. Wigner-Ville transform of the sum of 2 chirps. (see color plates)

Figure 8.8. Affine filter applied to a Wigner-Ville transform. (see color plates)

3.4 Interference Terms

Minimal interference terms are necessary for appropriate interpretation and discrimination of the signals being analyzed (Auger, 1995). The interference terms of the Wigner-Ville distribution are due to the WVD's quadratic structure. Interference terms occur in the case of multicomponent signals and can be represented mathematically with quadratic cross terms. While filtering of the interference terms is desirable, the use of linear filters can distort the resolution and concentration of the

---

[1]The figures in this chapter illustrating the Wigner-Ville transform were generated using the Time-Frequency Toolbox for MATLAB found at http://www-rocq.inria.fr/fractales/Software/TFTB.

auto component terms. However, the nonlinear center affine filter[2] as described in (Arce, 2000) can be applied to effectively filter out the cross terms while leaving the auto component terms relatively unaffected. For the previous example of the sum of 2 chirps, the center affine filter was applied to the Wigner-Ville transform. A window size of 13 was used in this example and the resulting image can been seen in Figure 8.8.

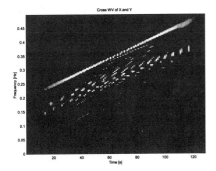 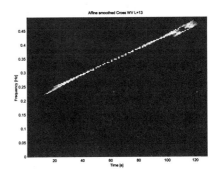

Figure 8.9. Cross Wigner-Ville of 2 chirp signals with a common component. (see color plates)

Figure 8.10. Cross Wigner-Ville Affine Filtered. (see color plates)

In another example, we take the cross Wigner-Ville transform of 2 signals which are each the sum of 2 chirps, but share a common component which is a chirp varying from [0.2-0.5]. In Figure 8.9 we can detect the common component, but once again the interference terms are strongly present. Application of the center affine filter to this figure produces the result in Figure 8.10 allowing the common frequency component to be readily discerned. In the next section, the use of the Wigner-Ville time-frequency representation (TFR) along with the center affine filtering method is used to analyze two protein families.

4. Application of Time-Frequency Analysis to Protein Families

4.1 Fibroblast Growth Factors

We illustrate the use of the Wigner-Ville time-frequency distribution using the same FGF proteins and EIIP representation that was discussed in Sections 2 and 3.2.

---

[2] A MATLAB implementation of the center affine filter can be found at http://www.ece.udel.edu/~bloch/downloads.

As stated previously, experiments have shown that the potential cell attachment sites of FGF's are between residues 46-48 and residues 88-90 and the characteristic frequency has been shown in the literature as 0.4512 (Fang, 1999).

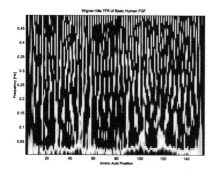
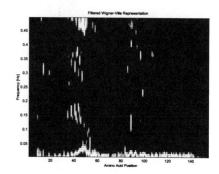

Figure 8.11. Wigner-Ville TFR of Basic Human FGF. (see color plates)

Figure 8.12. Activation sites of Basic Human FGF. (see color plates)

The discrete Wigner-Ville time-frequency representation of human basic FGF is shown in Figure 8.11. Cross terms make interpretation of this representation difficult. After application of the center affine filter as described in (Arce, 2000), one can see in Figure 8.12 that the bright regions in the lowest frequency range correspond to experimentally proven activation sites and the bright region around frequency 0.45 corresponds to the characteristic frequency component.

4.2   Homeodomain Proteins

Homeodomain proteins contain a 60-amino acid DNA binding domain found in numerous eukaryotic transcription factors. The homeodomain family is a useful system for studying sequence-structure and sequence-function relationships because several hundred sequences are known and the structures of several homeodomains have been determined (Clarke, 1995).

Application of the Wigner-Ville TFR to a homeodomain B6 protein that has been represented via the EIIP mapping results in the plot shown in Figure 8.13. There doesn't appear to be any clear signature in the plot which one could relate to active regions on the protein. However, if an alternative mapping is used, such as one measuring the hydrophobicity of the amino acids, a clear indication of the binding region is detected at amino acid positions 146-205 (Figure 8.14). This example illustrates the importance of the particular numerical representation being used to describe the protein family being studied (Bloch, 2001).

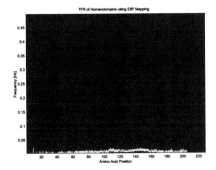

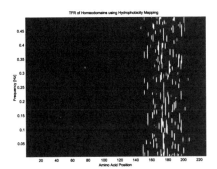

Figure 8.13. TFR of Homeodomain Proteins Using EIIP Mapping. (see color plates)

Figure 8.14. TFR of Homeodomain Proteins Using Hydrophobicity Mapping. (see color plates)

5. Conclusions

This chapter illustrates the usefulness of time-frequency signal processing for the analysis of protein sequence data. Time-frequency representations such as the Wigner-Ville distribution, when appropriately filtered for interference terms, provide frequency as well as spatial information and lead to the ability to identify biologically active sites of certain proteins. However, as illustrated in the homeodomain example, not all mappings are capable of correlating to biological properties. The selection of the numerical mapping is problem-specific since different mappings may highlight different structural properties of proteins which are directly related to a particular protein's function.

GenomeNet[3], a Japanese network of database and computational services for genome research, currently maintains a database of over 400 different numerical mappings of amino acids (Kawashima, 1999). Work is underway to extend the methods described in this chapter, and involves the applicability of these different numerical indices based on the protein family being analyzed.

---

[3] GenomeNet WWW Server: http://www.genome.ad.jp/dbget/aaindex.html.

References

Arce, G.R. and Hasan, S.R. (2000) "Elimination of Interference Terms of the Discrete Wigner Distribution Using Nonlinear Filtering." *IEEE Transactions on Signal Processing*, 48, 2321-2331.

Auger, F. and Flandrin, P. (1995) "Improving the Readability of Time-frequency and Time-scale Representations by the Reassignment Method." *IEEE Transactions on Signal Processing*, 43, 1068-1089.

Bloch, K.M. and Arce, G.R. (2001) "Time-Frequency Analysis of Protein Sequence Data." *2001 IEEE-EURASIP Workshop on Nonlinear Signal and Image Processing*.

Clarke, N.D. (1995) "Covariation of Residues in the Homeodomain Sequence Family." *Protein Science*, 4, 2269-2278.

Cosic, I. (1994) "Macromolecular Bioactivity: Is it Resonant Interaction Between Macromolecules? - Theory and Applications." *IEEE Transactions on Biomedical Engineering*, 41, 1101-1114.

Fang, Q. and Cosic, I. (1999) "Prediction of Active Sites of Fibroblast Growth Factors Using Continuous Wavelet Transforms and the Resonant Recognition Model." *Proceedings of The Inaugural Conference of the Victorian Chapter of the IEEE EMBS*, 211-214.

Hunter, L. (1993) "Molecular Biology for Computer Scientists." *Artificial Intelligence and Molecular Biology*, AAAI Press.

Kawashima, S., Ogata, H., and Kanehisa, M. (1999) "AAindex: Amino Acid Index Database." *Nucleic Acids Research*, 27, 368-369.

Pennisi, E. (2000) "Genomics Comes of Age." *Science*, 290, 2220-2221.

Qian, S. and Chen, D. (1996) *Joint Time-Frequency Analysis: Methods and Applications*, Prentice-Hall PTR.

Tomii, K. and Kanehisa, M. (1996) "Analysis of Amino Acids and Mutation Matrices for Sequence Comparison and Structure Prediction of Proteins." *Protein Engineering*, 9, 27-36.

Veljkovic, V., Cosic, I., Dimitrjevic, B. and Lalovic, D. (1985) "Is it Possible to Analyze DNA and Protein Sequences by the Methods of Digital Signal Processing?." *IEEE Transactions on Biomedical Engineering*, 32, 337-341.

Chapter 9

STATISTICS OF THE NUMBERS OF TRANSCRIPTS AND PROTEIN SEQUENCES ENCODED IN THE GENOME

Vladimir A. Kuznetsov
Laboratory of Integrative and Medical Biophysics, National Institute of Child Health and Human Development, NIH, Bethesda, Maryland, USA

1. Introduction

Recent achievements in genome sequencing, coupled with the ability of large-scale gene-expression technologies to simultaneously measure the numbers of copies of distinct mRNAs and the number of distinct proteins corresponding to thousands of genes in a given cell type have revolutionized biological research by allowing statistical analysis of data sets related to varying structural and functional properties of biological systems. These studies have provided a lot of information on the gene, proteins, genetic and biochemical networks and, potentially, allow the discovery of general regularities that integrate this information into organizational principles governing complex biological systems and their evolution.

A basic topic of any statistical inference of complex systems is the characterization of the distribution of object frequencies for a population (for example, the distribution of allele frequencies in a population (Huang & Wier, 2001) or the distribution of homologous families by family size ratios (Friedman & Hughes, 2001) based on statistical analysis of the samples. In this chapter we will analyze the class of skew distributions that appear in samples provided by large-scale gene expression experiments and by proteome data sets. The observed distributions have the following characteristic in common: there are few frequent, and many rare classes. Such distributions appearance is so frequent in many natural and artificial large-scale complex systems and in the phenomena in which they appear so diversely (physics, biology, economics, sociology, internet, linguistics, etc. (Yule, 1924; Simon and Van Wormer, 1963; Mandelbrot, 1982; Borodovsky & Gusein-Zade, 1989; Adami, 1998; Ramsden & Vohradsky, 1998; Jeong et al., 2000; Kuznetsov, 2001b; Newman et al., 2001), that one may be lead to conjecture that all these phenomena can be similar in terms of the structure of the underlying probability mechanisms.

The empirical probability distributions to which we shall refer specifically are: (A) the distribution of mRNA transcripts by their frequencies of occurrences in gene-expression databases; (B) the distribution of the numbers of promoters regulated by the regulatory DNA-binding proteins; (C) the distribution of identifiable protein clusters by the cluster sizes; (D) the distribution of the identifiable protein domains by their appearance values in the proteome . For such data sets, the true underlying probability distribution functions have not been previously quantitatively characterized. In particular, RNA-DNA hybridization experiments (Bishop et al., 1974) and many more recent large-scale gene-expression profiling studies of mRNA populations have demonstrated a very broad range of transcript copy numbers for genes expressed in a given cell type (0.1-30,000 copies per mammalian cell). Many thousands of expressed genes in eukaryotic cells are "rare" (less than 1 copy per cell) in the numbers of associated transcripts. However, the true distribution of gene expression levels (the numbers of mRNAs at the each possible level of occurrence) in a cell has not been previously identified due to experimental errors, undersampling, and non-reliable detection of the low abundant transcripts.

The determination of expressed genes at all biologically significant levels in eukaryotic cells is a challenging biological problem (Bishop et al., 1974; Velculescu et al., 1999). A more simple and practically important goal of genome projects is to systematically identify genes. Last year, two papers announced drafts of the human genome sequence (Venter et al., 2001; IHGSC, 2001), but estimates of the number of human genes continues to fluctuate. Current estimates center around 30,000-40,000, with occasional excursions to 100,000 or more (Pennisi, 2000; Hogenesch et al., 2001). One reason for the continuing ambiguity is that many genes are neither well-defined nor easily recognizable. These estimations are based on samples of four methods: cDNA cloning and EST (expression sequence tag) sequencing of mRNAs (Emmert-Buck et al., 2000; Ewing & Green, 2000), the Serial Analysis of Gene Expression (SAGE) method using short-nucleotide sequence *tags* that match different poliadenilated mRNAs (Velculescu et al., 1997; Velculescu et al., 1999), conserved coding exon and protein domain coding sequence identification by comparative genome analysis (IHGSC, 2001; Crollius et al., 2000), and computational gene prediction (Venter et al., 2001; IHGSC, 2001). These methods work better for large, highly expressed, evolutionarily conserved protein coding genes, and ambiguously predict other genes. For example, although the large-scale gene expression technologies have become increasingly powerful and are widely used, frequent errors in identification of hundreds and thousands of rarer transcripts create considerable uncertainty in fundamental questions such as the total number of genes expressed in an organism and the biological significance of rare transcripts. Many of the low-abundant transcripts may be essential for determining normal and pathological cell phenotypes (Guptasarma, 1995; Ohlsson et al., 2001). However, thousands of rarely-expressed genes have not been characterized and a rationale for the extreme number of the low abundant transcripts has remained unresolved.

Thus, an important issue for gene identification in the postgenome era is determining the true statistical distributions of the number of genes expressed at all possible expression levels both in a single cell and in a population of the cells. For better understanding the general statistical principal governing the biological complexity we also need to analyze the distributions of regulatory proteins in protein-DNA networks, proteins in protein clusters, and protein domains in proteomes.

In this chapter we will analyze the statistical distributions arising in large-scale gene-expression and protein data sets. This analysis allows us to identify the true underling probability functions and to estimate the number of evolutionarily conserved protein-coding genes in the human and mouse genomes, and the number of the protein-coding genes expressed in a human cell. We also will show how these distributions help in understanding of the general principal governing the functional adaptation of cells to environmental stress-factors and in understanding the biological complexity growth during evolution.

2.   Distributions of the Gene Expression Levels

2.1   Empirical Distributions

Gene expression within a cell is a complex process involving chromatin remodeling, transcription and export of RNA from the nucleus to cytoplasm where mRNA molecules are translated into proteins. Each of these steps is carried out by highly specialized set, elaborate machinery, typically consisting of tens to hundreds of components. Currently, large-scale gene-expression profiling methods (e.g., SAGE) (Velculescu et al., 1997; Velculescu et al., 1999), cDNA or oligonucleotide microarrays (Holstege et al., 1998; Jelinsky & Samson, 1999) involve making cDNA sequences to the less stable mRNA molecules and then using short-nucleotide sequence *tags* that match different mRNAs to quantify their relative abundance in the cell sample. These methods can measure mostly highly and moderately abundant transcripts from large numbers of cells (i.e., not a single cell). However, thousands of genes expressed at very low copy number (≤ 1 copy per cell) can not be reliably and unambiguously detected.

The complete gene expression profile for a given cell type is the list of all expressed genes, together with each gene's expression level defined as the average number of cytoplasmic mRNA transcripts per cell. For each gene-expression data set, we define the *library* as a list of sequenced tags that match mRNAs associated with genes, together with the number of occurrences of each specific tag. The statistics of expressed genes in such libraries can be characterized by *gene expression level probability functions* (GELPF), which for each possible gene expression level value gives the probability of that value occurring for a randomly-chosen gene. The empirical GELPF specifies the proportions of distinct tags which have 1, 2, etc. transcripts present in an associated mRNA sample (i.e., a normalized histogram of gene expression levels). Analysis of such empirical histograms can lead to models of the underlying GELPF in a cell and in a population of cells.

Let M be the size of the library, i.e. the total number of tags in it, and n_m be the observed number of distinct tags that have the expression level m (occurring m times) in a given library of size M. The discrete value n_m roughly reflects the number of expressed genes with expression level m in the cell sample due to experimental errors, non-unique tag-gene matching and incorrect annotation of genes (see below). Let J denote the observed expression level for the most abundant tag in the library. The $\sum_{m=1}^{J} n_m = N$ is the number of distinct tags in the library. The points $(m, g(m))$ for $m = 1, \ldots, J$, where $g(m) = n_m/N$, form the histogram corresponding to the empirical GELPF.

Analyzing more than 50 human cDNA and SAGE libraries presented by Cancer Genome Anatomy Project (CGAP) database (www. ncbinlm.nih.gov/CGAP; www.ncbi.nlm.nih.gov/SAGE) we observed that the empirical relative frequency distribution of expressed genes, constructed for all analyzed libraries exhibited remarkably similar monotonically-skewed shapes with a greater abundance of rare transcripts and more gaps among the higher-occurrence expression levels (Kuznetsov & Bonner, 1999; Emmert-Buck et al., 2000). Moreover, histograms for yeast SAGE libraries, as well as for mouse and for human SAGE or cDNA libraries also exhibit remarkably similar behavior (Figure 9.1, Table 9.1).

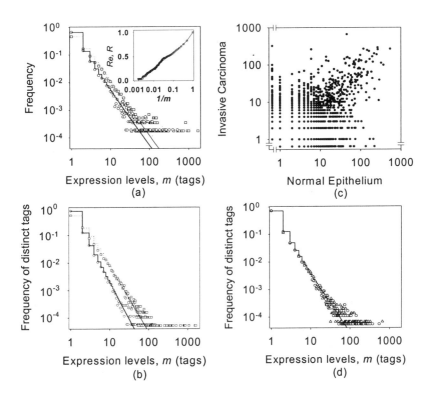

Figure 9.1. Fitting of the empirical frequency distributions of the gene expression levels. Log-log plots. (a) ○: G2/M-arrested yeast cell library with 20,096 SAGE tags; solid step-function line: GDP model; □ : three pooled yeast cell libraries with 59,494 SAGE tags (see Table 9.1); dotted step-function line GDP model for □-data. Insert plot.○: empirical cumulative fraction function R_e-values ($R_e = (\sum_{j=1}^{m} j \cdot n(j, M))/(NM)$, where $n(j, M)$ is the observed number of distinct tags represented exactly by j copies in a library with size $M; N$ is the total number of distinct tags in a library for the ○-histogram (main plot); solid line is corresponding theoretical model R ($R(m) = \sum_{j=1}^{m} j \cdot f(j))/\sum_{j=1}^{J} j \cdot f(j)$) computed by the GDP model $f(j)$(main plot). (Cumulative data reduces the apparent "noise" in the histogram data). (b) Log-log plot. □ : human normal brain cells library 154 with 81,516 SAGE tags; dotted step-function line : best-fit GDP model for □-data; ○: 8,936 SAGE tags sampled without replacement from library 154 (sub-library 154.1; Table 9.1); solid step-function line: best-fit GDP model for ○-data. (c) Correlation plot of genes expression values in normal human mammary epithelial cells (library 6360) and human breast invasive ductal carcinoma cells (library 17716); (d) Log-log plot: ○, △: empirical histograms for normal and carcinoma cells described on plot (c), respectively; solid line links fitted GDP model data for ○-data.

Table 9.1. Characterization of empirical frequency distributions of gene expression levels for human (H.s), mouse (Mo.) and yeast SAGE libraries. URLs of libraries for human and mouse tissues (*www.ncbinlm.nih.gov/UniLib*; *www.ncbi.nlm.nih.gov/SAGE*) and for yeast cells on G2/M-, S-, and log- phases of cell growth (*www.sagenet.org*)

Sample	M	N	p_1	J	J/M	$k \pm SE$	$b \pm SE$
H.s.154.1	8936	4590	0.76	181	0.030	1.36±0.023	-0.12±0.013
H.s.17716	39473	14978	0.73	655	0.017	1.09±0.001	-0.26±0.001
H.s. 6360	49167	18511	0.72	540	0.011	1.10±0.001	-0.23±0.001
H.s.154	81516	19137	0.53	1598	0.020	1.25±0.012	0.57±0.016
Mo. 19018	43274	17754	0.72	630	0.015	1.19±0,001	-0.17±0.001
Mo. 20427	64240	24796	0.73	425	0.007	1.14±0.001	-0.20±0.001
Yeast, G2/M	19527	5303	0.67	519	0.027	0.96±0.006	-0.20±0.006
Yeast, S	19871	5785	0.67	561	0.028	0.98±0.004	-0.20±0.004
Yeast, log	20096	5324	0.66	636	0.032	0.97±0.004	-0.17±0.004
Yeast, total	59494	11329	0.62	1716	0.029	0.94±0.008	-0.11±0.008

Several classes of distribution functions (Poisson, exponential, logarithmic series, power law Pareto-like (Johnson et al.,1992) were fit to the histograms. The model that best fit the empirical GELPF's was the Pareto-like function (Kuznetsov, 2001b):

$$f(m) = z^{-1}/(m + b)^{k+1}, \qquad (9.1)$$

where $f(m)$ is the probability that a distinct tag (representing a gene), randomly chosen from N distinct tags, occurs m times in the library. The function $f(m)$ involves two unknown parameters, k, and b, where $k > 0$, and $b > -1$; the normalization factor z is the generalized Riemann Zeta-function value: $z = \sum_{j=1}^{J}(j + b)^{-(k+1)}$, where J is the observed expression level for the most abundant tag in the library; consequently it increases with library size M and for large M value $J >> aM$, where $a > 0$. We will call Eq.9.1 the Generalized Discrete Pareto (GDP) model. The parameter k characterizes the skewness of the probability function; the parameter b characterizes the deviation of the GDP distribution from a simple power law (the GDP with $b = 0$, see, for example, dotted line on Figure 9.1a). Insert plot on Figure 9.1a illustrates that the fitted GDP model predicts well the empirical cumulative fraction function $R_e(m)$, which demonstrates that our model fits well over the entire range of experimental values.

2.2 Effect of sample size on the distribution shape

Similarly-sized libraries made using the same method from many different human tissues and cell lines have similar numbers of distinct gene tags and are characterized by empirical GELPFs with nearly equivalent parameters in their best-fit GDP models (Table 9.1). For example, the correlation plot in Figure 9.1c shows many differences between expression levels of a same distinct SAGE tag (representing mRNAs) observed in two libraries derived from normal human mammary epithelial cells and from breast invasive carcinoma tissue (library sizes 49,167 and 39,473 SAGE tags, respectively). The Pearson correlation coefficient, r, between the gene expression values for these two samples is 0.45. This is a moderate level of similarity of gene-expression profiles. However, our empirical GELPF and best-fit parameters of the GDP model for these libraries are very similar (Figure 9.1d; Table 9.1).

As the size of any library increases the shape of the empirical GELPF changes systematically: 1) the p_1, fraction of distinct tags represented by only one copy, becomes smaller, 2) the J increases proportionally to M; 3) the parameter b becomes smaller, and 4) the parameter k slowly increases (Figure 9.1b; Table 9.1). Although the yeast genome is less complex, yeast libraries behave similarly (Table 9.1; Figure 9.2a). We also found that for yeast and human libraries, all values of scaling parameter a ($a = J/M$) fall within narrow ranges (Table 9.1). These observations suggest that all studied cell types have a general underlying probability distrubution whose skewed form is dependent on the size of the sample.

Importantly, in self-similar (fractal) biological and physical systems, described by a simple power law, the parameter(s) are independent of the size of the system (Adami, 1998; Stanley et al, 1999; Jeong et al., 2000; Rzhetsky & Gomez, 2001), but not in our case. For example, Table 9.1 shows that the parameter b in the GDP model becomes larger as library size increases ($r = 0.9$ for 18 SAGE libraries, data not presented). Power law models, including Eq.9.1, also predict an unlimited increase in the number of species as the sample size approaches infinity (Appendix A), whereas the number of species (expressed genes) is a finite number. Thus, we must take into account the sample-size effect. We developed a new statistical distribution model, called the Binomial Differential (BD) distribution (Kuznetsov, 2001a; Kuznetsov, 2001b), which assumes explicitly that parameters are size-independent and a number of expressed genes is a finite number (see below). This model also assumes that each expressed gene has a finite probability of being observed in any given library obtained from a cell population, and each gene is statistically independently expressed. Although transcription events of specific genes may be correlated in a given cell, most transcription events within a cell population are expected to be statistically independent events. These assumptions are consistent with observations (Chelly et al., 1989; Ko, 1992; Ross et al., 1994; Newlands et al., 1998). In the next section we briefly describe our new model.

3.   Probability distribution and an estimator of the total number of expressed genes

Let us assume that N_t genes $1, 2, ..., N_t$ are expressed with M_t associated transcripts in total in the cells of a large cell population. Also assume that these genes are expressed independently with respective probabilities $q_1, q_2, ..., q_{N_t}$, where Pr(a random transcript corresponds to gene i) $= q_i$.

Let the random variable s_i denote the number of transcripts in a random library of size M. Note $\sum_{i=1}^{N_t} s_i = M$. When $M \ll M_t$, sampling with replacement is an acceptable model of library construction. This follows a multinomial distribution (Johnson et al.,1992) . The joint probability of observing $s_1 = y_1$ mRNA transcripts of gene 1, $s_2 = y_2$ mRNA transcripts of gene 2, ..., $s_{N_t} = y_{N_t}$ mRNA transcripts in a given library with size M is defined by the probability function $f(y_1, \ldots, y_{Nt}; M) := \Pr[s_1 = y_1, \ldots, s_{N_t} = y_{Nt}]$, where

$$f(y_1, \ldots, y_{Nt}; M) := \frac{M!}{\prod_{j=1}^{N_t} y_j!} \prod_{j=1}^{Nt} q_j^{y_j}. \tag{9.2}$$

The function f has the unknown parameters $q_1, q_2, ..., q_{Nt}$, and N_t, together with the constraints $\sum_{i=1}^{N_t} q_i = 1$ and $\sum_{j=1}^{N_t} y_j = M$.

Let p_m denote the probability that a randomly chosen distinct gene is represented by m associated transcripts in the library for $m = 1, 2, \ldots$. Based the joint probability function f, we estimated the number of genes at each expression level m and taken for all sampled transcript, when M is large enough, we obtain the probability function of gene expression levels, p_m, in terms of the number of transcripts M and the estimated number of genes $N(M)$ in a sample of size M as follows (Kuznetsov, 2001a; Kuznetsov, 2001b):

$$p_m \approx h(m) := (-1)^{m+1} \frac{1}{N} \frac{M!}{m!(M-m)!} \frac{d^m N}{dM^m}, \tag{9.3}$$

where $m = 1, 2, \ldots$ and $N(M)$ is the differentiated function of M. We will call the function $h(m)$ the binomial differential (BD) function. Taking $m = 1$ in Eq.9.2, we obtain the ordinal differential equation:

$$\frac{dN}{dM} = p_1 \frac{N}{M} \tag{9.4}$$

with $N(1) = 1$. We call the function $N(M)$ defined by Eq.9.4 the population "logarithmic growth" (LG) model. p_1 is a decreasing function of M (Kuznetsov, 2001b). We have used the empirical approximation:

$$p_1 = \frac{1 + (1/d)^c}{1 + (M/d)^c}, \tag{9.5}$$

where the c and d are positive constants. Using an explicit specification of p_1 allows us to fit the BD and LG models to empirical histograms. With p_1 defined by Eq.9.5, Eq.9.4 has an exact solution for $N(M)$ in the limit as $M \to \infty$:

$$N(M) = \left(M^c \frac{1 + 1/d^c}{1 + (M/d)^c} \right)^{\frac{1 + 1/d^c}{c}} \tag{9.6}$$

with

$$\lim_{M \to \infty} N(M) = N_t = (1 + d^c)^{\frac{1 + 1/d^c}{c}}. \tag{9.7}$$

N_t is "the gene richness" estimator of the number of expressed genes in a large population of cells. Using Eq.9.2 with the fitted parameters d and c provides a mean of computing $p_1, p_2, \ldots$ at a given library size M. Note, unlike the fixed GDP models, the BD probability function depends on the number of distinct genes, N, and the library size, M; it also yields the finite value N_t for the total number of genes as $M \to \infty$. The probability function p_m has a skewed form, and is approximated by the power law form ($p_m \sim m^{-2}$; Lotka-Zipf law, http://linkage.rockefeller.edu/wli/zipf), which describes many other large-scale, complex phenomena such as income, word occurrence in text, numbers of citations to a journal article, biological genera by number of species, etc. (Yule, 1924; Simon and Van Wormer, 1963; Mandelbrot, 1982).

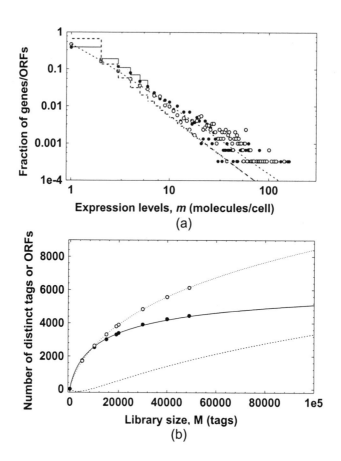

Figure 9.2. Analysis of the GELPF for yeast transcriptome. (a) Log-log plot. Dotted step-function line: best-fit GDP model for original G2/M phase-arrested yeast cell library (see Figure 9.1a). Solid step-function line: the fraction of genes/ORFs estimated by the BD model for a single yeast cell; • : data simulated from fitted GDP model for N =3,009 ORFs; ○: relative frequency of 3,009 genes/ORFs in a single log-phase yeast cell, estimated from GeneChip data (www.hsph.harvard/geneexpression) (Jelinsky et al., 2000). Dashed line links the fitted GDP model with $k = 0.86 \pm 0.01$, $b = 0.37 \pm 0.003$ for ○-data. Histogram for ○-data was constructed as follows: for each ORF/gene, the scaled hybridization intensity signal value, I, in the yeast GeneChip database (Jelinsky et al., 2000), was converted using empirical formula $m = (I - 20)/165$ to the number of mRNA molecules per single yeast cell ○-data. (b) Growth curves for the SAGE library. ○ : number of "true" distinct tags of sub-libraries from pooled yeast library of 47,393 "true" tags; dashed line: LG model with $d = 20,000 \pm 1,946$; $c = 0.356 \pm 0.02$ for ○-data; •: number of genes/ORFs observed in these sub-libraries; LG model with $d = 6,575 \pm 185$, $c = 0.579 \pm 0.01$ for •-data; short-dashed line: the number of redundant "true" tags.

4. Determination of the number of expressed genes and GELPF in a single cell

4.1 The number of expressed genes and GELPF in a single yeast cell

Without removing experimental errors in SAGE libraries one can not obtain an accurate estimate of the total number of expressed genes N_t and GELPF . We selected only SAGE tags whose location on the organism's chromosome map coincided with protein-coding gene or open reading frame (ORF) regions. Our gene richness estimator (Eq.9.7) could then from this data set estimate N_t and reconstruct the true underlining GELPF, even when the SAGE library matches a fraction of all genes. Figure 9.2 illustrates our approach using the Tag Location database (http:genome-www.stanford.edu/Saccharomyces), which contains ~8,500 distinct SAGE tags that match ~4,700 of ~6,200 yeast genes or open reading frames (ORFs). An ORF is a DNA sequence which is (potentially) translatable into protein.

First, since almost all yeast protein-coding genes/ORFs and their location on chromosomes are known, we obtained a fraction the "true" distinct tags (and their expression levels) in a yeast SAGE library by eliminating "the sequencing error tags" (Velculescu et al., 1997) that fail to match yeast genome, and erroneous tags that fail to match known 3'NLaIII genes/ORFs regions and adjacent 3'end regions presented in the chromosome Tag Location database. Second, by random sampling "true" tags, we constructed population growth curves for the numbers of "true" distinct tags and for the corresponding numbers of genes/ORFs tabulated in the Tag Location database (Figure 9.2b). The LG model (Eq.9.6) fits the size-dependent data both for "true" tags and data for genes/ORFs. In the case of "true" distinct tags (○, Figure 9.2b), our estimator (Eq.9.7) predicts a very large value: $25,103 \pm 2,000$ distinct "true" tags in a large yeast cell population. For genes/ORFs (●, Figure 9.2b), a reasonable estimate (see, for example, Cantor & Smith, 1999; Johnson, 2000) of the total number of expressed genes, $N_t = 7,025 \pm 200$, was obtained. Because a pooled library has been used, we can conclude that all or almost all yeast genes are expressed in growing normal yeast cell population, i.e. $N_t \approx G$, where G is the total number of genes in the entire genome.

At any given time, a cell generates only a subset of the possible mRNA transcripts encoded by N_{cell} genes in the genome. Using the estimated parameters $c = 0.579$ and $d = 6,580$ in the LG function (Eq.9.6) and an estimate $M_{cell} = 15,000$ of the number of mRNAs per yeast cell (Velculescu et al., 1997), Eq.9.7 predicts $N_{cell} = 3,009$ ORFs per cell.

This estimate is consistent with our estimate for a single yeast cell in the G2/M phase-arrested state (2,936 genes/ORFs;Kuznetsov, 2001b) and ~10% smaller than a published estimate of the number of expressed genes/ORFs in a log-phase yeast cell (3,298 genes/ORFs;Velculescu et al., 1997). Note, the latter estimate has a positive bias because it was obtained without elimination of SAGE tags that do not match to non-coding regions of DNA and without correction of the estimate on SAGE tag redundancy.

The GELPF for a single yeast cell (Figure 9.2a) was estimated for corrected data with both the BD (Eq.9.3) and GDP models (Eq.9.1) (Kuznetsov, 2001b). To validate an GELPF model based on SAGE data, we also used independent data obtained by Affymetrix GeneChip technology (Jelinsky & Samson, 1999) which detects more, rarely-expressed yeast genes than the SAGE method. Figure 9.2a shows a histogram constructed for the normalized hybridization signals converted to gene expression values for 3,000 more-highly expressed genes/ORFs representing ~16,000 transcripts in a yeast log-phase cell. Figure 9.2a shows that the frequency distribution for the GeneChip microarray data follows the GDP model and is consistent with the GELPF for corrected SAGE data. Similar skewed form of frequency distributions were also observed in 30 other microarray samples (untreated and treated) from yeast cells (Holstege et al., 1998; Jelinsky & Samson, 1999; Jelinsky et al., 2000).

4.2 Estimate of the number of expressed genes and the GELPF in a human cell

After validation with the yeast database, we can apply the BD model to a large human cell SAGE transcriptome (Velculescu et al., 1999) in order to predict the GELPF for a single human cell and for large populations of the same cell type. Figure 9.3a shows that the GDP model (step-function 1) fits well to an empirical histogram of distinct tags for a pooled library of colon cancer cells (with library size 591,717 tags) (Velculescu et al., 1999). To obtain the "true" distinct tags for human colon cancer transcriptome data set, we first selected all 69,381 distinct tags and then discarded the tags that did not match UniGenes associated with "known" genes in the UniGene database (Lash et al., 2000). We also discarded those tags that are classified as "sequence errors" in the yeast database. We obtained 19,777 "true" tags forming a new library with 411,361 tags. The number of mRNAs in a typical human cell, M_{cell}, is $\sim 300,000$ copies per cell (Velculescu et al., 1999). Taking 300,000 tags randomly from 411,361 tags, we obtained 17,690 "true"

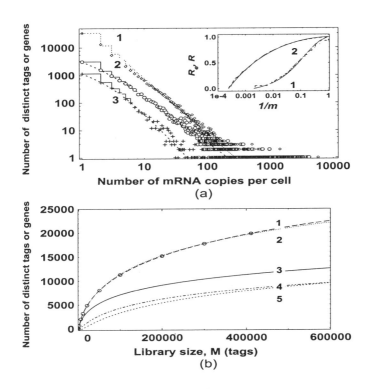

(a)

(b)

Figure 9.3. Analysis of the GELPF for human and yeast transcriptomes. (a) Log-log plot. ◇ : the numbers of 69,381 distinct tags for the human colon cancer transcriptome SAGE library of size 591,717 tags; dotted step-function line : best-fit GDP models with $k = 0.98 \pm 0.005, b = 0.405 \pm 0.007$ for ◇-data. Solid step-function line in curve-sets 1 and 2 : the number of genes by the BD model for a single yeast cell and a single human cell, respectively; dashed line in curve-sets 1 and 2 : the best-fit the GDP model to these data points for the yeast (+) and the human single cell (○; with $k = 0.75 \pm 0.089$, $b = 1.07 \pm 0.25$), respectively. Histograms 1 and 2 were generated from fitted GDP model for 3,009 genes in a single yeast cell and 10,337 genes in a single human cell,respectively. Insert plot. Dotted lines : the cumulative fraction plots computed from the ○- and +- histograms, respectively; solid lines : the cumulative fraction function, R, from best-fit GDP model (for dashed lines on main plot). (b) Growth curves for tag subsets of human library. ○ : the number of "true" distinct tags in sub-libraries from the human colon cancer cell lines transcriptome library of 411,391 "true" tags; curve 1 (dashed line) : best-fit LG model (with $d = 46,232 \pm 200, c = 0.319 \pm 001$) for ○-data; curve 2 (dotted line) : best-fit of $N_g + N_r$ (see Appendix B) to ○-data; curve 3 (solid line) the number of different genes N_g; curve 4 (– – –) : the number of redundant distinct tags N_r; curve 5 (short-dashed line) : the number of redundant distinct tags for yeast data (see also shot-dashed line in Figure 9.2b).

distinct tags which represent the content of a single colon cancer cell
(Figure 9.3a, GDP model: step-function line 2). Figure 9.3b shows that
Monte-Carlo generated growth curve for "true" tags data is fitted by the
LG model. However, this fit yields the value $N_t = 72,884$ as $M \to \infty$
(with $c = 0.319 \pm 0.010$; $d = 46,232 \pm 200$). This estimate is about
twice as big as the recent estimates (~26,000-35,000 genes) provided by
the Human Genome Sequencing Projects for the entire human genome
(Venter et al., 2001; IHGSC, 2001). This implies that the total number
of "true" tags that match the same gene dramatically grows when the
SAGE library size becomes bigger.

Let $N_g(M)$ denote the true number of expressed genes in a library
of size M. If $N(M)$ is the total number of "true" distinct tags in the
same library, then $N_r = N - N_g$ is the number of distinct tags that
redundantly represent genes. Introducing the function $N_r(M)$ allows
us to deal with the tags-to-gene and tag-to-genes multiple matching
problem (Velculescu et al., 1999; Lash et al., 2000; Caron et al., 2001).
Figure 9.3b shows the number of expressed genes N_g (curve 3), the
number of redundant tags N_r (curve 4), and the sum of $N_g + N_r$ (curve
2) with increasing sample size. Figure 9.3b also shows that the number of
redundant tags occurring in a human library (curve 4) is larger than the
number of redundant tags extrapolated for the same-size yeast library
(curve 5) (see also Figure 9.2b). The latter difference is expected due
to the higher complexity of the human genome. The best-fit estimate
of N_g (curve 3, Figure 9.3b) predicts $N_{cell} = 10,336$ expressed genes for
a single cell (at $M_{cell} = 300,000$ tags); $N_t = 31,238$ expressed genes
are predicted by the LG model for a large colon cancer cell population
($M \to \infty$). Note that our procedure for selection of "true" tags tend to
be *conservative*, because some of the UniGene clusters may be erroneous
itself, and UniGene database is large, but not a complete catalogue of
all human transcript, i.e. does not cover entire human genome(Lash et
al., 2000). Therefore, some SAGE tags which we discarded may still
represent real transcripts, as was shown many times (for example, Croix
et al., 2000).

Comparison of GELPFs for yeast and human single cells.
Figure 9.3a shows the probability function for the number of expressed
genes for a single colon cancer human cell (step-function 2) and for a
single yeast cell (step-function 3) estimated by the BD model. These
functions were computed as described in Appendix B. To show the variability of the frequencies of different genes at the different expression
levels, we also sampled expression levels from the fit GDP model N_{cell}
times ($N_{cell} = 10,336$) (o, Figure 9.3a), just as we did for yeast data.

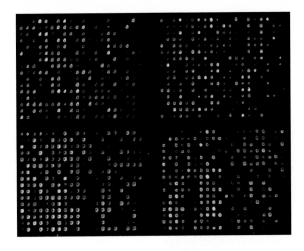

Figure 1.2 (see page 4)

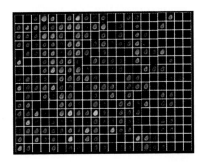

Figure 1.4 (see page 6)

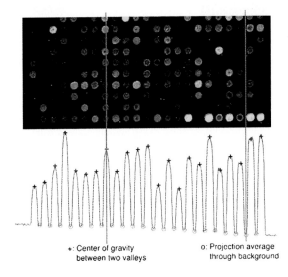

+: Center of gravity
between two valleys

o: Projection average
through background

Figure 1.5 (see page 7)

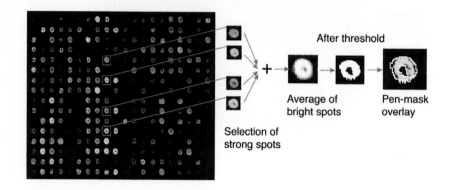

After threshold

Average of bright spots

Pen-mask overlay

Selection of strong spots

Figure 1.7 (see page 10)

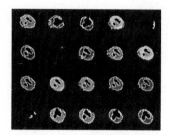

Figure 1.8 (see page 11)

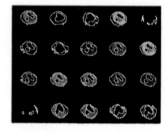

Figure 1.9 (see page 11)

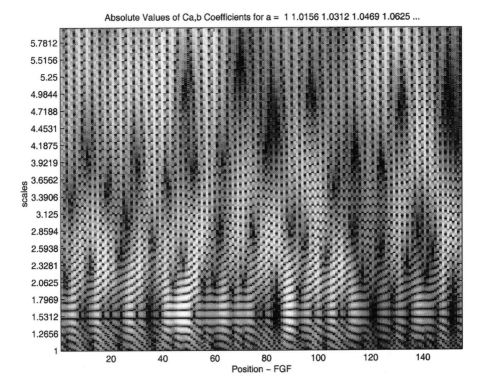

Figure 8.6 (see page 119)

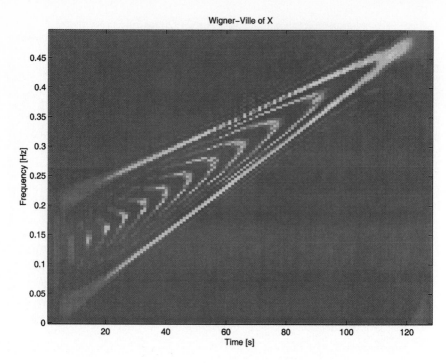

Figure 8.7 (see page 120)

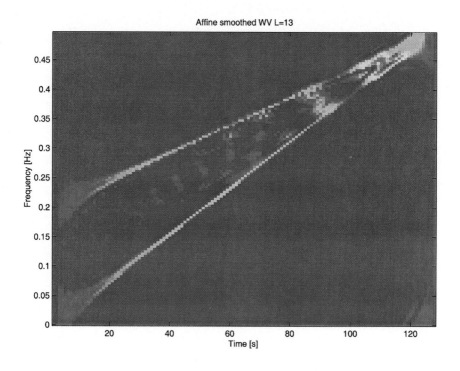

Figure 8.8 (see page 120)

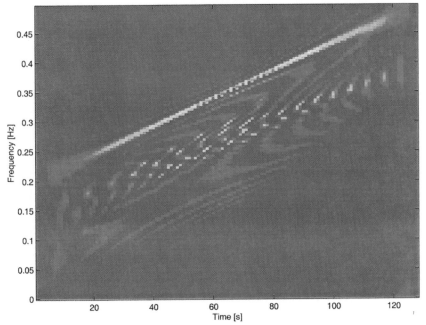

Figure 8.9 (see page 121)

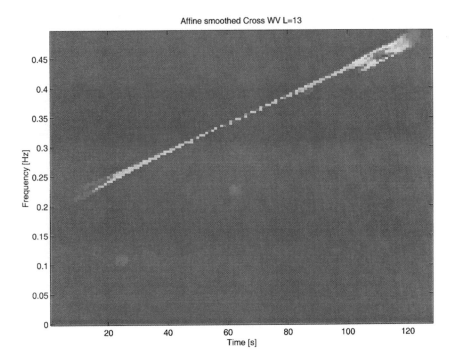

Figure 8.10 (see page 121)

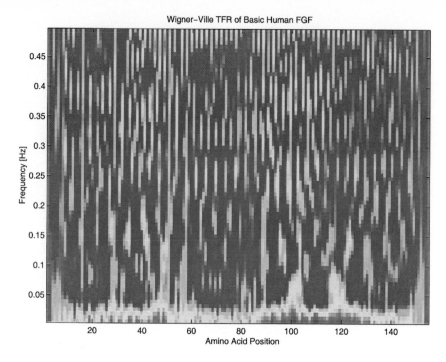

Figure 8.11 (see page 122)

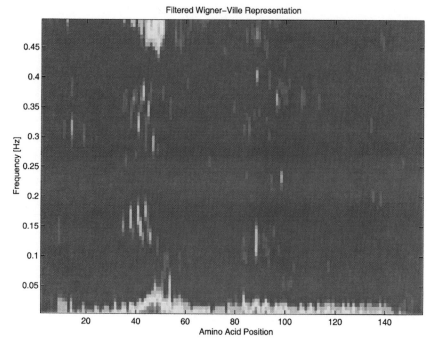

Figure 8.12 (see page 122)

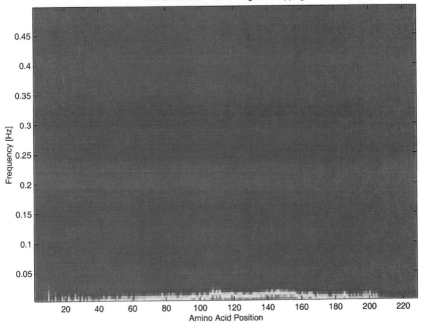

Figure 8.13 (see page 123)

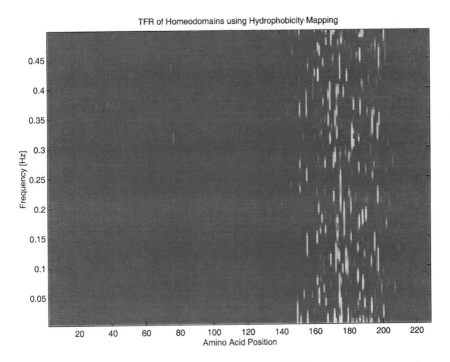

Figure 8.14 (see page 123)

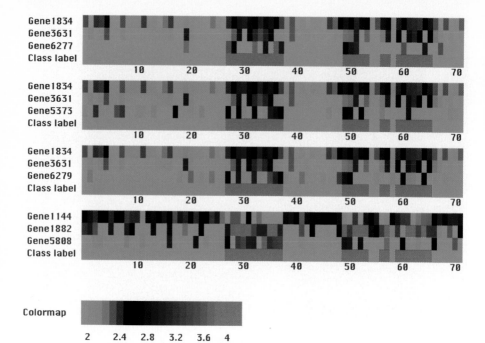

Figure 10.8 (see page 195)

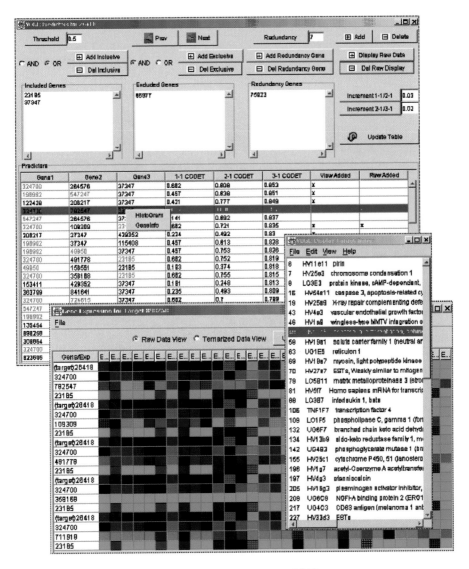

Figure 13.1 (see page 237)

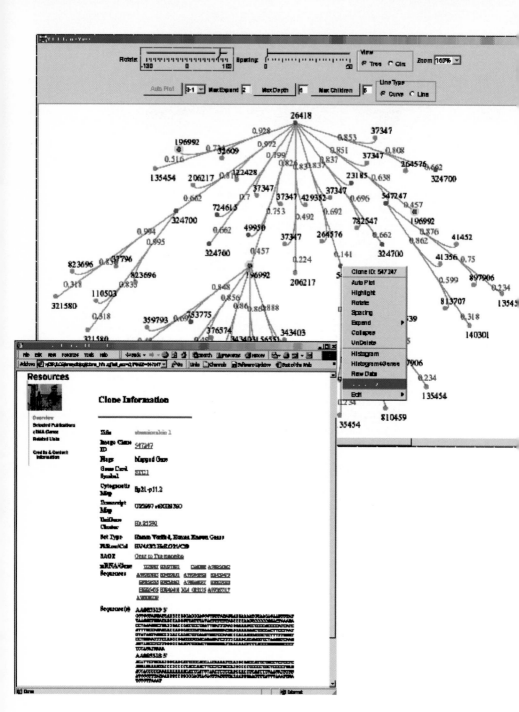

Figure 13.2 (see page 238)

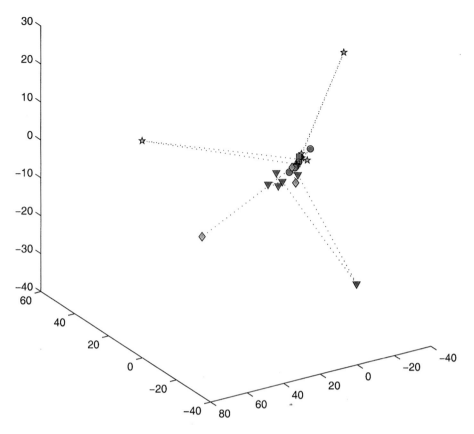

Figure 14.5 (see page 251)

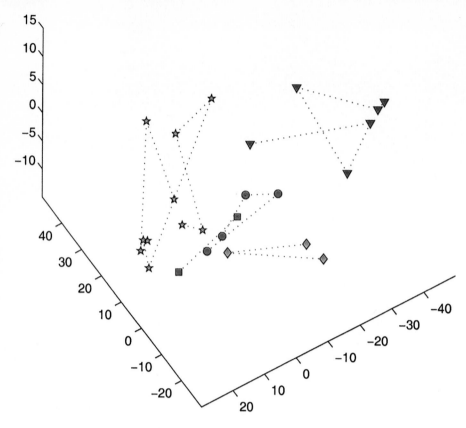

Figure 14.6 (see page 252)

The shapes of the yeast and human GELPFs are quite similar (Figure 9.3a). However, 38% and 31% of active genes are represented by a single mRNA copy per cell for yeast and human cells, respectively. The GDP model parameter values are significantly different ($k = 1.51 \pm 0.002$ and $b = 2.07 \pm 0.009$ for a single yeast cell, and $k = 0.747 \pm 0.089$ and $b = 1.074 \pm 0.250$ for a single human cell, respectively) as well as the cumulative fraction of mRNAs plots (insert plot on Figure 9.3a). On average, there are five mRNAs per gene in a single yeast cell and 30 mRNAs per gene in a single human cell.

5. Global Transcription Response to Damaging Factors

The GELPF can be used to characterize global changes in transcriptional profiles that take place when cells are exposed to reactive chemical agents, such that virtually every molecule in a cell is at risk of being altered in some way. We have inspected data at http:/www.hsph.harvard.edu/geneexpression which represents 30 mRNA profiles of ~6,200 yeast genes/ORFs sampled from normal cells at the different phases of cell growth and cells treated with clinically relevant chemical and physical damaging factors. These data sets have been obtained by GeneChip microarrays technology. Despite the differences of chemical factors and differences in transcription profiles in cells on different phases of cell growth, the responses of cells on diverse exposures look very uniform. We found that global response of transcription machinery in yeast cells is a non-linear function of damaging level (Figure 9.4a). Interestingly, the relationships between total number of expressed genes in treated cells and the levels of cell damaging is opposite to mRNAs response (Figure 9.4b).

Figure 9.4c shows the typical changes of the empirical frequency distribution and of the shape of fit GDP model in the case of response *S. cerevisiae* genes upon exposure to a mildly toxic dose of a alkylating agent, methyl methanesulfonate (MMS). Inserted plot in Figures 9.4c shows that the number of low-expressed genes per treated cell is increased, but expression levels of highly-abundant transcripts are decreased. Note that for all our analysis the hybridization signals corresponding to ≥ 0.5 mRNA molecule per cell were considered. Correlation plot on Figure 9.4d shows that gene-expression profiles for pairs of the samples obtained from normal cells are highly reproducible ($r_{12} = 0.91$). Figure 9.4d also shows that damaging factor mostly amplifies a transcription of many hundreds of rarely-transcribed genes and reduces an abundance of transcripts of highly- and moderately- active genes. At 60 min after initiation of the treatment, the total number of expressed genes per cell was increased by ~ 600 (20%) genes; however, the total number of mRNAs was decreased by $\sim 2,000$ molecules (12%). We obtained similar relationships in normal and MMS-treated cells sampled on the G1-, S- and G2- phase of cell growth and in cells sampled on log-phase of cell growth and then treated with several distinct damaging agents (data not presented). Jelinsky *et al.* (1999; 2000) shows that MMS exposure represses groups of genes involved in RNA and nucleotide synthesis, in the synthesis and assembly of ribosomal proteins and, simultaneously, MMS activates protein-degradation processes.

Non-linear amplification and/or induction of the expression of hundreds of very low-transcribed genes at any phase of cell growth in a quantitatively predictable manner is a big surprise. How does the cell regulate the massive production of these rare mRNAs in such low, but reasonably determined numbers per cell? We could assume that in a normal yeast cell the limited amount of specific transcription factors and other less specific regulatory elements of transcription machinery may exist. Most of these regulator molecules could be pre-occupied during much time with the transcription of genes coding the highly-abundant and moderately-abundant proteins (i.e., ribosomal genes). Stress factors or damaging agents could dramatically reduce de novo transcription and de novo protein synthesis (Jelinsky & Samson, 1999; Jelinsky et al., 2000), lead to temporal increasing of free transcription factors in cell nucleus and increase a possibility of the interaction of transcription factors with promoter regions of many normally rarely-expressed genes.

Interestingly, the observed global gene response to different stress conditions in yeast cell populations demonstrates the properties of finite-size self-organized critical (SOC) systems (diverse class of random cellular automata models) (Kauffman, 1993; Adami, 1998). Such systems, which based on simple logical rules, assume that a finite number of different components (i.e., genes) locally interact in the system, where a noise can propagate through the entire system and leads, at some conditions, to catastrophically big events (Adami, 1998). Power law-like distributions associated with SOC systems demonstrate size-dependent effects, robustness to perturbations, changes in the system composition and in network re-organization in response to random environment.

6.   Stochastic and Descriptive Models of Gene Expression Process

Analysis of human transcriptome has allowed us to estimate the cumulative numbers of both expressed protein-coding genes as well as er-

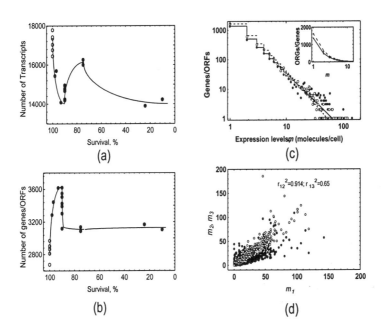

Figure 9.4. Global transcriptional response of yeast cells to damaging agent MMS. Relationships between the number of mRNA transcript per cell (a), the number of expressed genes per cell (b) and the percent survival (as determined by colony-forming activity test). The percent of survival for several data points did not measured directly, but it was estimated by the dose-dependence or time-dependence relationships for a same damaging factor. o, • :untreated and treated cells, respectively. (c): Log-phase yeast cells were exposed to 0.1% MMS for one hour and resulted in the 90% survival (sa determined by the colony-forming ability). Log-log plot. o : normal cell sample (#zero2) with 2,970 genes/ORFs representing 16,762 mRNAs/cell; dotted step-function line : best-fit GDP model at $k = 0.86 \pm 0.001$ and $b = 0.37 \pm 0.003$ for o-data; • : 3,548 genes/ORFs representing 14,733 mRNAs/cell; solid step-function line: best-fit GDP model at $k = 1.26 \pm 0.003$ and $b = 1.0 \pm 0.009$ for o-data. Insert plot: Log-linear plots for the left-side of the distributions presented on major plot at $m < 15$. (d) o, • : Correlation plots of the expression value of pairs (m_1, m_2) in normal cell samples (#zero1, #zero3;$r_{12}^2 = 0.91$) and of (m_1, m_3) in pair of normal and treated cell samples (#zero2, #MMS;$r_{13}^2 = 0.65$), respectively.

roneous and redundant sequences. After eliminating the erroneous tags and redundant tags, we estimated that there are ~10,000 protein-coding genes expressed in any random single human colon cancer cell. This estimate is ~5.5 times smaller than the previous estimate of 56,000 expressed genes per cell made for the same human SAGE transcriptome (Velculescu et al., 1999). Thus, an effective process for removing both erroneous tags and redundant multiple matching tags from SAGE libraries is critical for obtaining an accurate profile of expressed genes and for more accurately estimating the number of genes in human cells and in other organisms. After removing the errors and redundancy, GELPF for yeast shows that ~1,200 genes/ORFs are represented on average by a single mRNA molecule per cell (Figure 9.2a). The population growth curve for genes/ORFs (Figure 9.2b) indicates that ~3,800 additional genes/ORFs (~55% of all yeast genes) are expressed at less then one transcript per cell. These estimates are consistent with our estimates based on data found in the large-scale yeast oligonucleotide microarray databases (Figure 9.4c). Similar statistical analysis for human transcriptome data indicates that ~70% of all protein-coding human genes are expressed with less than 1 transcript per cell (Figure 9.3a). These results strongly support our assumption that a major fraction of expressed genes in an eukaryotic cell are transcribed sporadically and randomly.

This point is supported by many observations. Activating gene expression is a complex, multi-step probability process that must ultimately load the transcription machinery onto a gene. This can only occur when appropriate chromatin states allow specific transcription factors to access the relevant sequences within a promoter. Once bound, these factors and associated transcriptional cofactors form a complex that positions RNA polymerase so transcription can begin. The transcription of different genes in different species is a fast process with a rate of ~1-2 kb/min (Femino et al., 2000; Jackson, 2000). During proliferation, a mammalian cell can produce ~ 10^7 copies of the specific protein. An example of this is the transcription initiation factor eIF-4A (Jackson, 2000). A single copy gene can sustain this level of protein synthesis if it transcribed continuously with one to two initiations per minute (assuming transcription rate of 2 kb/min, translation rate of 250 amino acids per minute at 1 initiation per 0.4 minutes and average mRNA life-time and cell cycle parameters taken in (Jackson, 2000). However, even in proliferating mammalian cells, many of genes are transcribed infrequently, typically less than once per hour (Jackson, 2000). It is clear from the corrected probability distribution of gene expression levels in both proliferative and non-proliferative cells (Figure 9.1, Figure 9.3; Table 9.1) that the most mRNAs presented at < 2 copies per cell need only to be transcribed very seldom (i.e. > 1 molecule per 5 hours at a typical half-life ($\tau_{1/2} = 5\ h^{-1}$) for human mRNAs (Jackson, 2000). Mechanistic aspects of such pulse transcription process in a cell may be due to (1) temporal modulation of supercoiling effects of chromosomes, (2) temporal inactivation of regulatory repressors (i.e. the promoter methylation), and (3) the limited amount of specific transcription factors and other regulatory elements of transcription machinery which could be preoccupied for a long time with the transcription of genes coding for abundant proteins.

There is growing evidence that initiation of transcription of at least a major fraction of expressed genes has been observed to occur sporadically and randomly both in time and location on chromosomes in a variety of cell systems. Much work in support of this mechanism has been done in cell cultures in which gene activity was determined on separated cells rather than as the entire cell tissue population. It was reported (Weintraub, 1988; Walters et al., 1995) that enhancers act to increase the probability but not the rate of transcription. This random on/off model of initiation/suppression of transcription is supported by the results in (Ko, 1992; McAdams, 1999). It was also shown that luciferase reporter promoter genes in transfected cells shuttle on and off in individual cells (Ross et al., 1994; Hume, 2000), theorizing that transcription might be a random event occurring infrequently and in short pulses. Newlands et al. (1998) and Sano et al. (2001) have shown striking examples of such stochastic gene expression patterns.

Based on statistical analysis of GELPFs in different cell tissue types, states of the same-type cells and cell types from different species, we suggested that the existence of such random transcription process implies that all or almost all protein-coding genes in a genome should have a small but positive probability to be transcribed in any given cell during any fixed time-interval. This suggestion is consistent with the observation of many low abundance transcripts of various tissue-specific genes in human cells of different type, such as fibroblasts, lymphocytes, etc. (Chelly et al., 1989). Although not all cells of a population would have a copy of a specific transcript at a given moment, we would expect to see all these genes expressed, at least at a low level, in a sufficiently large cell population at any point in time. That is, ergodicity holds. This point is supported by data in GeneChip database (Jelinsky et al., 2000): we observed that only 250 ORFs ($\sim$150 of them are "questionable" or "hypothetical" ORFs) of $\sim$6,200 genes/ORFs were not expressed in 6 samples from normal growing yeast cells. In mammalian cells, a significant fraction of genes are established with silencing (transcripts are not observed); the silent state of a gene can be inherited, but later reactivated involving the stochastic, all-or-non mechanism at the level of a single cell (Sutherland, et al., 2000). Hybridization experiments show that in proliferated HeLa cells at least 50,000 different transcriptional units in non-repeated DNA are transcribed (Jackson, 2000). That number overestimates recent estimates of the number of genes in the entire human genome reported by the Human Genome Sequencing Projects (26,000-35,000 genes; Venter et al., 2001; IHGSC, 2001). By our conservative estimates, presented above, ~31,200 genes are active in human colon cell type. Based on this data, we might suggest that all or almost all genes of entire human genome are transcribed in distinct human cell types, i.e. we postulate a formula: $G \approx N_{ti}$, where G, N_{ti} are the total number of expressed genes in the entire human genome and in i-th tissue type.

Random and seldom on/off transcription events must be more general in genes and gene clusters with low probability of transcription. Physically, such random "basal" levels of gene transcription events might reflect nonlinear on/off-triggering of the independent "gene regulatory complexes" to internal or external fluctuations including thermal molecular motion. Generally, movement of proteins within the nucleus by passive, randomly directed diffusion and other dynamical properties of proteins in the nucleus are associated with stochastic mechanisms of gene expression (Misteli, 2001). Low probability transcription events for many genes could be mediated by its own specific transcripts (Thieffry et al. 1998). In recent years, a diverse class of nuclear small regulatory RNAs (often denoted riboregulators) was found, which involved in cellular responses to stress conditions, and which act as global (negative) regulators by affecting the expression of many genes (Eddy, 2001). Many of these non-coding RNAs are cis-anti-sense RNAs that overlap coding genes on the other genomic strand. In our analysis of yeast transcriptome , we found that in a cell ~500 distinct tags matched anti-sense sequences matches totally ~1000 genome regions. A significant fraction of RNA transcripts associated with this large number of SAGE tags might be involved in negative control of transcription of protein-coding genes. On the another hand, fast metabolism of nuclear non-coding RNA molecules provides more mobility of enhancer and promoter regions, as well as many protein molecules involved in transcription process and, finally, permits a spatio-temporal conditions for the transcription of particular groups genes which previously were non-activated (Jackson, 2000). Thus, we suspect that in response to environmental changes, various stress conditions and local fluctuation of molecular composition the initiation of transcription process and regulation post-transcription events for many rarely-expressed genes could be under dynamical control of small non-coding genes, many of them represent anti-sense RNA transcripts. Such autoregulation tends to keep the low-expressed gene "half-on" sporadically providing the mechanism of low expression for many genes.

Because the distribution of the life-times of "switch-on" and "switch-off" states for genes in a single cell appears to have a long right tail (McAdams, 1999), one of the two alleles of the same locus for a given low-expressed gene might be present in the same state for a long period of time. If this is the case, a natural clonal selection process in a population of the cells could select the clone(s) with monoallelic gene expression , i.e. phenomenon in which only a single copy, or allele, of a given gene of diploid organisms appears (Hollander, 1999; Ohlsson et al., 2001; Cook et al., 1998). Asymmetric transcription of "long-term controlled" rarely-expressed genes might be also limited by non-uniformly distributed concentration of components of the preinitiation complex and some specific transcription factors. These concentration limits might also lead to monoallelic gene expression phenomena.

7. Probability Distributions of the Numbers of Putative Proteins by Their Cluster Sizes and DNA-binding Proteins by the Regulated Promoters

The probability distribution of expressed genes at the level of cytoplasmic mRNAs in a cell effects the probability distribution of the synthesizing proteins. We expect certain similarity in (self-)organization of the gene expression processes at the protein level, to those that we observed for the mRNA transcription level. Recently, Drs. Ramsden and Vohradsky (Ramsden & Vohradsky, 1998; Vohradsky & Ramsden, 2001) have found that the rates of occurrence of observed highly abundant proteins of prokaryotic organisms follow a skew distribution, which we assume could be fit by our GDP and BD models[1].

---

[1] To analyse empirical distributions of the rates of proteins synthesis Drs. Ramsden and Vohradsky (Ramsden & Vohradsky, 1998) used a model proposed by Mandelbrot for the rank-frequency distribution of words in text sequences Mandelbrot, 1982. However, the model fit systematically only a portion of the range that can be approximated by the model. The results indicated that occurrences of most probable "protein" words as well as most rarely observed "protein" words are very different from those predicted by the Mandelbtot model for natural languages. Note also, that the linguistic value of Mandelbrot model has been doubted for a long time. Mandelbrot (Mandelbrot, 1982) and others (Li, 1992) have shown that even randomly generated "text" (with words of different length) exhibits power-law behavior closed to that of natural languages.

The proteins clustered by the similarity of the protein sequences shows another interesting example of the skew distribution in proteomic. Li *et al.* (2001) applied the clustering algorithm, which they developed, to count the number of the protein clusters in a population of 40,580 putative 'proteins' found in the International Protein Index database (http://www.ensemble.org/IPI). Log-log plot in Figure 9.5a shows that empirical frequency distribution of the constructed protein clusters versus the cluster sizes contains two distinct distributions. The frequency distribution of the fist highly diverse subpopulation of the clusters (contained ~97% of 40,580 proteins), fitted well by the GDP model (at $k = 1.65 \pm 0.006$; $b = -0.32 \pm 0.04$; $J = 71$). The second subpopulation of the clusters contains a few large-size clusters (i.e. zinc finger proteins cluster is the largest, with 479 proteins). The negative value of parameter b, and large fraction of "singleton" protein sequences ($p_1 = 0.87$), indicates the large number of "erroneous" small-size clusters and to undersampling of the data set. We might expect that many of observed 25,237 "singleton" proteins might be clustered when a population of well-defined proteins will be larger and when clustering algorithms become more accurate.

Analysis of the empirical frequency distribution of the numbers of 97 identified proteins regulating a given number of promoters in *E.coli* provides interesting insight into the structure of the entire regulatory system of initiation of transcription (Thieffry et al. 1998). On a log-log plot, this data give us an idea of the complexity of transcriptional regulation in *E. coli.* by proteins (Figure 9.5b). This figure shows that GDP model sufficiently well approximates the empirical frequency distribution at $k = 1.45 \pm 0.053$; $b = 1.91 \pm 0.17$. Figure 9.5a reflects also how complex the initiation of transcription for a given promoter can be. For example, the last right-hand point of Figure 9.5b shows that there is a single common protein in *E.coli* proteome which in potentially involved in regulation of 71 distinct promoters; however, 54 of 97 regulating proteins (55%) interact with only 1 to 2 promoters and, on average, one of these DNA-binding proteins controls 4 promoters.

8.   Protein Domain Statistics in Proteomes

8.1   Statistical Analysis of Proteome Complexity

Basic evolutionary processes are probably most effectively understood in terms of distinct protein domain coding DNA sequences and thus in terms of associated protein domains, rather than in terms of genes as the primary evolving units (Rubin et al., 2000; Li et al., 2001; Koonin

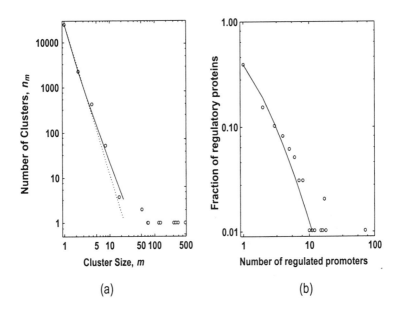

Figure 9.5. Skew probability distributions in proteomic. (a) Probability distribution of putative protein clusters by the cluster sizes. The 40,580 putative proteins have been selected and clustered by a measure of proportionality of identical amino acids in aligned region between paired sequences (Li et al., 2001) at the level of similarity $I \geq 30\%$ proposed by the authors. Log-log plot: o : a number of protein clusters; solid line and dotted line: fitted GDP ($b = -0.32 \pm 0.004$; $k = 1.65 \pm 0.006$) and simple power law (at $\kappa = 2.33 \pm 0.03$) for o-data at $J \leq 71$, respectively. The 2-d, 3-d and 4-th empirical points on the left side of the plot represent averages of 3, 5 and 10 observations, respectively. (b) Relationships between a fraction of regulatory proteins and the number of regulated promoters in *E. Coli* (Thieffry et al. 1998). Log-log plot: o : data; solid line: best-fit GDP model at $b = 1.91 \pm 0.17$ and $k = 1.45 \pm 0.053$ for o -data.

et al., 2000; Venter et al., 2001; IHGSC, 2001). Protein domains generally correspond to locally folded segments of proteins that serve as active binding sites governing selective interactions of a protein with other specific molecules. For many different proteins, these domains can be found alone or in combination with other or the same domains so as to combinatorially form the functional parts of the proteins that make up the proteome. Currently, about 4,700 "identifiable" domains that collectively occurred in more than 600,000 putative (predicted) proteins of 6 eukaryotes and more than 40 prokaryotes were described in the Interpro database at the end of 2001 (www.ebi.ac.uk/interpro/). Current protein domain and protein family databases are incomplete and have errors and redundancies. It is obvious that many distinct domain coding DNA sequences and thus the associated protein domains have been (1) highly conserved in evolution and (2) shared among different organisms (Rubin et al., 2000; Li et al., 2001; Venter et al., 2001; IHGSC, 2001). For example, about 50% of the 1401 Interpro protein domains of *Saccharamyces cerevisiae, Drosophila melanogaster, Caenorhabditis elegans* are shared among these organisms. Only about 7% of Interpro domains are unique to the human proteome in comparison to these organisms (IHGSC, 2001). During evolution, proteomes have become more complex through such processes as gene duplication, mutation, and shuffling in vertical descent within the same species organisms (paralogous processes) and in horizontal displacement of genes between different species (orthologous processes) (Rubin et al., 2000; Koonin et al., 2000; Friedman & Hughes, 2001). It appears that a domain that has appeared in a specie has a significant chance to be conserved in evolution. Rather, more combinations of domains arise as protein complexity grows. Thus, most of the biological functions appear to be provided by a limited set of shared protein domains which can be traced back to a common ancestor species (Rubin et al., 2000; Venter et al., 2001; IHGSC, 2001). This suggests that basic evolutionary processes could be fruitfully understood in terms of protein domains.

How can we quantitatively describe proteome complexity? The frequencies of occurrence of distinct domains are very different even in the same eukaryotic proteome (Rubin et al., 2000): most domains are found only once (though the same domain is present in many other organisms) in a given proteome; however, a small percentage of domains occur very often. A given protein contains from one to many dozen domains, and may include the same domain several times. Moreover, a given domain may appear in many distinct proteins in the same proteome and in the proteomes of many distinct species. Generally, possible relationships between protein domains and proteins may be presented by a bipartite

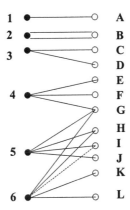

Figure 9.6. A bipartite graph network of six domains (labeled 1 through 6) and 12 proteins (labeled **A** through **L**), with edges joining each protein to the domains in which it is included. Dashed line indicates an appearance of domain 6 in protein **I**.

graph. Figure 9.6 shows a small bipartite graph with six domains (labeled 1 to 6) and 12 proteins (labeled **A** through **L**), with edges joining each protein to the domains it includes. Let the *proteome domain profile* for a specific organism be the list of all distinct protein domains inherent to the organism, together with the numbers of occurrences of each of the domains in the proteome of the organism. Each protein that makes up the proteome is represented just once in counting domain occurrences. There is a fundamental statistical description of a proteome domain profile which consists of the distribution of the domain occurrence values in the profile.

Let X denote the number of occurrences of a random distinct domain in the profile. We may define the *domain occurrence probability function* (DOPF) $f(m) := P(X = m)$ for $m = 1, 2, \ldots$. The DOPF gives the probability of each possible domain occurrence value 1,2, etc. arising for a random domain in the proteome domain profile.

Currently, the proteome domain profiles for studied organisms, even those having fully sequenced genomes, are incomplete and can only be approximated by various inference-based methodologies such as analysis of the genome sequences of the organism. Let n_m be the observed number of distinct domains which have the occurrence value m (occurring m times) in a given sample from the proteome. Let J denote the observed occurrence value for the most frequent domain in the sample proteome. The $\sum_{m=1}^{J} n_m = N$ is the number of distinct domains in the sample. The points $(m, g(m))$ for $m = 1, \ldots, J$, where $g(m) = n_m/N$, form the histogram corresponding to the *empirical* DOPF. Thus, the

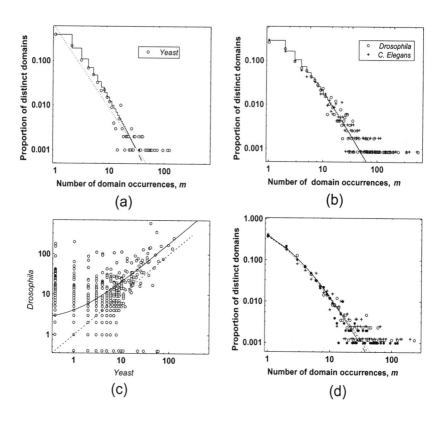

Figure 9.7. Comparative analysis of proteome complexity in yeast, worm and fly. (a) • : Empirical frequency distribution of domain occurrence values in yeast proteome sample; solid step-function line: the GDP model for o-data; dotted line: best-fit Riemann-Zipf model ($b = 0; k > 0$) for •-data. (b) Empirical frequency distribution of domain occurrence values in fly (o) and worm (+) proteome samples; solid step-function line: the GDP model for o-data; (c) Correlation plot: co-occurrences of protein domains in yeast and fly proteomes. Solid line: the regression the model $m_{fly} = 2.23 + 1.85 \cdot m_{yest}$, where m_{fly} and m_{yest} are the occurrence values for fly and yeast proteome samples, respectively; dotted line is a diagonal. (d) Empirical frequency distribution of domain occurrence values in yeast proteome sample (•) and same-size sub-samples randomly-chosen from fly (o) and worm (+) proteome samples; solid line, dashed line and dotted line fit points calculated by the best-fit GDP model for •-, o-, and +-data, respectively; sub-samples from worm and fly data sets containing by 4,745 domain occurrences.

Table 9.2. Characteristics of protein domain families. N_t : the total number of genes/ORFs; A: the number of identifiable proteins corresponding to domain families; $A_{m>1}$: the number of identifiable proteins contained more than one InterPro domains; N: the number of distinct InterPro domains; M: the total number of domain occurrences per proteome (Rubin et al., 2000). μ: average number of appearances of a given domain in proteome ("average diameter" of domains-proteins network); α: average number of domains per protein; p_1: the fraction of domain occurred ones; k, b : parameters of the GDP model; κ : the slope parameter of simple power law; Ψ_{GDP}, Ψ_{PL} : model selection criterion (Kuznetsov, 2001b) for the GDP and the simple power law, respectively

	Fly	*Worm*	*Yeast*	*Fly$_s$*	*Worm$_s$*
N_t	13603	18424	6241	-	-
A	7419	8356	3056	-	-
$A_{m>1}$	2130	2261	672	-	-
N	1177	1144	1000	847	787
M	11569	12822	4547	4547	4547
α	9.8	11.2	4.4	5.4	5.8
μ	1.56	1.53	1.49	-	-
J	579	545	125	236	199
p_1	0.26	0.29	0.39	0.38	0.42
k	1.24±0.02	1.01±0.001	1.54±0.019	1.41±0.008	1.0±0.007
b	3.0±0.016	1.95±0.007	2.10±0.063	2.00±0.03	0.91±0.02
Ψ_{GDP}	4.7	5.7	5.5	5.3	5.5
κ	0.69	0.65	0.82	0.46	0.53
Ψ_{PL}	0.8	1.4	0.03	3.6	4.7

empirical DOPF specifies the fractions of distinct domains that occur $1, 2, \ldots$ times in the sample. The empirical DOPF is just the normalized histogram of domain occurrence values (Figure 9.7a,b).

We analyzed the empirical DOPF for the proteomes of a number of fully-sequenced organisms. The data we used was obtained from the Interpro database (http://www.ebi.ac.uk/interpro). We examined how the empirical DOPFs change with the increasing complexity of eukaryotic organisms including *S. cerevisiae*, *C. elegans*, *Drosophila melanogaster*, *Homo Sapience*, *Guillardia Theta* and also several bacteria (*L. Lactis*, *E. Coli*, etc*).

We observed that for each proteome, the histograms of domain occurrence values exhibited remarkably similar monotonically-skewed shapes with a greater abundance of bigger domain occurrence values and more gaps among these higher-occurrence values. Figures 9.7a,b show the empirical probability functions (histograms) of the domain occurrence values in the proteome samples of *S. cerevisiae*, *C. Elegans* and *Drosophila melanogaster*. The histograms for the fly and for the worm are most similar, reflecting their greater complexity and proteome size. Note, "the protein sets" used were samples of the proteomes, not the complete proteome data sets. In particular, we have used a list of 1401 InterPro domains (Rubin et al., 2000). We observed 1,000 of these domains in the yeast protein set, 1,144 in the worm protein set, and 1,177 in the fly protein set.

All the histograms we studied can be approximated by the generalized discrete Pareto probability (GDP) function (Kuznetsov, 2001b):

$$f(m) = z^{-1}/(m+b)^{k+1},$$

where $f(m)$ is the probability that a domain, randomly chosen from the N distinct domains, occurs m times in the sample of proteome. The function f involves two unknown parameters, k, and b, where $k > 0$, and $b > -1$; the normalization factor z is the generalized Riemann Zeta-function value: $z = \sum_{j=1}^{J}(j+b)^{-(k+1)}$.

Let M be the total number of occurrences of domains in the sample proteome. We will call M the *connectivity number*. This number reflects the complexity of the proteome: M is larger for multicellular eukaryotic organisms (the fly and the worm has the connectivity numbers 12,822 and 11,569, respectively) than for single-cell organisms (M is only 4,547 for the yeast). Interestingly, that the mean of domain occurrence value in a set of distinct domains, M/N, is also larger for more complex organisms. Moreover, M positively correlates with the occurrence value J of the most frequent domain in the sample proteome, and negatively correlates with a frequency of domains occurred only once, p_1, in sample proteome.

New domain usage within a proteome results from gene duplication, shuffling, and/or adding or deleting sequences that code for domains. This results in new proteins from old parts. A large proportion of proteins in different organisms share the same domains, even over widely divergent evolutionary paths. Importantly, a domain occurring once in some species, has a high probability to be used in other species multiple times in different proteins. To specify this rule we analyzed the occurrence of domain-coding sequences that have been found in tiny 551-kilobase eukaryotic genome of the cryptomonad *G. theta*, in the domain-coding sequences for more complex organism yeast. We counted the number of occurrences of all 211 distinct domains which have been found in *G. theta*. Table 3 shows the frequency distributions of the *G. theta* domains, identified (or missing) in the yeast proteome. Only 9% (19 of all 211 distinct *G. theta* domains) are missing in the yeast proteome; 18 of these 19 domains occurring in the *G. theta* proteome, only once or twice. Most of the *G. theta* domains have been multiplied in the yeast proteome. We observed a positive correlation ($r = 0.61$; $p < 0.001$)

Yeast	G. Theta 1	2	3	4	5	6	9	10	11	12	Total
0	14	4							1		19
1	44	2			1						47
2	50	5									55
3	14	3	2			1					20
4	8	2	1	1							12
5	10										10
6	1	1	1								3
7	1	3	1								5
8	3	1									4
9	2	1					1				4
10	1										1
11	1										1
12	1			1							2
13	1	1									2
14	2		2							1	5
16	1					1					2
17	1										1
18			1								1
19	1										1
20	1	1									2
21		1									1
22		1									1
33	1						1				2
35		1									1
37		1									1
40			1								1
54-115				1		2	2	1		1	7
	158	28	9	3	1	4	4	1	1	2	221

Table 9.3. Co-occurrence distribution of protein domains, which have been found in cryptomonad organism *G. theta*, in the sample proteome of *C. Serevisiae*. For all the domains that occurred *s* times ($1 \leq s \leq 12$) in *G. theta*, we counted the number of times these domains occurred in the sample proteome for the yeast.

between the *G. theta* domain occurrence values for these two species. These observations suggest that gene loss (or protein-coding sequence loss) is under *negative* selection, while the processes that add a new gene are under *positive* selection during organism complexity growth.

The Pearson correlation coefficient, r, between the domain occurrence values for pair of organisms provides an estimate of their evolutionary similarity. Figure 9.7c shows that a greater similarity is observed between the worm and the fly ($r = 0.66$) and that yeast is more closely related to the fly ($r = 0.58$) than to the worm ($r = 0.47$).

The changes of the parameter values of the GDP model for the different species (see Figures 9.7a,b) strongly correlate with the changes of major characteristics of empirical DOPF: M, p_1, and J. This suggests the existence of a size-dependence for the parameters of the distribution.[2] In particular, the slope parameter k, negatively correlates with mean number of times that a random domain occurs within the proteome and positively correlates with the fraction of single-domain proteins in the proteome. Thus the slope parameter, k, might be used to characterize the evolutionary trend toward using more domains (selected by evolution) in the proteome as well as trend toward using already existing specific domain(s) (by reusing the new multi-domain protein(s) or recombining domains). The values of parameters k and b in the GDP model may also reflect the evolutionary branching order of species (e.g., yeast (first), worm (second), fly (third) and human (fourth) (Friedman & Hughes, 2001).

Several theoretical models for the growth of multi-domain proteins during evolution have been suggested in (Koonin et al., 2000; Rzhetsky & Gomez, 2001; Friedman & Hughes, 2001). Koonin *et al.* (2000) for example, showed that multi-domain proteins frequently are created through the (genetic) fusion of additional domains at one or both ends of a protein. We considered a more flexible evolutionary process in which single domain insertion occurs randomly within multi-domain proteins. To be precise, we simulated the reverse process (i.e., going back one evolutionary step to a simpler proteome) modeled by the random removal of one domain copy from the proteome. We applied this random removal process to both the worm and fly proteomes separately until

---

[2] Our goodness of fit analysis of frequency distributions of domain occurrence levels in eukaryotic proteomes shows that data do not follow a simple power law (dotted line on Figure 9.1a,b; Table 9.2) or the Zipf law (rank-frequency distribution form of power law). Recent attempts to use the linear regression models to fit such data sets in log-log plot appear to have led authors to a strong bias in parameter estimations and to incorrect conclusions that protein domain networks in proteomes and in genomes (Rzhetsky & Gomez, 2001; Wuchty, 2001) follow scale-free statistical prorperties, i.e. size-independency of the distributions.

we obtained proteomes with the same total number of domains (i.e., $M = 4,547$) as observed for yeast (Figure 9.7d). The empirical DOPF of the domain occurrence values for these two sub-samples closely approximate the yeast DOPF (Figure 9.7, Table 9.2). The parameters of the GDP model, fly (at $M = 4,547$) for the simulated "fly ancestor" and for yeast were very similar. Greater differences between the DOPF for the simulated "worm ancestor", worm (at $M = 4,547$) and the observed DOPF for yeast were observed. In both simulations, a longer right tail of the DOPF for the simulated data (compared to the observed DOPF for the yeast) was observed. These deviations probably reflect a natural selection process that preserves various non-random co-occurrences for some sets of domains. Such co-occurrences are not properly presented by random removal of domains.

Figure 9.8 shows that human protein domain complexity is greater than the other organisms with fully sequenced genomes; for example, the connectivity number for humans, *E. coli* and *N. meningitidis* is 36,082, 3,649 and 1,693, respectively. On the other hand, a fraction of domains that occurred once in these organisms is 0.27, 0.52, and 0.66, respectively. The slope parameter, k, of the DOPFs is the smallest in the case of human sample proteome. Figure 9.8 also shows that random sub-samples taken from the human proteome have domains occurrence profiles that have quite similar empirical DOPFs to the DOPFs for different species having smaller complexity.

Thus, the process of evolutionary growth of the proteome as measured by its domain network complexity can be characterized by the DOPF across species. Our reverse evolution simulation experiments based on random removal of domains fit sufficiently well to claim that such randomness is likely the predominant processes in the evolution of the proteome (Figure 9.7 and Figure 9.8) These experiments assume that the likelihood of addition of any domain to the proteome is equally likely and not dependent on the time of its occurrence and the location within a protein. Missing some older domains in the course of evolution also presents but with the slower rate (see, for example, Table 9.3) than for adding a new domain or using old one. The major non-random process would probably consist of removing or adding domain pairs and more larger protein families or domain clusters domains. This would clearly introduce co-occurrence events (which certainly exist in evolution) in our model.

In bulk, the process of proteome complexity growth looks like the *cumulative advantage dynamics* of the system of independent elements (protein domains) with associated counters: new domains are occasion-

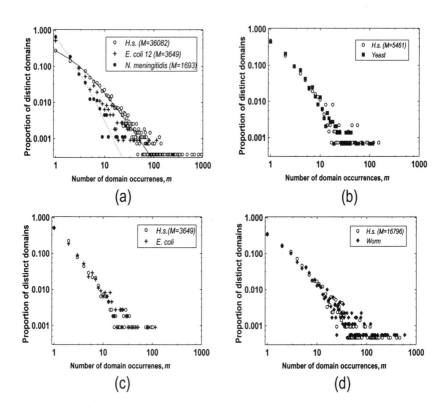

Figure 9.8.   Comparative analysis of proteome complexity in human vs *E.coli, yeast, worm*
and *N. meningitis-B*. (a)∘, +, • : Empirical frequency distributions of domain occurrence
values in the human, *E.coli,* and *N. meningitis-B* proteome samples, respectively; ∘-data
set is connected by line. All empirical DOPFs fit well by the GDP model. Empirical frequency
distribution of domain occurrence values in human proteome sub-samples (∘) vs yeast (b;■),
E. coli (c; +) and worm (d; ♦) proteome samples. M is the connectivity number. To construct
the histograms we have used data from the fourth release (on December, 11, 2001) of the
InterPro database.

ally added while the counters are incremented at a rate depending on their count values.

Note that our model can be also formulated in probability terms. (1) In each random time-point in the course of evolution, the proteome connectivity number is increased by one domain entry. (2) Any new entry (repeated gene duplication, loss, gene translocation, as well as the *de novo* origin of genes and the horizontal transferred gene, etc.) occurs in the course of evolution at approximately the same probability. (3) The probability of entry of a given domain which has already presented in a proteome is defined by the proportion of this domain in the proteome. Such probabilistic model, excepting domain loss process, leads to the DOPF in form of the BD probability distribution (Kuznetsov, 2001b) (see next subsection).

Our assumptions are weaker than Koonin *et al.* (2000). However since the majority of randomly selected addition sites within the entire proteome are at ends (that is the majority of proteins contain only one or two domains (Rubin et al., 2000; see also Table 9.2), the left side of the DOPF should not be too sensitive to differences between the two specific models. The right tail of the frequency distribution might be much longer for our model. Within the framework of our model one could specify and test different mechanisms for evolution of the genome and proteome complexity and explore changes in DOPF's for specific models of biological evolution, including Yule's (Yule, 1924), Fisher's (Fisher, 1930) and more recent models. In doing this analysis, we can suggest that at the proteome level, selection pressures do not produce a drastic global effect on the functional form of the distribution of evolutionarily conservative protein domains, but do effect the shape of the distribution, accounting for the increase in the sizes of some gene families and re-location of some gene clusters.

8.2 Prediction of the Numbers of Protein-Coding Genes in the Genome and of Protein Domains in the Entire Proteome

Although the exact numbers of genes in the human and mouse genomes remains to be determined, vertebrate genomes clearly contain more genes than those of *Drosophila* and *C. elegans* (Venter et al., 2001; IHGSC, 2001). In Figure 9.9, we show that remarkable linear relationship between the number of evolutionarily conserved protein-coding genes, G, and the proteome connectivity number (the total number of observed InterPro domains), M, in complete proteome samples of about $65 \pm 3.4\%$ ($n = 20$) of assumed complete proteome. Eighteen of these

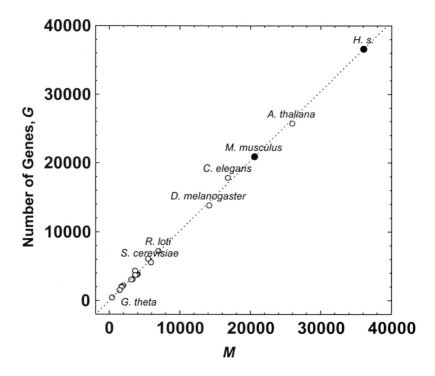

Figure 9.9. Relationships between the number of evolutionarily conserved protein-coding genes in the genome, G, and the observed proteome connectivity number, M, for *H. sapiens*, *M. musculus* and the 18 organisms with fully-sequenced genomes: *A. thaliana, , C. elegans, M. melanogaster, R. loti, P. aeruginosa, S. cerevisiae, C. acetobutylicum, V. cholerae, M. tuberculosis, C. crescentus, E. coli K-12, Synechocystis sp. (PCC 6803), D. radiodurans, L. lactis, N. meningitidis-A, N. meningitidis-B, M. leprae, G. theta.* ○ : data points for the 18 fully-sequenced genomes, ●: the predicted points for mouse and human. The regression line $G=1.015M$ was estimated using the 18 data points of the fully-sequenced genomes. (We analysed data of InterPro database, http//:www.ebi.ac.uk/interpro).

20 data points are for organisms with fully-sequenced genomes, except for the two prediction points for mouse and human shown as solid circles. According to the plot, 36,600 and 20,900 evolutionarily conserved protein-coding genes are in human and in mouse genomes, respectively.

The linearity of this graph is surprising. It may be that we are seeing the early part of a saturation curve of proteome evolution or did not have sufficiently representative statistical samples. Indeed our extrapolated human data point may be an over-estimate of the number of genes because the human and mouse proteomes available to us are not yet complete and the results must be considered preliminary. However, even if the average of these proteomes represents 60-65% of the total, it seems unlikely that predictions will be changed substantially with additional data. The slope in Figure 9.9 is 1.015. It suggests that in proteome size growth, *occasionally* appearing *new entry* of a domain sequence encoded in the genome of the organism leads, on average, to a new functional gene copy. This is consistent with our model of the protein complexity growth (see above) and with our observations that for all analyzed prokaryotic and eukaryotic organisms the average values of the connectivity number, α ($\alpha = M/A$; or "the average diameter" of proteome network) is approximately constant, or more precisely, ranges in a very narrow interval ($1.27 - 1.83$ with mean value 1.5 ± 0.139, ($n = 20$)). Although the human, *A. thaliana* and mouse genomes show a higher proportion of recent gene duplications than other known eukaryotic organisms (Douglas at al., 2001; Venter et al., 2001), a-value for these organisms is still only 1.83, 1.46, 1.82, respectively. It implies a high proportion of single-gene and two-member families in both genomes.

Figure 9.10 shows that as the number of evolutionarily conserved protein-coding genes increases the relationships between the number of distinct protein domains, N, the fractions of rarely observed domains (p_1, p_2, and p_3) and the occurrence value of the more frequent domain change in the reciprocate and predictable manner. For example, the p_1, fraction of distinct domains represented by only one entry in proteome, becomes smaller, but the occurrence value of mostly common domain in proteome, J, increases proportionally to the number of the genes in genomes ($J/G \gg a$, where a is a nearly constant parameter for all organisms studied here, equal to 0.027 with the standard errors ± 0.008 ($n = 20$)). Increasing the number of genes also leads to monotonically decreasing the slope of probability function. Generally, we can consider such the size-dependent evolution process in terms of the BD model (Kuznetsov, 2001b) if in the Eqs. (9.3-9.7) the variable N is, formally, the number of distinct domains and the variable M to re-sign as G.

Figure 9.10 shows that our BD model fit well all observed data, except the data point for *A. thaliana*. Only two fitted parameters of the model ($c = 0.486 \pm 0.0260$, $d = 4,500 \pm 300$) allows us to estimate the DOPF for a given organism with known number of genes. (Note: Our estimates are relatively robust and consistent with data in Figure 9.9. When we excluded the data point for mouse, human and *A. thaliana* from our curve-fitting analysis, we obtained the numerically similar solution of the BD model with the parameter values closely to the values presented above). Moreover, Eq.9.7 predicts existence of ˜$5,400 \pm 500$ domains in the entire proteome world (see also Figure 9.10a).

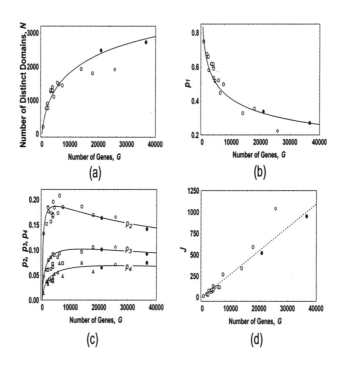

Figure 9.10. Relationships between the number of evolutionarily conserved protein-coding genes in the 20 genomes and some characteristics of the DOPF. (a) The number of distinct domains in the proteomes, (b) a fraction of distinct domains that occurred once, (c) fractions of distinct domains that occurred two and three times in a given proteome. o : 17 data points for fully-sequenced genomes (see Figure 9.9), except for the three data points for mouse and human (solid circles), and *A. thaliana* ($\Diamond$); lines in the all plots: best-fit curves providing by the BD model (at c=0.486±0.0260, d=4,500±400) to o- and •-data points. (d) Observed occurrence values for the most frequent domain in a given proteome. Regression line $J(G)=0.028G$ fitted to oand •-data points. In all plots, assignments are uniformed.

Thus, as the proteome increases due to evolution, each independent domain, alone or in combination with other domains, forms the repertoire of proteins of the proteome. This process may be considered mostly as a random birth and death process with advantage accumulation of domains in more complex proteomes. New domains and new use of already present domains likely are appearing in proteomes for different species relatively independently with both accumulating with similar probability. Generally, our analysis suggests that the increase in gene number in evolution occurred predominantly as a result of independent and random events: adding of single domains, either by mutation, gene conversion, reverse transcription, transposition or by occasional horizontal transferring of genes, or by combinatorial genome shuffling; multiple independent gene duplications and occasional gene block duplications became more probable in higher eukaryotic organisms.[3] These forms of genome adaptations allow the populations of unicellular and multicellular organisms ensuring survival in unpredictably changing environments. This can appear as the several overlapping Pareto-like probability distributions with very long left tail of the empirical DOPF. Figures 9.7, 9.8 and 9.10 also demonstrate a general rule of growing of the organism's complexity: to solve adaptation problems more complex organisms preferentially increase using the assembly of the "old" protein domains and blocks into new combinations. This conclusion is consistent with the results of recent analysis by Hughes and Friedman (Hughes et al., 2001; Friedman & Hughes, 2001), who suggested that the increase in the gene numbers in vertebrates may occur as a result of multiple independent gene duplication, as well as occasional duplication of chromosomal blocks.

---

[3] Evolutionary biologists have hypothesized that gene duplication, giving rise to multi-gene families, has played an important role in evolution (Ohno, 1970; Vision et al., 2000). In particular, a role for duplication of complete genomes in adaptive evolution through the single round or two rounds genome duplication by polyploidization (the 1R and 2R hypotheses) have been widely cited. In particular, genomic duplication in *A. thaliana* has been argued (Vision et al., 2000). Consistent with such hypotheses, there has been pronounced non-monotonicity in the shape of the empirical PODF or the peak(s) at the domain occurrence value 2 and/or 4 on the PODF for some organisms. However, for all of the 20 analyzed prokaryotic and eukaryotic organisms we did not observe peak(s) on the PODF (see for example Figure 9.7 and Figure 9.8). For the vast majority of the organisms we did not observe significant increment in proportion of the domain occurrence value 2 and/or 4 on the empirical PODFs (Figure 9.7, Figure 9.8, Figure 9.10). Only *A. thaliana, D. melanogaster* and a few bacteria organisms demonstrate pronounced negative deviation of the observed proportion p_1-value, and some positive deviation of the observed proportion p_2 values from corresponding values predicted by our probability model (Figure 9.10). Thus, our analysis shows inconsistencies of observations with the one round or two rounds of polyploidization hypotheses for most of the analyzed organisms.

The molecular basis of observed universality of the BD model in the genome-related phenomena is not well understood at present, but it is at least fundamentally important in that this universality is related to stochastic and probabilistic processes in the course of evolution.

9.   Conclusion

The goal of this study was to characterize the statistical distributions appearing in different genome-related phenomena. We observed such distributions for transcript copy numbers in a single eukaryotic cell, for sizes of protein clusters, for protein domain occurrence values in bacteria and eukaryotic proteomes, and for numbers of regulatory DNA-binding proteins in the prokaryotic genome. In all these cases, the data follows a Pareto-like distribution. We also found that the form of the distribution systematically depends on the size of the sample, the physiological state of the cells, the cell type and the complexity of the organism. It allows us to reveal several general rules of cell response to damaging stress factors and of proteome complexity growth in the course of evolution.

We developed a stochastic model of finite population growth that leads to a size-dependent Pareto-like distribution of the number of occurrences of species in the population. This model is called the Binomial Differential (BD) distribution. We presented evidence that the empirical frequency distributions of gene expression levels for all analyzed yeast, mouse and human cell types are well described by this model. Our distribution models could explain this universality by a stochastic mechanism.

We also developed a new methodology to remove major experimental errors from gene-expression data sets generated by the SAGE method. The Binomial Differential model allows us to estimate the total number of expressed genes in a single cell and in a population of given cells even when transcriptome databases are incomplete and have errors.

Mechanistic models of stochastic gene-expression processes in adaptability, phenotypic and genetic variability in a same-type cell population are discussed. Further identification of the distribution of gene expression levels in different cells appears to be important in understanding the relationships between stochastic and more probabilistic mechanisms that govern expression of genes associated with different cell types, normal and pathological processes in multi-cellular organisms. Statistical analysis of the global gene-expression data could also help to understand how the stochastic variability of gene expression in a single cell might lead to changes of the genotype repertoire in developing tissues and in the entire organism (Till et al., 1964; Ross et al., 1994; Hollander, 1999; Hume, 2000; Shulman & Wu, 1999; Ohlsson et al., 2001).

Our ergodicity hypothesis of an initiation of the gene-expression process suggests that a random and sporadic "basal" transcription mechanism exists for all protein-coding genes in all or almost all eukaryotic cell types. This mechanism may provide a basic level of phenotypic diversity and adaptability in cell populations. For example, sporadic expression of a large number of rarely-transcribed genes can provide apparent monoallelic expression phenomena in drawing cell populations and can explain the phenomenon of "no-phenotype" gene knockouts (cases where loss of function of a gene has no visible effect on the organism). It is possible that the rarely expressed genes might evolve by a weak selection process. Many of these genes are of no immediate benefit to the genome, but could be significant during natural selection.

Based on the analysis of "identifiable" protein domain occurrences in the sample proteomes of several fully-sequenced organisms, we developed a simple probabilistic model that describes the proteome complexity of an organism and shown that this description is consistent as the proteome grows in evolution. Our model is the Binomial Differential probability function of domain occurrence values in a proteome. We found that such a function is common for all studied organisms with fully-sequenced genomes and that the form of this distribution changes systematically as the proteome complexity grows. Fitting and extrapolation of the BD model predicts ~5,400 distinct protein domains in the entire proteome. We also discovered that the rate of growth of the number of protein-coding genes in evolving genomes appears to be a linear function of the total number of domains in the corresponding proteome samples. It allows us to estimate the number of protein-coding genes in incompletely sequenced genomes.

Finally, ~ 31,200 expressed genes were estimated based on the distribution of transcript numbers in the human colon cancer transcriptome and ~ 36,600 evolutionarily conserved protein-coding genes were predicted for the entire human genome based on the analysis of the numbers of protein domains in proteomes of fully-sequenced genome organisms.

Thus, discovery of universal size-dependent relationships in distributions of the genes by their expression levels in cells of different organisms and in distributions of proteins by their domains in evolution of eukaryotic organisms may help us understand the basic mechanics governing the growth of genome and proteome complexities during evolution and to quantitatively evaluate the relationship between natural selection and random events in the evolution process. Obviously, our assumptions and predictions were based on the sample analysis and on the simple statistical models: further development of theory and obtaining new and more complete and reliable data is required.

The Pareto-like skewed distributions observed in genomic and proteomic phenomena could be associated with fundamental properties of the self-organized critical system models. Such cellular automata models demonstrate the prominent size-dependent effect and are relatively robust to changes in the composition of their distinct components, to damaging of their sub-systems, to dynamics of network in the system. It will be important to study more deeply the relationships between properties of SOC systems and empirical statistical distributions presented here. Promising results in this direction have been recently obtained by introducing so-called Probabilistic Boolean Networks, the probabilistic generalization of the well-known Boolean network models, which, in particular, allows the analysis of small networks in large-scale genetic systems (Shmulevich et al., 2002).

It is likely that our statistical approach and models could be applicable for analysis of many other evolving large-scale systems (for example, in business, linguistics, sociology, informatics, internet, physics, etc.).

Acknowledgments

I thank Robert Bonner and Gary Knott for useful discussions and critical comments.

Appendix A: Infinity Limit for Population Growth Associated with the Generalized Pareto Probability Distribution

Here we will estimate relationships between N and M for the Generalized Pareto distribution when M is large enough and parameters of the distribution are constants (i.e. size-independent). For convenience let us fist find these relationships for a continued form of the Generalized Pareto density function:

$$f(m) = ks\frac{(m_o + b)^k}{(m + b)^{k+1}}, \tag{9.8}$$

where $m \subset [m_o, aM], s = 1/(1 - ((m_o + b)/(aM + b))^k)$, where m_o, k, b are constants. $m_o > 0; k > 0, b > -1; a_o < a < 1, a_o > 0$. Function f could be estimated as follows:

$$f = n_m/N, \tag{9.9}$$

where n_m is the observed number of distinct items (tags or genes) which have the occurrence level m in a given sample of size M. Then, using estimation

$$M = \int_{m_o}^{aM} n_m m \, dm, \qquad (9.10)$$

we can obtain a formula

$$N = M \cdot A(M), \qquad (9.11)$$

where

$$A(M) = [ks(m_o + b)^k \int_{m_o}^{aM} \frac{m}{(m+b)^{k+1}} dm]^{-1}. \qquad (9.12)$$

Taking the integral we find that if $M \to \infty$ than $N \sim M^k$ at $0 < k < 1$; $N \sim M/\log M$ at $k = 1$; and $N \sim M$ at $k > 1$. Thus, in all cases for parameter k we have $N \to \infty$ when $M \to \infty$.

Using a discrete analog of the Pareto probability function (see Eq.9.1) and $M = \sum_{m=1}^{aM} m \cdot n_m = N \sum_{m=1}^{aM} m \frac{1}{(m+b)^{k+1}}$, where $N = \sum_{m=1}^{aM} n_m = \sum_{m=1}^{aM} (1/(m+b)^{k+1})$ and $k > 0$; $b > -1$, $a_o < a < 1$, $a_o > 0$, we can obtain a formula:

$$N = M \cdot A_d(M), \qquad (9.13)$$

where

$$A_d(M) = \frac{\sum_{m=1}^{aM} \frac{1}{(m+b)^{k+1}}}{\sum_{m=1}^{aM} \frac{m}{(m+b)^{k+1}}}.$$

In the case of the Lotka-Zipf law ($b = 0, k = 1$), we have:

$$A_d(M) = \frac{\sum_{m=1}^{aM} \frac{1}{m^2}}{\sum_{m=1}^{aM} \frac{1}{m}} \approx \frac{\pi^2/6}{\ln(aM) + \gamma},$$

where $\gamma = 0.5772...$ is the Euler constant. Then,

$$\lim_{M \to \infty} N \approx \frac{(\pi^2/6)M}{\log M} \to \infty.$$

In the other specific case $b = 0, k = 2$ we have:

$$\lim_{M \to \infty} N \approx (1 + \pi^2/15)M \to \infty.$$

One could check that Eq.(9.13) has a similar growth to infinity for N as we have shown for Eq.(9.11), when $M \to \infty$.

Appendix B: Population Growth Curve for the Number of Human Expressed Genes

We model size-dependences N_r and N_g for human gene-expression data as follows:

$$dN_g/dM = p_{1g}N_g/M, \tag{9.14}$$

$$dN_r/dM = p_1 N_g/(M + a), \tag{9.15}$$

where $N_g(1) = 1, N_r(1) = 0, p_{1g} = (1+(1/d_g)^c)/(1+(M/d_g)^c), p_1 = (1+(1/d)^c)/(1 + (M/d)^c); c, d, d_g, a$ are positive constants. The parameter a reflects the lag in accumulation of the redundant distinct tags in a growing library (see also the analogous curve on Figure 9.2b). The parameters $c = 0.319$ and $d = 46,232$ were determined above by fitting Eq.9.6 to "true" distinct tag data points. The parameter values $a = 8,890 \pm 1430$ and $d_g = 17,532 \pm 400$ were estimated by fitting the function $N_g(M) + N_r(M)$ (curve 2, Figure 9.3b) to the same data. Eq.9.14 was used to compute the number of expressed genes in a single human cell and in a large population of these cells. Eq.9.14 was combined with Eqs.9.4-9.5 and Eq.9.1 for a specified human cell type sample to estimate the numbers of expressed genes at each expression level in a single human cell (curve 2, Figure 9.3a). This computation method is the same as we used for yeast data (Kuznetsov, 2001b).

References

C. Adami, (1998) *Introduction to artificial life*, Springer-Verlag, New-York.

J. O. Bishop, J. G. Morton, M. Rosbash, and M. Richardson, (1974) Three classes in Hela cell messenger RNA. *Nature*, 250, 199-204.

M. Yu. Borodovsky, S.M. Gusein-Zade (1989) A general rule for ranged series of codon frequencies in different genomes. *J. Biomolecular Structure and Dynamics*, 6, 1001-1012.

C.R. Cantor, C. L. Smith (1999) *Genomics*, J. Willey & Sons, Inc. New York.

J. Chelly, J.-P. Concordet, J.-C. Kaplan, A. Kahn (1989) Illegitimate transcription: Transcription of any gene in cell type. *Proc. Natl. Acad. Sci. U.S.A.*, 86, 2617-2621.

J.-J. Chen, J. D. Rowley & S. M. Wang (2000) Generation of longer cDNA fragments from serial analysis of gene expression tags for gene identification. *Proc. Natl. Acad. Sci. U.S.A.*, 97, 349-353.

D. L. Cook, A.N. Gerber and S.T. Tatscott (1998). Modeling stochastic gene expression: Implications for haploinsufficiency, *Proc. Natl. Acad. Sci. U.S.A.*, 95, 15641-15646.

Caron, H., et al. (2001) The human transcriptome map: clustering of highly expressed genes in chromosomal domains. *Science*, 291, 1289-1292.

B.S. Croix et al. (2000) Genes expressed in human tumor endothelium. *Science*, 289, 1197-1202.

R. Crollius, et al. (2000) Estimate of human gene number provided by genomewide analysis using *Tetraodon nigroviridis* DNA sequence. *Nature Genetics*, 25, 235-238.

S. Douglas et al.(2001) The highly reduced genome of an enslaved aldal nucleus. *Nature*, 410, 1091-1096.

S.R. Eddy (2001) Non-coding RNA genes and the modern RNA world. *Nature Rev. Genetics*, 2, 919-928.

M. R. Emmert-Buck, *et al.* (2000) Molecular profiling of clinical tissue specimens: Feasibility and applications. *Am. J. Pathol.*, 156, 1109-1115.

B. Ewing, P. Green (2000) Analysis of expressed sequence tags indicates 35,000 human genes. *Nature Genetics 25*, 232-234.

A.M. Femino, F.S. Fay, K. Fogarty and R.H. Singer (1998) Visualization of single RNA transcripts in situ. *Science*, 280, 585-590.

R.A. Fisher (1930). *The genetical theory of natural selection.* Oxford: Clarendon Press.

R. Friedman and A.L. Hughes (2001) Pattern and timing of gene duplication in animal genomes. *Genome Res.*, 11, 1842-1847.

P. Guptasarma (1995) Does replication-induced transcription regulate synthesis of the myriad low number proteins of Escherichia coli? *BioAssays*, 17, 987-997.

J.B. Hogenesch, et al. (2001) A comparison of the Celera and Ensemble predicted gene sets reveals little overlap in novel genes. *Cell,* 106, 413-415.

G.A. Hollander. On the stochastic regulation of interleukin-2 transcription. *Seminars in Immunology,* 11, 357-367.

F. C. P. Holstege, et al. (1998) Dissecting the regulatory circuitry of a eukaryotic genome. *Cell,* 95, 717-728.

Huang S.-P. and Weir B. S. (2001) Estimating the total number of alleles using a sample coverage method. *Genetics,* 159, 1365-1373.

Hughes A.L., da Silva J., Freadman R. (2001) Ancient genome duplications did not structure the human Hox-bearing chromosomes. *Genome Res.,* 11, 771-780.

D. A. Hume (2000) Probability in transcriptional regulation and implications for leukocyte differentiation and inducible gene expression. *Blood,* 96, 7, 2323-2328.

International Human Genome Sequencing Consortium (2001) Initial sequencing and analysis of the human genome. *Nature,* 409, 860-921.

D.A. Jackson, A. Pombo and F. Iborra (2000) The balance sheet for transcription: an analysis of nuclear RNA metabolism in mammalian cells. *FASEB J.,* 14, 242-254.

S. A. Jelinsky and L. D. Samson (1999) Global response of *Saccharomyces cerevisiae* to alkylating agent. *Proc. Natl. Acad. Sci. U.S.A.,* 96, 1486-1491.

S. A. Jelinsky, P. Estep, G.M. Church, and L. D. Samson (2000) Regulatory networks revealed by transcriptional profiling of damaged *Saccharomyces cerevisiae* cells: Rpn4 links base excision repair with proteasomes. *Molec. and Cell. Biology,* 20, 8157-8167.

H. Jeong, B. Tombor, R. Albert, Z.N. Ottval, A.-L. Barabasi (2000) The large-scale organization of metabolic networks. *Nature* 407, 651-654.

M. Johnson (2000) The yeast genome: on the road to the gold age. *Current Opinion in Genetics and Development,* 10, 617-623.

N. L. Johnson, S. Kotz, A. W. Kemp, *Univariate Discrete Distributions.* John Wiley & Sons, Inc., New-York, 1992.

S. A. Kauffman (1993) The origins of Order: Self-Organization and Selection in Evolution. Oxford University Press, New-York.

M.S.H. Ko (1992) Induction mechanism of a single gene molecule: stochastic or deterministic. *BioAssays,* 14, 341-346.

E. Koonin, L. Aravind & A. S. Kondrashov (2000). The impact of comparative genomics on our understanding of evolution. *Cell,* 101, 573-576.

V. A. Kuznetsov & R.F. Bonner (1999) Statistical tools for analysis of gene expression distributions with missing data. In: 3rd Annual Conference on Computational Genomics. Nov.18-21. Baltimore, MD:The Institute for Genomic Research, p.26.

V. A. Kuznetsov (2000) The genes number game in growing sample. *J. Comput. Biol.*, **7**, 642.

V. A. Kuznetsov (2001) Analysis of stochastic processes of gene expression in a single cell. In: *2001 IEEE-EURASIP Workshop on Nonlinear Signals and Image Processing*, University of Delaware, Baltimore, MD, USA, June, 2001.

V. A. Kuznetsov (2001) Distribution associated with stochastic processes of gene expression in a single eukaryotic cell. *EURASIP J. on Applied Signal Processing*, 4, 285-296.

A. S. Lash, *et al.* (2000) SAGEmap: A public gene expression resource. *Genome Res.*, 10, 1051-1060, 2000.

W. Li (1992), Random texts exhibit Zipf's-law-like word frequency distribution, *IEEE Transactions on Information Theory* , 38, 1842-1845.

W. Li (1999) Statistical properties of open reading frames in complete genome sequences. *Computers & Chemistry*, 23, 283-301.

W.-H. Li, Z. Gu, H.Wang and A. Nekrutenko (2001) Evolutionary analyses of the human genome. *Nature*, 409, 847-849.

B. Mandelbrot (1982). Fractal Geometry in Nature. New York:Freeman.

H.H. McAdams and A. Arkin (1999) It's a noisy business! Genetic regulation at the nanomolar scale. *Trends in Genetics*, 15, 65-69.

T. Misteli (2001) Protein dynamics: Implications for nuclear architecture and gene expression. *Science*, 291, 843-847.

S. Newlands, *et al.* (1998) Transcription occurs in pulses in muscle fibers. *Genes Dev.*, 12, 2748-2758, 1998.

M. E. J. Newman, S.H. Strogatz and D.J. Watts (2001) Physical Rev. E., 64, 026118-1-02618-17.

E. Pennisi (2000) And the gene number is...? *Science*, 288, 1146-1147.

R. Ohlsson, A. Paldi, and J.A. Marshall Graves (2001) Did genomic imprinting and X chromosome inactivation arise from stochastic expression? *Trends in Genetics*, 17, 136-141.

S. Ohno (1970) *Evolution by gene duplication.* Springer Verlag, New York.

A. Pombo et al. (2000) Specialized transcription factories within mammalian nuclei. *Critical Reviews in Eukaryotic Gene Expression*, 10, 21-29.

J. J. Ramsden, J. Vohradsky (1998) Zipf-like behavior in prokaryotic protein expression. *Phys. Review. E.*, 58, 7777-7780.

I. L. Ross, C. M. Browne, and D. A. Hume (1994) Transcription of individual genes in eukaryotic cells occurs randomly and infrequently. *Immunol. Cell. Biol.* 72, 177-185.

G. M. Rubin et al. (2000) Comparative genomics of the Eukaryotes. *Science*, 287, 2204-2215.

A. Rzhetsky and S. M. Gomez (2001) Birth of scale-free molecular networks and the number of distinct DNA and protein domains per genome. *Bioinformatics*, 17, 988-996.

Y. Sano et al. (2001) Random monoallelic expression of three genes clustered within 60 kb of mouse t complex genomic DNA. *Genome Res.*, 11, 1833-1841.

I. Shmulevich, E.R. Dougherty, S. Kim, and W. Zhang (2002) Probabilistic Boolean networks: a rule-based uncertainty model for gene regulatory networks *Bioinformatics*, 18 (in press).

M.J. Shulman & G.E. Wu, (1999) Hypothesis:genes which function in a stochastic linage commitment process are subject to monoallelic expression. *Seminars in Immunology*, 11,369-371.

H.A. Simon & T.A. Van Wormer (1963). Some Monte-Carlo estimates of the Yule distribution, *Behavior Science*, 8, 203-210.

H.E. Stanley, *et al.* (1999) Scaling features of noncoding DNA. *Phys. Review. E*, 273, 1-18.

H.G. Sutherland, *et al.* (2000) Reactivation of heritably silenced gene expression in mice. *Mammalian Genome*, 11, 347-355.

D. Thieffry, A.M. Huerta, E. Perez-Rueda and J. Collado-Vides (1998) From specific gene regulation to genomic networks: a global analysis of transcriptional regulation in *Escherichia coli*. *BioEssays*, 20, 433-440.

J.E. Till, E.A. McCulloch, L. Siminovish (1964) A stochastic model of stem cell proliferation, based on the growth of spleen colony-forming cells. *Proc. Natl. Acad. Sci. U.S.A.*, 51, 29-38.

V. E. Velculescu, *et al.* (1997) Characterization of yeast transcriptome. *Cell*, 88, 243-251.

V. E. Velculescu, *et al.* (1999) Analysis of human transcriptomes. *Nat. Genet.*, 23, 387-388.

J.C. Venter, J.C., *et al.* (2001) The sequence of the human genome. *Science*, 291, 1304-1351.

T.J. Vision, D.G. Brown and S.D. Tanksley (2000) The origins of genome duplications in *Arabidopsis*. *Science*, 290, 2114-2117.

J. Vohradsky and J.J. Ramsden. (2001) Genome resource utilization during prokaryotic development. *FASEB J.* (express article 10.1096/fj.00-0889fje).

M.C. Walters, *et al.*(1995) Enhancers increase the probability but not the level of gene expression. *Proc. Natl. Acad. Sci. U.S.A.*, 92, 7125-7129.

H. Weintraub (1988) Formation of stable transcription complexes as assayed by analysis of individual templates. *Proc. Natl. Acad. Sci. U.S.A.*, 85, 5819-5823.

S. Wuchty (2001) Scale-free behavior in protein domain networks. *Molec. Biol. Evol.*, 18, 1694-1702.

G.U. Yule (1924) A mathematical theory of evolution, based on the conclusions of Dr. J.C. Willis, F.R.S. *Philosophical Transactions of the Royal Society of London. Ser. B.*, 213, 21-87.

Chapter 10

NORMALIZED MAXIMUM LIKELIHOOD MODELS FOR BOOLEAN REGRESSION WITH APPLICATION TO PREDICTION AND CLASSIFICATION IN GENOMICS

Ioan Tabus, Jorma Rissanen, and Jaakko Astola

Institute of Signal Processing, Tampere University of Technology, Tampere, Finland

1. Introduction

We discuss the NML models for Boolean model classes. Such a model for the linear regression problem was introduced and analyzed recently, (Rissanen, 2000). We restate the classification problem as a modeling problem in terms of a class of parametric models, for which the maximum likelihood parameter estimates can be easily computed. We first review the NML model for Bernoulli strings as the solution of a minmax optimization problem. We then introduce two model classes for the case where the binary strings to be modeled are observed jointly with several other binary strings (regression variables). We derive the NML model for both classes, provide fast evaluation procedures, and we review the connection with the MDL principle.

The concept of gene expression was introduced four decades ago with the discovery of the messenger RNA, when the theory of genetic regulation of protein synthesis was described (Jacob and Monod, 1961). Each human cell contains approximately three billion base pairs, which encode about 30 000 to 40 000 genes. In any particular cell only a small fraction of these genes is being actively transcribed. Genes carry the essential information for protein synthesis. Cell metabolism, in both healthy and malignant states, is governed by complex interactions between genes. A gene may have a varying level of expression in the cell at a certain time. Many important biological processes can be probed by measuring the gene transcription levels, and models of regulatory pathways can be inferred from these measurements (Russel *et al.*, 2000).

The availability of cDNA microarrays (Schena *et al.*, 1995) makes it possible to measure simultaneously the expression levels for thousands of genes. The huge amount of data provided by cDNA microarray measurements is explored in order to answer fundamental questions about gene functions and their inter-dependencies, and hopefully to provide answers to questions like "what is the type of the disease affecting the cells and what are the best ways to treat it". Large-scale studies of gene expression are expected to provide the transition from 'structural' to 'functional' genomics (Hieter and Boguski, 1997), where knowing the complete sequence of a genome is only the first step in understanding how it works.

As a first step in deciphering complex biological interactions at a biomolecular level it is necessary to study the microarray data in order to provide sets of possible rules for how the expressions of certain genes determine the expression level of a target gene. Gene expression data obtained in microarray experiments may often be discretized as binary or ternary data, the values 1,0,-1 carrying the meanings of overexpressed, normal, and repressed, respectively, which are the descriptors needed to define regulatory pathways (Kim and Dougherty, 2000).

The prediction and classification problems addressed here refer to processing of the binary or ternary valued cDNA microarray data in order to find groups of genes, or factors, which are very likely to determine the activity of a target gene (or a class label), as opposed to the oversimplified model in which one gene only influences a single target gene (or class label). Previous solutions to related problems were given by using perceptron predictors, see (Kim and Dougherty, 2000; Kim *et al.*, 2000), which were shown to provide more reliable information than the determination of dependencies based solely on the correlation coefficient linking single genes.

In our approach to prediction we look for flexible classes of models with good predictive properties, but then we also must consider the complexity of the models, the right balance between the two modeling aspects being set by the MDL principle.

One possible way to study the classification problem is in terms of Boolean regression models. Suppose that the data available is in a matrix X, where the entry $x(i,j) \in \{0,1\}$ is a binary (quantized) gene expression, the row index $i \in \{1,\ldots,N\}$ identifies the gene, and the column index $j \in \{1,\ldots,n\}$ identifies the "patient". We denote by $\underline{x}_j$ the jth column of the matrix X. Furthermore, a class label y_j is known for all patients (e.g. $y_j = 0$ or $y_j = 1$ for the j'th patient having disease type A, or type B, respectively). Our goal is to build Boolean models $\hat{y}_j = f(x_{i_1,j}, \ldots, x_{i_k,j})$ and to select the set of informative genes,

$\{i_1, i_2, \ldots, i_k\}$. One natural requirement for a good model is to get a small sum of absolute errors

$$\sum_{j=1}^{n} |y_j - \hat{y}_j|. \tag{10.0}$$

We show that this minimization actually results as a consequence of the NML formalism.

2.   The NML model for Bernoulli strings

In this section we assume that a Bernoulli variable Y with $P(Y = 0) = \theta$ is observed repeatedly n times, generating the string $y^n = y_1, \ldots y_n$. We look for a distribution $q(y^n)$ over all strings of length n, such that the ideal codelength $\log \frac{1}{q(y^n)}$ assigned to a particular string y^n by this distribution, is as close as possible to the ideal codelength $\log \frac{1}{P(y^n; \hat{\theta}(y^n))}$ obtainable with the Bernoulli models. In the coding scenario, the decoder is allowed to use a predefined distribution, $q(\cdot)$, but he cannot use the distribution $P(\cdot; \hat{\theta}(y^n))$ because he does not have $\hat{\theta}(y^n)$ available. The latter is the most advantageous distribution in the family $P(y^n; \theta)$ for the string y^n, since it maximizes $P(y^n; \hat{\theta}(y^n))$ and therefore minimizes the ideal codelength $\log \frac{1}{P(y^n; \hat{\theta}(y^n))}$. We use the term 'ideal' code length because we drop the unessential requirement that a real code length must be an integer-valued quantity. The distribution $q(y^n)$ is selected such that the "regret" of using $q(y^n)$ instead of $P(y^n; \hat{\theta}(y^n))$, namely,

$$\log \frac{1}{q(y^n)} - \log \frac{1}{P(y^n; \hat{\theta}(y^n))} = \log \frac{P(y^n; \hat{\theta}(y^n))}{q(y^n)}, \tag{10.1}$$

is minimized for the worst case y^n; i.e.

$$\min_{q} \max_{y^n} \log \frac{P(y^n; \hat{\theta}(y^n))}{q(y^n)} \tag{10.2}$$

Theorem 1 *(Shtarkov, 1987) The minimizing distribution is given by*

$$q(y^n) = \frac{P(y^n; \hat{\theta}(y^n))}{C_n}, \tag{10.3}$$

where

$$C_n = \sum_{m=0}^{n} \binom{n}{m} \left(\frac{m}{n}\right)^m \left(1 - \frac{m}{n}\right)^{n-m}. \tag{10.4}$$

Proof

Fix y^n, having m zeros. Then, if $P(Y = 0) = \theta$, we have by independence $P(y^n; \theta) = (\theta)^m (1 - \theta)^{n-m}$, and the ML estimate of θ satisfies $\frac{m}{\theta} - \frac{n-m}{1-\theta} = 0$, which gives $\hat{\theta}(y^n) = \frac{m}{n}$ and $P(y^n; \hat{\theta}(y^n)) = \left(\frac{m}{n}\right)^m \left(1 - \frac{m}{n}\right)^{n-m}$. The constant C_n in (10.5) clearly normalizes $P(y^n; \hat{\theta}(y^n))$ to a distribution $q(y^n)$ such that $\sum_{y^n \in \{0,1\}^n} q(y^n) = 1$. To show that this distribution is minmax, we consider any different distribution $p_n(y^n)$, and take a string z^n such that $p_n(z^n) < q(z^n)$ (such a z^n exists since $p_n(\cdot)$ and $q(\cdot)$ must sum to unity and they cannot be equal for all strings. We have $\max_{y^n} \log \frac{P(y^n; \hat{\theta}(y^n))}{p_n(y^n)} \geq \log \frac{P(z^n; \hat{\theta}(z^n))}{p_n(z^n)} > \log \frac{P(z^n; \hat{\theta}(z^n))}{q(z^n)} = \log C_n = \max_{y^n} \log \frac{P(y^n; \hat{\theta}(y^n))}{q(y^n)}$, which shows $q(\cdot)$ to be minmax. The minimizing distribution in (10.17) is also called the NML (universal) model (Barron et al., 1998).

Another strong optimality property of the NML models was recently proved in (Rissanen, 2001), where the following minmax problem was formulated: find the (universal) distribution which minimizes the average regret

$$\min_q \max_g E_g \log \frac{P(Y^n; \hat{\theta}(Y^n))}{q(Y^n)} = \min_q \max_g \sum_{y^n} g(y^n) \log \frac{P(y^n; \hat{\theta}(y^n))}{q(y^n)} \quad (10.5)$$

where $g(\cdot)$, the generating distribution of the data, and $q(\cdot)$ run through any sets that include the NML model.

Theorem 2 *(Rissanen, 2001) The minimizing distribution $q(\cdot)$ and the worst case data generating distribution for the minmax problem (10.6) is given by (10.4) and (10.5).*

Proof
We have

$$\min_q \max_g E_g \log \frac{P(Y^n; \hat{\theta}(Y^n))}{q(Y^n)} = \min_q \max_g D(g||q) - D(g||\hat{q}) + \log C_n,$$

where $\hat{q}(y^n) = \frac{P(y^n; \hat{\theta}(y^n))}{C_n}$ and $D(g||q)$ is the relative entropy, or Kullback-Leibler distance between distributions g and q. Since only the first term depends on q, the minimizing q is $q = g$, which makes $D(g||q) = 0$, and the maximizing g is $g = \hat{q}$, which makes $D(g||\hat{q}) = 0$, and finally $g = q = \hat{q}$ proves the claim.

Note how the minmax problem generalizes Shannon's noiseless coding theorem, which states that if the data have been generated by a single distribution g, then the ideal coding distribution that minimizes the Kullback-Leibler distance $E_g \log(g(X^n)/q(X^n))$ has to mimic g. Note too that in the optimization problem above no assumption was made that the data generating mechanism is included in the class of models, which is of utmost relevance for practical problems in which nothing is known about any 'true' data generating mechanism.

3. The NML model for Boolean regression

We consider a binary random variable Y, which is observed jointly with a binary regressor vector $\underline{X} \in \mathcal{B}^k$. In a useful model class a carefully selected Boolean function $f : \mathcal{B}^k \to \{0, 1\}$ should provide a reasonable prediction $f(\underline{X})$ of Y, in the sense that the absolute error $\mathcal{E} = |Y - f(\underline{X})|$ has a high probability of being 0. Since $\mathcal{E}, Y, f(\underline{X})$ are binary-valued we have $\mathcal{E} = |Y - f(\underline{X})| = Y \oplus f(\underline{X})$, which also implies $Y = f(\underline{X}) \oplus \mathcal{E}$, where $\oplus$ is modulo 2 sum.

To formalize this we consider a model defined as follows:

$$Y = f(\underline{X}) \oplus \mathcal{E} = \begin{cases} f(\underline{X}) & if \quad \mathcal{E} = 0 \\ \overline{f(\underline{X})} & if \quad \mathcal{E} = 1 \end{cases} \tag{10.6}$$

where $f(\cdot)$ is a Boolean function and the error $\mathcal{E}$ is independently drawn from a Bernoulli source with parameter θ; i.e., $P(\mathcal{E} = 0) = \theta$ and $P(\mathcal{E} = 1) = 1 - \theta$, or for short

$$P(\mathcal{E} = j) = \theta^{1-j}(1 - \theta)^j, \text{ for } j \in \{0, 1\}. \tag{10.7}$$

Denote by $\underline{b}_i \in \{0, 1\}^k$ the vector having as entries the bits in the binary representation of integer i, i.e., $\underline{b}_0 = [0, \dots, 0, 0]$, $\underline{b}_1 = [0, \dots, 0, 1]$, etc. Further, define by (10.7) and (10.8) the conditional probability of $y \in \{0, 1\}$ for code $\underline{b}_i \in \{0, 1\}^k$,

$$P(Y = y | \underline{X} = \underline{b}_i) = \theta^{1-y \oplus f(\underline{b}_i)}(1 - \theta)^{y \oplus f(\underline{b}_i)} \overset{def}{=} P(y; f, \underline{b}_i, \theta). \tag{10.8}$$

The Boolean regression problem considered is to find the optimal universal model (in a minmax sense to be specified shortly) for the following class of models:

$$\mathcal{M}(\theta, k, f) = \{P(y; f, \underline{b}_i, \theta) = \theta^{(1-y \oplus f(\underline{b}_i))}(1 - \theta)^{(y \oplus f(\underline{b}_i))}\} \tag{10.9}$$

where $\theta \in [0, 1]$, $y \in \{0, 1\}$, $\underline{b}_i \in \{0, 1\}^k$. A more flexible class of models can be obtained by allowing the parameter θ to depend on the actually observed regressor vector $\underline{b}_i$. This leads us to define a second class of models, $\mathcal{M}(\Theta, k, f)$, as follows:

$$\mathcal{M}(\Theta, k, f) = \{P(y; f, \underline{b}_i, \theta_i) = \theta_i^{(1-y \oplus f(\underline{b}_i))}(1-\theta_i)^{(y \oplus f(\underline{b}_i))}\} \quad (10.10)$$

where $i = 0, \ldots, 2^k - 1$, $\Theta = \{\theta_0, \ldots, \theta_{2^k-1}\}$ is the set of parameters, θ_i is a parameter constrained to $[0,1]$, and k is the structure parameter.

3.1   The NML model for the Boolean class $\mathcal{M}(\Theta, k, f)$

When the sequence $y^n = y_1 \ldots y_n$ and another sequence of binary regressor vectors $\underline{b}^n = \underline{b}_{i_1}, \ldots, \underline{b}_{i_n}$ are observed, a member of $\mathcal{M}(\Theta, k, f)$ assigns to the sequence y^n the following probability

$$P(y^n; \Theta, k, f, \underline{b}^n) = \prod_{j=1}^{n} \theta_{i_j}^{(1-y_j \oplus f(\underline{b}_{i_j}))}(1-\theta_{i_j})^{(y_j \oplus f(\underline{b}_{i_j}))}. \quad (10.11)$$

By grouping the factors we get

$$P(y^n; \Theta, k, f, \underline{b}^n) = \prod_{\ell | \underline{b}_\ell \in \underline{b}^n} \theta_\ell^{\sum_{j | i_j = \ell}(1-y_j \oplus f(\underline{b}_\ell))}(1-\theta_\ell)^{\sum_{j | i_j = \ell}(y_j \oplus f(\underline{b}_\ell))} \quad (10.12)$$

which shows that we can carry out the maximization with respect to the θ_ℓ-parameters independently for each regression vector in the following way. Consider the subsequence $z^{m_\ell} = z_1 z_2 \ldots z_{m_\ell}$ formed with the observed values y_i that correspond to the regressor vector $\underline{b}_\ell$; denote by b_ℓ the prediction at regressor $\underline{b}_\ell$ (i.e. $b_\ell = f(\underline{b}_\ell) \in \{0,1\}$), and by $m_{\ell 0}$ and $m_{\ell 1} = \sum_{j=1}^{m_\ell} z_j$ the number of zeros and ones, respectively, in the sequence z^{m_ℓ}. In the rest of this subsection, we write for notational simplicity $m_\ell, m_{\ell_0}, m_{\ell_1}$ as m, m_0, m_1, respectively, for a fixed regressor vector, $\underline{b}_\ell$. The likelihood is maximized independently at every $\underline{b}_\ell$:

$$\max_{\substack{\theta_\ell \in [0,1] \\ b_\ell \in \{0,1\}}} P(z^m; \theta_\ell, b_\ell) = \max_{\substack{\theta_\ell \in [0,1] \\ b_\ell \in \{0,1\}}} \theta_\ell^{\sum_{j=1}^{m}(1-z_j \oplus b_\ell)}(1-\theta_\ell)^{\sum_{j=1}^{m} z_j \oplus b_\ell}$$

When $b_\ell = 0$, the extremum of $\log P(z^m; \theta_\ell, b_\ell = 0) = (m - m_1) \log \theta_\ell + m_1 \log(1 - \theta_\ell)$ is attained for $\frac{m - m_1}{\theta_\ell} - \frac{m_1}{1 - \theta_\ell} = 0$, which gives $\theta_\ell = \frac{m - m_1}{m}$ and $P_{max, b_\ell = 0} = \left(\frac{m - m_1}{m}\right)^{m - m_1} \left(\frac{m_1}{m}\right)^{m_1}$. The second derivative is negative:

$$(\log P(z^m; \theta_\ell, b_\ell = 0))'' = -\frac{m_0}{\theta_\ell^2} - \frac{m_1}{(1 - \theta_\ell)^2} < 0 \quad (10.13)$$

which ensures that the extremum is a maximum.

When $b_\ell = 1$, the maximum of $P(z^m; \theta_\ell, b_\ell = 1) = \theta_\ell^{m_1}(1 - \theta_\ell)^{m-m_1}$ is attained for $\frac{m_1}{\theta_\ell} - \frac{m-m_1}{1-\theta_\ell} = 0$, which gives $\theta_\ell = \frac{m_1}{m}$ and $P_{max,b_\ell=1} = \left(\frac{m_1}{m}\right)^{m_1}\left(\frac{m-m_1}{m}\right)^{m-m_1} = P_{max,b_\ell=0}$. The maximum likelihood estimate is not unique: either $b_\ell = 0$ and $\theta_\ell = \frac{m_0}{m}$, or $b_\ell = 1$ and $\theta_\ell = \frac{m_1}{m}$ maximizes to

$$
\begin{aligned}
P(z^m; \hat{\theta}_\ell(z^m), \hat{f}_{y^n}(\underline{b}_\ell)) &= \max_{\theta_\ell \in [0,1], b_\ell \in \{0,1\}} P(z^m; \theta_\ell, b_\ell) \\
&= \left(\frac{m_1}{m}\right)^{m_1}\left(\frac{m-m_1}{m}\right)^{m-m_1} \quad (10.14)
\end{aligned}
$$

For the entire string y^n, the maximised likelihood is by independence

$$
P(y^n; \hat{\Theta}(y^n), k, \hat{f}_{y^n}, \underline{b}^n) = \prod_{\underline{b}_\ell | \underline{b}_\ell \in \underline{b}^n} \left(\frac{m_{\ell_1}}{m_\ell}\right)^{m_{\ell_1}}\left(\frac{m_\ell - m_{\ell_1}}{m_\ell}\right)^{m_\ell - m_{\ell_1}} \quad (10.15)
$$

The normalization constant of the NML model when the regressor string $\underline{b}^n$ was observed, is the following

$$
\begin{aligned}
\sum_{y^n \in \{0,1\}^n} P(y^n; \hat{\Theta}(y^n), k, \hat{f}_{y^n}, \underline{b}^n) &= \prod_{\ell | \underline{b}_\ell \in \underline{b}^n} \sum_{z^{m_\ell} \in \{0,1\}^{m_\ell}} P(z^{m_\ell}; \hat{\theta}_\ell(z^{m_\ell}), \hat{f}(\underline{b}_\ell)) \\
&= \prod_{\ell | \underline{b}_\ell \in \underline{b}^n} C_{m_\ell},
\end{aligned}
$$

where m_ℓ is the total number of times $\underline{b}_\ell$ appears in the sequence $\underline{b}^n = \underline{b}_{i_1}, \dots, \underline{b}_{i_n}$, and C_i was introduced in (10.5).

Theorem 3 *The universal NML model is*

$$
q(y^n) = \frac{P(y^n; \hat{\Theta}(y^n), k, \hat{f}_{y^n}, \underline{b}^n)}{\prod_{\ell | \underline{b}_\ell \in \underline{b}^n} C_{m_\ell}}. \quad (10.16)
$$

Proof The proof is the same as that of Theorem 1.
$\square$

Remark The maximum likelihood alone for the optimum Boolean regressor function does not give any preference to $b_\ell = f(\underline{b}_\ell)$ being either 0 or 1. However, we may further require the value of $f(\underline{b}_\ell)$ to be a good prediction of y, and the minimization of the mean square estimate, which is the conditional expectation, $f(\underline{b}_\ell) = 1$ if $E(y|\underline{b}_\ell) > 0.5$ and $f(\underline{b}_\ell) = 0$ if $E(y|\underline{b}_\ell) < 0.5$ will fix the choice.

Since $E(y|\underline{b}_\ell) > 0.5$ is equivalent to $m_1 > m_0$, we define with this selection of the Boolean function uniquely the ML estimates to be $\hat{\theta}_\ell(z^m) = \max(\frac{m_0}{m}, \frac{m_1}{m}) \geq 0.5$ and $\hat{f}_{y^n}(\underline{b}_\ell) = 1_{m_1 > m_0}$, where 1_x is the selector function, being 1 whenever the condition x is true, and 0 otherwise.

3.2   The NML model for the Boolean class $\mathcal{M}(\theta, k, f)$

When the errors $\mathcal{E}(t)$ are assumed to be Bernoulli distributed $\mathcal{B}(\theta)$ with the same parameter for all the regressor vectors $\underline{b}_\ell$, we get the class of models $\mathcal{M}(\theta, k, f)$ defined in (10.10). When the sequence $y^n = y_1 \ldots y_n$ and the sequence of binary regressor vectors $\underline{b}^n = \underline{b}_{i_1}, \ldots, \underline{b}_{i_n}$ are observed, a member of the class $\mathcal{M}(\theta, k, f)$ assigns to the sequence y^n the following probability

$$
\begin{aligned}
P(y^n; \theta, k, f, \underline{b}^n) &= \prod_{j=1}^{n} \theta^{(1-y_j \oplus f(\underline{b}_{i_j}))}(1-\theta)^{(y_j \oplus f(\underline{b}_{i_j}))} \\
&= \theta^{\sum_{j=1}^{n}(1-y_j \oplus f(\underline{b}_{i_j}))}(1-\theta)^{\sum_{j=1}^{n}(y_j \oplus f(\underline{b}_{i_j}))} \\
&= \theta^{n_0}(1-\theta)^{n-n_0}, \quad\quad\quad (10.17)
\end{aligned}
$$

where n_0 is the number of zeros in the sequence $\{\varepsilon_j = y_j \oplus f(\underline{b}_{i_j})\}_{j=1}^{n}$. The ML estimate of the model parameters,

$$
(\hat{\theta}(y^n), \hat{f}_{y^n}) = \arg \max_{\theta, f} P(y^n; \theta, k, f, \underline{b}^n), \quad\quad (10.18)
$$

can be obtained in two stages, first by maximizing with respect to f,

$$
\max_f P(y^n; \theta, k, f, \underline{b}^n), \qu\quad\quad (10.19)
$$

and observing that the optimal $f(\cdot)$ does not depend on θ. For a fixed $\theta > 0.5$, the function $P(y^n; \theta, k, f, \underline{b}^n) = \theta^{n_0}(1-\theta)^{n-n_0}$ increases monotonically with n_0, and (10.20) is maximized by maximizing n_0, or, equivalently, by minimizing $n - n_0$

$$
\begin{aligned}
\min_f(n - n_0) &= \min_f \sum_{j=1}^{n} |y_j - f(\underline{b}_{i_j})| \\
&= \min_f \sum_{\ell=0}^{2^n-1} \sum_{j|\underline{b}_{i_j}=\underline{b}_\ell} (1 - y_j)f(\underline{b}_\ell) + y_j(1 - f(\underline{b}_\ell)) \\
&= \min_f \sum_{\ell=0}^{2^n-1} m_{\ell_0} f(\underline{b}_\ell) + m_{\ell_1}(1 - f(\underline{b}_\ell)). \quad (10.20)
\end{aligned}
$$

Equation (10.21) shows that f is optimum both for the mean absolute error (MAE) and the mean square error (MSE) as criteria. It can also be seen that the assignment of $f(\underline{b}_{\ell})$ depends solely on m_{ℓ_0}, m_{ℓ_1}, and the solution is given by

$$\hat{f}_{y^n}(\underline{b}_{\ell}) = \begin{cases} 0 & if \quad m_{\ell_0} \geq m_{\ell_1} \\ 1 & if \quad m_{\ell_0} < m_{\ell_1} \end{cases}, \tag{10.21}$$

which can readily be computed from the data set. Denote by $n_0^*(y^n)$ the number of zeros in the sequence $\{\varepsilon_j = y_j \oplus \hat{f}_{y^n}(\underline{b}_{i_j})\}_{j=1}^n$. To completely solve the ML estimation problem we have to find

$$\max_{\theta} P(y^n; \theta, k, \hat{f}_{y^n}, \underline{b}^n), \tag{10.22}$$

for which the maximizing parameter is $\hat{\theta}(y^n) = \frac{n_0^*(y^n)}{n}$. Therefore

$$P(y^n; \hat{\theta}(y^n), k, \hat{f}_{y^n}, \underline{b}^n) = \left(\frac{n_0^*(y^n)}{n}\right)^{n_0^*(y^n)} \left(1 - \frac{n_0^*(y^n)}{n}\right)^{n - n_0^*(y^n)} \tag{10.23}$$

We need to define a distribution $q(y^n)$ over all possible sequences y^n, which is the best in the minmax sense

$$\min_q \max_{y^n} \frac{P(y^n; \hat{\theta}(y^n), k, \hat{f}_{y^n}, \underline{b}^n)}{q(y^n)}. \tag{10.24}$$

The NML model, then, is

$$\begin{aligned} q(y^n) &= \frac{P(y^n; \hat{\theta}(y^n), k, \hat{f}_{y^n}, \underline{b}^n)}{C_n(k, \underline{b}^n)} \\ &= \frac{\left(\frac{n_0^*(y^n)}{n}\right)^{n_0^*(y^n)} \left(1 - \frac{n_0^*(y^n)}{n}\right)^{n - n_0^*(y^n)}}{\sum_{x^n} \left(\frac{n_0^*(x^n)}{n}\right)^{n_0^*(x^n)} \left(1 - \frac{n_0^*(x^n)}{n}\right)^{n - n_0^*(x^n)}}, \end{aligned} \tag{10.25}$$

where

$$C_n(k, \underline{b}^n) = \sum_{y^n} \left(\frac{n_0^*(y^n)}{n}\right)^{n_0^*(y^n)} \left(1 - \frac{n_0^*(y^n)}{n}\right)^{n - n_0^*(y^n)} \tag{10.26}$$

We remark that n_0^* depends on y^n through $\hat{f}_{y^n}$ in a complicated manner. When $k = 0$, the normalization factor is $C_n(0, \underline{b}^n) = C_n$, given in (10.5).

Alternative expressions for the coefficient $C_n(k, \underline{b}^n)$ provide a faster evaluation. We need to specify the distinct elements in the set $\{\underline{b}_{\ell} | \underline{b}_{\ell} \in$

$\underline{b}^n\}$ as $\{\underline{b}_{j_1}, \ldots, \underline{b}_{j_K}\}$, and denote by z^q the subsequence of y^n observed when the regressor vector is $\underline{b}_{j_q}$. Let n_q be the length of the subsequence z^q having m_q zeros.

For the optimal Boolean function $\hat{f}_{y^n}(\underline{b}_{j_q})$ in (10.22), the number of zeros in the sequence of errors, evaluated for the regressor vector $\underline{b}_{j_q}$, is $\max(m_q, n_q - m_q)$. Therefore the overall number of zeros in the sequence of errors is $n_0^*(y^n) = \sum_{q=1}^{K} \max(m_q, n_q - m_q)$, and the Bernoulli parameter is $\hat{\theta}_{y^n} = \frac{n_0^*(y^n)}{n} = \frac{1}{n} \sum_{q=1}^{K} \max(m_q, n_q - m_q)$.

We may write the conditional probability of the string y^n when the optimal Boolean function $\hat{f}_{y^n}(\cdot)$ is used and the numbers of zeros in the subsequences $\{z^q\}$ are $m_1, \ldots, m_K$, as follows:

$$P(y^n; \hat{f}_{y^n}, m_1, \ldots, m_K) =$$

$$= \left(\frac{\sum_{q=1}^{K} \max(m_q, n_q - m_q)}{n} \right)^{\sum_{q=1}^{K} \max(m_q, n_q - m_q)} \times$$

$$\times \left(\frac{\sum_{q=1}^{K} \min(m_q, n_q - m_q)}{n} \right)^{\sum_{q=1}^{K} \min(m_q, n_q - m_q)} . \quad (10.27)$$

There are $\prod_{q=1}^{K} \binom{n_q}{m_q}$ ways to generate sequences y^n having the given numbers $m_1, \ldots, m_K$ of zeros in the subsequences $z^1, \ldots, z^K$, respectively, and all of them have the same probability, given by (10.28). In order to get the sum $C_n(k, \underline{b}^n)$ we group the terms in (10.27) as follows

$$C_n(k, \underline{b}^n) = \sum_{m_1=0}^{n_1} \binom{n_1}{m_1} \sum_{m_2=0}^{n_2} \binom{n_2}{m_2} \cdots \sum_{m_K=0}^{n_K} \binom{n_K}{m_K} \times$$

$$\times \hat{\theta}^{\sum_{q=1}^{K} \max(m_q, n_q - m_q)} (1 - \hat{\theta})^{\sum_{q=1}^{K} \min(m_q, n_q - m_q)}, \quad (10.28)$$

where $\hat{\theta} = \frac{1}{n} \sum_{q=1}^{K} \max(m_q, n_q - m_q)$. While the expression for $C_n(k, \underline{b}^n)$ in (10.27) involves summation of 2^n terms, the expression (10.29) requires addition of only $\prod_{q=1}^{K} n_q$ terms.

A fast algorithm for computation of $C_n(k, \underline{b}^n)$ Observe that (10.29) can be alternatively expressed as

$$C_n(k, \underline{b}^n) = \sum_{n_1^*=0}^{n} \left(\frac{n_1^*}{n} \right)^{n_1^*} \left(1 - \frac{n_1^*}{n} \right)^{n - n_1^*} S_{K, n_1, \ldots, n_K}(n_1^*) (10.29)$$

where $S_{K, n_1, \ldots, n_K}(n_1^*)$ is the number of sequences y^n having n_1^* ones in the residual sequence. We note that $n_1^* = \sum_{q=1}^{K} \min(m_q, n_q - m_q)$ and the numbers $S_{K,n_1,\ldots,n_K}(n_1^*)$ can be easily computed, recursively in K. Denote first

$$h_\ell(m) = \begin{cases} 0 & if & m > \frac{n_\ell}{2} \\ \binom{n_\ell}{m} & if & m = \frac{n_\ell}{2} \\ 2\binom{n_\ell}{m} & else \end{cases}, \qquad (10.30)$$

which is the number of sequences of n_ℓ bits, having either m bits set to 1, or $n_\ell - m$ bits set to 1, for $0 \le m \le \frac{n_\ell}{2}$. By combining each of the $S_{K-1,n_1,\ldots,n_{K-1}}(n_1^* - m_K)$ sequences having $n_1^* - m_K$ ones in the residual sequence, with each of the $h_K(m_K)$ sequences having either m_K bits set to 1, or $n_K - m_K$ bits set to 1, we get sequences having $(n_1^* - m_K) + \min(m_K, n_K - m_K) = n_1^*$ bits of 1 in their residual sequence. Therefore the following recurrence relation holds:

$$S_{K,n_1,\ldots,n_K}(n_1^*) = \sum_{m_K=0}^{n_K} h_K(m_K) S_{K-1,n_1,\ldots,n_{K-1}}(n_1^* - m_K) \quad (10.31)$$

where, by convention, $S_{K-1,n_1,\ldots,n_{K-1}}(n_1^* - m_K) = 0$ for negative arguments, $n_1^* - m_K < 0$.

We note that the recurrence is simply a convolution sum, $S_{K,n_1,\ldots,n_K} = h_K \otimes S_{K-1,n_1,\ldots,n_{K-1}}$, and from here we conclude that

$$S_{K,n_1,\ldots,n_K} = h_1 \otimes h_2 \otimes \ldots \otimes h_K. \qquad (10.32)$$

We can easily see that $S_{K_1,n_1,\ldots,n_{K_1}}(i) = 0$ for $i > \frac{\sum_q^{K_1} n_q}{2}$, due to the fact that the optimal residual sequence cannot have more than $\frac{\sum_q^{K_1} n_q}{2}$ ones. Also, from (10.31) we note that only $\frac{1}{2^K}\prod_{q=1}^K n_q$ terms have to be added when evaluating all convolution sums (10.33). Therefore the computation of $C_n(k, \underline{b}^n)$ by using (10.30) is about 2^K faster than when using (10.29).

3.3   A two part code for the Boolean class $\mathcal{M}(\Theta, k, f)$

We consider a coding problem where the decoder has available the sequence of regression vectors $\underline{b}^n$, and the set $\{\underline{b}_{j_1}, \ldots, \underline{b}_{j_K}\}$, with the numbers $\{n_q\}$ (recall that we denote by z^q the subsequence of y^n observed when the regressor vector is $\underline{b}_{j_q}$, and by n_q the length of the subsequence z^q having m_q zeros).

In order to specify the parameters $\{\hat{\theta}_q = \frac{m_q}{n_q}\}$ it is necessary to encode the values $\{m_q\}$, which can be done with

$$Cost(\hat{\Theta}) = \sum_{q=1}^{K} \log_2 n_q \tag{10.33}$$

bits, while the cost of encoding y^n, conditional on $\hat{\Theta}$, is

$$Cost(y^n|\hat{\Theta}) = -\log_2 \prod_{q=1}^{K} \hat{\theta}_q^{m_q} (1 - \hat{\theta}_q)^{n_q - m_q} \tag{10.34}$$

bits, leading to a overall description length of $Cost(\hat{\Theta}) + Cost(y^n|\hat{\Theta})$.

3.4 A two part code for the Boolean class $\mathcal{M}(\theta, k, f)$

The decoder has again available the sequence of regressor vectors $\underline{b}^n$. In order to specify the parameter $\hat{\theta}(y^n) = \frac{n_0^*(y^n)}{n}$ it is necessary to encode the value n_0^*, which can be done with

$$Cost(\hat{\theta}(y^n)) = \log_2 n \text{ bits.} \tag{10.35}$$

The ML estimates $\{\hat{f}(\underline{b}_q)\}$ need one bit each. Therefore the Boolean function for the K distinct regressor vectors has the cost $Cost(\hat{f}(y^n)) = K$. The cost of encoding y^n, conditional on $(\hat{\theta}, \hat{f}(y^n))$, is

$$Cost(y^n|\hat{\theta}(y^n), \hat{f}(y^n)) = -\log_2 \hat{\theta}(y^n)^{n_0^*(y^n)} (1 - \hat{\theta}(y^n))^{n - n_0^*(y^n)} \tag{10.36}$$

bits, and the overall description length is

$$\begin{aligned} Cost(y^n|\mathcal{M}(\theta, k, f)) &= Cost(\hat{\theta}(y^n)) + Cost(\hat{f}(y^n)) \\ &\quad + Cost(y^n|\hat{\theta}(y^n), \hat{f}(y^n)) \\ &= \log_2 n + K - \log_2 \prod_{q=1}^{K} \hat{\theta}(y^n)^{n_0^*(y^n)} (1 - \hat{\theta}(y^n))^{n - n_0^*(y^n)}. \end{aligned}$$

4. Experimental results

We illustrate the classification based on the NML model for Boolean regression models using the microarray DNA data Leukemia (ALL/AML) of (Golub et al., 1999), publicly available at http://www-genome.wi.mit.edu/MPR/data_set_ALL_AML.html.

The microarray contains 6817 human genes, sampled from 72 cases of cancer, of which 47 are of ALL type and 25 of AML type. The data is preprocessed as recommended in (Golub *et al.*, 1999) and (Dudoit *et al.*, 2000): first the gene values are truncated below at 100 and above at 16000; secondly, genes having the ratio of the maximum over the minimum less than 5, or the difference between the maximum and the minimum less than 500 are excluded; and finally, the logarithm to the base 10 is applied to the preprocessed gene expression values. The resulting data matrix $\tilde{X}$ has 3571 rows and 72 columns.

We design a two level quantizer by applying the LBG algorithm (Linde *et al.*, 1980) and the resulting decision threshold is 2.6455, when all the entries in the matrix $\tilde{X}$ are used as a training set (but we note that no information about the true classes is used during the quantization stage). The entries in the matrix $\tilde{X}$ are quantized to binary values, resulting in the binary matrix X.

4.1 The NML model for the Boolean regression models with $k = 1$

Table 10.1 lists the best 10 genes as ranked by the NML model for the Boolean class $\mathcal{M}(\theta, 1, f)$. The corresponding codelengths are in the range $26 - 44$ bits, well below 70 bits, which is the codelength assigned by the NML model for the Bernoulli class.

Figure 10.1 shows the gene expression values for the gene which minimizes the codelength assigned by the NML model of the Boolean class $\mathcal{M}(\theta, 1, f)$. We note that the binary profile for the quantized gene expression is quite robust to small changes in the value of the threshold, and also it correlates very well with the true class profile. In Figure 10.2 we compare the codelength assigned by the NML models for each of the classes $\mathcal{M}(\Theta, 1, f)$, and $\mathcal{M}(\theta, 1, f)$. The ranking of the best 6 genes is the same by both of the models.

4.2 The NML model for the Boolean regression models with $k = 2$

We consider here the Boolean classes of regression models using $k = 2$ regressors. There are $3570 \times 3571 = 12\,748\,470$ possible pairs of genes. We computed the codelength assigned by the NML model of the Boolean class $\mathcal{M}(\theta, 2, f)$ for all possible pairs of genes. In Figure 10.3 we show the codelength obtained by the best 100 pairs of genes, as assigned by NML model of the Boolean class $\mathcal{M}(\theta, 2, f)$. We also show for these gene pairs the codelength assigned by the NML model of the Boolean class $\mathcal{M}(\Theta, 2, f)$. The codelengths assigned to each gene pair by the two models are very close most of the time, although significant differences are observed for some gene pairs. Although the model $\mathcal{M}(\Theta, 2, f)$ is more flexible, it is sometimes able to explain the data in fewer bits, but the model $\mathcal{M}(\theta, 2, f)$ is better in most of the cases.

In Figure 10.4 we compare the NML model and the two part code for the class of models $\mathcal{M}(\theta, 2, f)$, for the best 100 pairs of genes. We observe that for each pair of genes the codelength associated by the simple two part code described in Section 3.4 is about 4 bits longer than the NML codelength. However, this difference is almost constant, suggesting that a first pre-selection according to the faster two-part code could be used to obtain a smaller pool of gene pairs out of the 12 748 470 possible pairs, and then the more precise NML model could be used for the final selection of the best gene pairs. Figure 10.5 shows a higher discrepancy between the codelengths assigned by NML model and the two part code for the class of models $\mathcal{M}(\Theta, 2, f)$. Finally, Figure 10.5 shows for the best 120 pairs of genes x_1x_2 the codelength assigned by the NML model of the class $\mathcal{M}(\theta, 2, f)$ together with the codelength assigned by the NML model of the class $\mathcal{M}(\theta, 1, f)$ (only the codelength for the best gene, x_1 or x_2, is shown). We conclude that the NML model $\mathcal{M}(\theta, 2, f)$ for each of the 120 pairs of genes explains the data better than their subnested NML models $\mathcal{M}(\theta, 1, f)$.

4.3 The NML model for the Boolean regression models with $k = 3$

In Figure 10.7 we show the best 100 triplets of genes $(x_1x_2x_3)$, ranked by the codelength assigned by the NML model of the class $\mathcal{M}(\theta, 3, f)$, together with the codelength assigned by the NML model of the class $\mathcal{M}(\theta, 2, f)$ (the codelength of the best out of three possible pairs (x_1x_2), (x_1x_3), (x_2x_3) is shown for each triplet). All listed models with three genes are better than the sub-nested models having only two genes.

4.4 Extension of the classification for unseen cases of the Boolean regressors

The Boolean regressors observed in the training set may not span all the 2^k possible binary vectors. If a binary vector $\underline{b}_q$ is not in the training set, the decision $f^*(\underline{b}_q)$ for classification remains undecided during the training stage. We pick the value of $f^*(\underline{b}_q)$ by the nearest neighbor voting, taking the majority vote of the neighbors $\underline{b}_\ell$ situated at Hamming distance 1, for which $f^*(\underline{b}_\ell)$ was decided during the training stage. Denote by $\mathcal{N}_1(\underline{b}_q) = \{\underline{b}_\ell : \underline{b}_\ell \in \underline{b}^n, w_H(\underline{b}_\ell \oplus \underline{b}_q) = 1, f^*(\underline{b}_\ell) \text{ is decided}\}$ the set of decided neighbors of $\underline{b}_q$ at Hamming distance 1. The voting decision is then

$$f^*(\underline{b}_q) = \begin{cases} 1 & if \quad \sum_{\underline{b}_\ell \in \mathcal{N}_1(\underline{b}_q)} f^*(\underline{b}_\ell) > \sum_{\underline{b}_\ell \in \mathcal{N}_1(\underline{b}_q)} (1 - f^*(\underline{b}_\ell)) \\ 0 & if \quad \sum_{\underline{b}_\ell \in \mathcal{N}_1(\underline{b}_q)} f^*(\underline{b}_\ell) < \sum_{\underline{b}_\ell \in \mathcal{N}_1(\underline{b}_q)} (1 - f^*(\underline{b}_\ell)) \end{cases} \quad (10.37)$$

If after voting there still remains a tie ($\sum_{\underline{b}_\ell \in \mathcal{N}_1(\underline{b}_q)} f^*(\underline{b}_\ell) = \sum_{\underline{b}_\ell \in \mathcal{N}_1(\underline{b}_q)} (1 - f^*(\underline{b}_\ell))$), we take the majority vote of the neighbors at Hamming distance 2, and continue, if necessary, until a clear decision is reached.

4.5 Estimation of classification errors achieved with Boolean regression models with $k = 3$

The Leukemia data set from (Golub *et al.*, 1999) was considered recently in a study that compared several classification methods (Dudoit *et al.*, 2000). The evaluation of the performance is based there on the classification error as estimated in a cross-validation 2:1 experiment. In order to compare our classification results with these results, we estimate the classification error in the same way, namely by dividing the 72 patient set at random into a training set of $n_T = 48$ patients and a test set of $n_s = 24$ patients, finding the optimal predictor $f^*(\cdot)$ over the training set, classifying the test set by use of the predictor $f^*(\cdot)$ (the extension for cases unseen in the training set is done as in the previous subsection), and counting the number of classification errors produced over the test set. The random split is repeated a number of $n_r = 10000$ times, and the estimated classification error is computed as the percentage of the total number of errors observed in the ($n_r \cdot n_s$) test classifications. We mention for comparison that the best classification methods tested in (Dudoit *et al.*, 2000) have classification errors higher than 1%. As we see in Table 10.3 and Figure 10.8 there are several predictors with three genes, achieving classification error rates as low as 0.004%. Note the remarkable agreement in ranking of the gene triplets by the NML codelength assigned by the NML model of the class $\mathcal{M}(\theta, 3, f)$ and the estimated classification error rates.

As for the genes involved in the optimal predictors of Table 10.3, we note that five genes belong to the set of 50 "informative" genes selected in (Golub *et al.*, 1999), namely $M23197, M84526, M27891, M83652, X95735$.

5.   Conclusions

Boolean regression classes of models are powerful modeling tools having associated NML models which can be easily computed and used in MDL inference, in particular for factor selection. Comparing the MDL methods based on the two-part codes with those based on the NML models, we note that the former is faster to evaluate, but the latter provides a significantly shorter codelength and hence a better description of the data. When analyzing the gene expression data, speed may be a major concern, since one has to test $\binom{n}{k}$ possible groupings of k genes, with n in the order of thousands and k usually less than 10. The two-part codes may then be used for pre-screening of the gene groupings, to remove the obviously poor performers, and then the NML model could be applied to obtain the final selection from a smaller pool of candidates. The running time for all our experiments reported here is in the order of tens of minutes.

The use of the MDL principle for classification with the class of Boolean models provides an effective classification method as demonstrated with the important cancer classification example based on gene expression data. The NML model for the class $\mathcal{M}(\theta, k, f)$ was used for the selection of informative feature genes. When using the sets of feature genes, selected by NML model, we achieved classification error rates significantly lower than those reported recently for the same data set.

References

Barron, A., Rissanen, J., Bin, Y. (1998) The minimum description length principle in coding and modeling. *IEEE Trans. on Information Theory, Special commemorative issue: Information Theory 1948-1998*, 44:6, 2743–2760.

Dudoit, S., Fridlyand, J., Speed, T.P. (2000) *Comparison of Discrimination Methods for the Classification of Tumors Using Gene Expression Data*. Dept. of Statistics University of California, Berkeley, Technical Report 576.

Golub, T.R., Slonim, D.K., Tamayo, P., Huard, C., Gaasenbeek, M., Mesirov, J.P., Coller, H., Loh, M.L., Downing, J.R., Caligiuri, M.A., Bloomfield,C.D., Lander, E.S. (1999) Molecular Classification of Cancer: Class Discovery and Class Prediction by Gene Expression Monitoring. *Science*, 286, 531–537.

Kim, S., Dougherty, E.R. (2000) Coefficient of determination in nonlinear signal processing. *Signal Processing*, 80, 2219–2235.

Hieter, P., Boguski, M. (1997) Functional genomics: it's all how you read it. *Science* 278, 601-602.

Jacob, F., Monod, J. (1961) Genetic regulatory mechanisms in the synthesis of proteins. *Journal of Molecular Biology* 3, 318-356.

Kim, S., Dougherty, E.R., Chen, Y., Sivakumar, K., Meltzer, P., Trent, J.M., Bitnner, M. (2000). Multivariate measurement of gene expression relationships. *Genomics*, 67, 201–209.

Linde, Y., Buzo, A., Gray, R.M. (1980) An algorithm for vector quantization design. *IEEE Transactions on Communications*, 28, 84–95.

Rissanen, J. (1978) Modelling by shortest data description. *Automatica*, 14, 465–471.

Rissanen, J. (1984) Universal coding, information, prediction and estimation. *IEEE Trans. on Information Theory*, 30, 629–636.

Rissanen, J. (1986) Stochastic complexity and modeling. *Ann. Statist.*, 14, 1080-1100.

Rissanen, J. (2000) MDL Denoising. *IEEE Trans. on Information Theory*, IT-46:7, 2537–2543.

Rissanen, J. (2001) Strong optimality of the normalized ML models as universal codes and information in data. *IEEE Trans. on Information Theory*, IT-47:5, 1712–1717.

Russel, P.J. (2000) *Fundamentals of genetics.* 2nd edition, San Francisco: Addison Wesly Longman Inc.

Schena, M., Shalon, D., Davis, R.W., Brown, P.O. (1995) Quantitative monitoring of gene expression patterns with a complementary DNA microarray. *Science* 270, 467-470.

Shtarkov, Yu.M. (1987) Universal sequential coding of single messages. *Translated from Problems of Information Transmission*, 23:3, 3–17.

Tabus, I., Astola, J. (2001) On the Use of MDL Principle in Gene Expression Prediction. *Journal of Applied Signal Processing*, 2001:4, 297-303.

Tabus, I., Astola, J. (2000) *MDL Optimal Design for Gene Expression Prediction from Microarray Measurements.* Tampere University of Technology, Technical Report,ISBN.952-15-0529-X.

Table 10.1.   The best 10 genes for predicting the class label with the NML model of the class $\mathcal{M}(\theta, 1, f)$.

Codelength	Gene Index	Gene accession number
26.7	2288	M84526
30.6	1834	M23197
30.6	3252	U46499
34.2	6855	M31523
34.2	760	D88422
37.5	6376	M83652
40.6	1685	M11722
43.5	6378	M83667
43.5	2128	M63379
43.6	1882	M27891

Table 10.2.   The best 14 pairs of genes for predicting the class label with the NML model of the class $\mathcal{M}(\theta, 2, f)$.

Codelength	Pair of Genes		Gene accession numbers	
18.3	2288	5714	M84526	HG1496-HT1496
19.2	1834	2288	M23197	M84526
19.2	2288	4079	M84526	X05409
19.3	2288	3631	M84526	U70063
19.3	2288	5808	M84526	HG2981-HT3127
19.3	1807	2288	M21551	M84526
19.3	1834	6277	M23197	M30703
19.3	1834	5373	M23197	S76638
19.3	1834	6279	M23197	X97748
19.4	758	4342	D88270	X59871
19.4	1144	1882	J05243	M27891
23.2	73	1834	AB000584	M23197
23.2	1834	5714	M23197	HG1496-HT1496
23.2	1834	5949	M23197	M29610

Table 10.3. The best 18 triplets of genes for predicting the class label according to the NML model of the class $\mathcal{M}(\theta, 3, f)$.

Code length	Classification error [%]	Triplet of Genes			Gene accession numbers		
6.9	0.912	1834	2288	5714	M23197	M84526	HG1496-HT1496
7.9	**0.010**	**1834**	**3631**	**6277**	**M23197**	**U70063**	**M30703**
7.9	0.891	758	4250	4342	D88270	X53586	X59871
8.0	0.652	2288	4847	6376	M84526	X95735	M83652
8.7	**0.008**	**1834**	**3631**	**5373**	**M23197**	**U70063**	**S76638**
8.7	**0.007**	**1834**	**3631**	**6279**	**M23197**	**U70063**	**X97748**
8.7	0.910	1144	1217	1882	J05243	L06132	M27891
8.8	0.649	302	2288	6376	D25328	M84526	M83652
8.8	0.055	1144	1834	1882	J05243	M23197	M27891
8.8	0.063	1834	1882	6049	M23197	M27891	U89922
8.8	**0.004**	**1144**	**1882**	**5808**	**J05243**	**M27891**	**HG2981-HT3127**
8.8	0.584	2288	3932	6376	M84526	U90549	M83652
8.9	0.558	2288	5518	6376	M84526	X95808	M83652
8.9	0.560	1399	2288	6376	L21936	M84526	M83652
8.9	0.620	1241	2288	6376	L07758	M84526	M83652
8.9	0.605	2288	3660	6376	M84526	U72342	M83652
8.9	0.582	2288	4399	6376	M84526	X63753	M83652
8.9	0.556	2288	4424	6376	M84526	X65867	M83652

Figure 10.1.   Gene expressions: quantized and unquantized expression levels of gene
2288, M84526 (DFD component of complement (adipsin)).

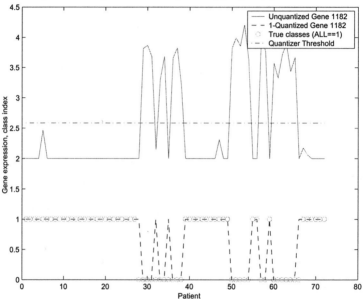

Figure 10.2.   The 100 best genes by the NML model of the class $\mathcal{M}(\theta, 1, f)$ (sorted in
increasing order of codelength). The NML codelength assigned by the class $\mathcal{M}(\Theta, 1, f)$ is
also shown for each gene. By comparison, the NML model of the Bernoulli class $\mathcal{B}(\theta)$ assigns
to the string y^n the codelength 70.57 bits.

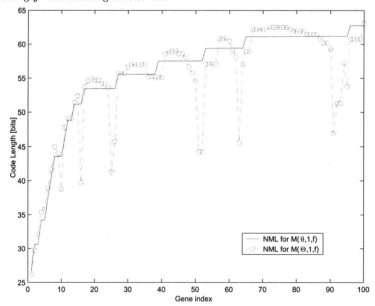

Figure 10.3. The 100 best pairs of two genes by NML model of the class $\mathcal{M}(\theta, 2, f)$ (sorted in increasing order of codelength). The NML codelength assigned by the class $\mathcal{M}(\Theta, 2, f)$ is also shown for each pair.

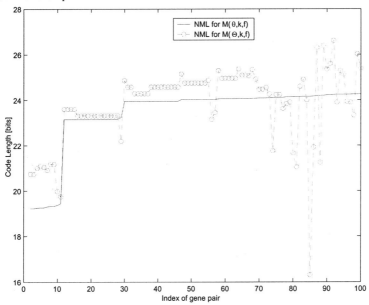

Figure 10.4. The 100 best pairs of two genes by the NML model of the class $\mathcal{M}(\theta, 2, f)$ (sorted in increasing order). The two-part code codelength assigned by the class $\mathcal{M}(\theta, 2, f)$ is shown to rank the gene pairs in a similar manner.

Figure 10.5. The 100 best pairs of two genes by the NML model of the class $\mathcal{M}(\Theta, 2, f)$ (sorted in increasing order of codelength). The two-part codelength assigned by the class $\mathcal{M}(\Theta, 2, f)$ is also shown for each pair.

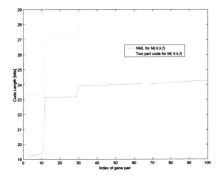

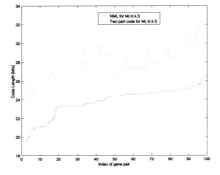

Figure 10.6.   The 120 best pairs of genes, by the NML model of the class $\mathcal{M}(\theta, 2, f)$ (sorted in increasing order of codelength). The minimum of the NML codelength assigned by the class $\mathcal{M}(\theta, 1, f)$ for each of the genes in the pair is also shown.

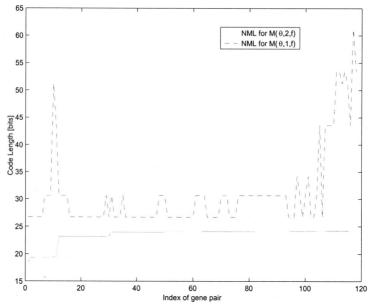

Figure 10.7.   The 100 best triplets of genes by the NML model of the class $\mathcal{M}(\theta, 3, f)$ (sorted in increasing order of codelength). The minimum of the NML codelength assigned by the class $\mathcal{M}(\theta, 2, f)$ for each pair of genes in the triplet is also shown.

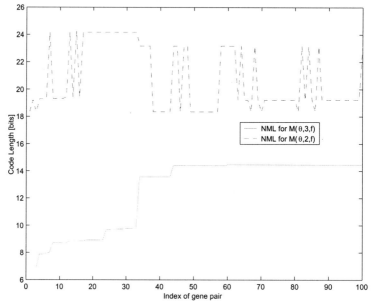

Figure 10.8. The four triplets of genes with the best classification error rates, 0.01%, 0.008%, 0.007%, and 0.004%, respectively, ranked as second, fifth, sixth and eleventh according to the NML model of the class $\mathcal{M}(\theta, 3, f)$. The original values (non-quantized) of the genes are shown, pseudo-colored using the colormap indicated at the bottom of the figure. The class labels are represented as follows: the ALL class with green and the AML class with red. The predictor based on the genes $x_1 = 1834, x_2 = 3631$ and $x_3 = 6277$ has the Boolean function $ALL = f(x_1, x_2, x_3) = \overline{x}_1 \overline{x}_2 \overline{x}_3$. The predictor based on the genes $x_1 = 1834, x_2 = 3631$ and $x_3 = 5373$ has the Boolean function $ALL = f(x_1, x_2, x_3) = \overline{x}_1 \overline{x}_2 \overline{x}_3$. The predictor based on the genes $x_1 = 1834, x_2 = 3631$ and $x_3 = 6279$ has the Boolean function $ALL = f(x_1, x_2, x_3) = \overline{x}_1 \overline{x}_2 \overline{x}_3$. The predictor based on the genes $x_1 = 1144, x_2 = 1882$ and $x_3 = 5808$ has the Boolean function $ALL = f(x_1, x_2, x_3) = x_1 \overline{x}_2 \vee \overline{x}_2 \overline{x}_3$. (see color plates)

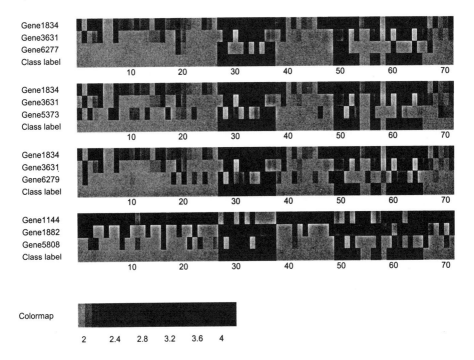

Figure 10.9.   The best 100 triplets of genes by the NML model of the class $\mathcal{M}(\theta, 3, f)$, and the estimated classification error of the corresponding predictors.

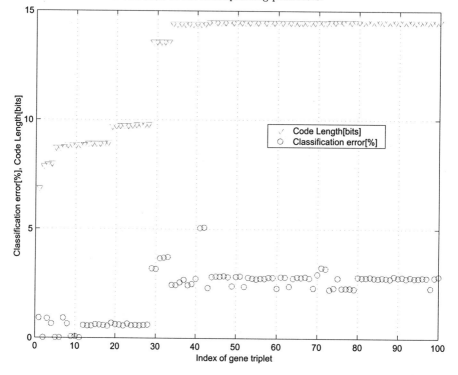

Chapter 11

INFERENCE OF GENETIC REGULATORY NETWORKS VIA BEST-FIT EXTENSIONS

Ilya Shmulevich[1], Antti Saarinen[2], Olli Yli-Harja[2], and Jaakko Astola[2]

[1] *Cancer Genomics Laboratory, Department of Pathology*
The University of Texas M. D. Anderson Cancer Center, Houston, Texas, USA
[2] *Institute of Signal Processing, Tampere University of Technology, Tampere, Finland*

1. Introduction

One of the most important breakthroughs in recent years in molecular biology is microarray technology, which allows monitoring of gene expression at the transcript level for thousands of genes in parallel (Schena *et al.*, 1995; Celis *et al.*, 2000). Even though mRNA is not the final product of a gene, armed with the knowledge of gene transcript levels in various cell types, under different developmental stages (Wen *et al.*, 1998), and under a variety of conditions, such as in response to specific stimuli (Iyer *et al.*, 1999; DeRisi *et al.*, 1997), scientists can gain a deeper understanding of the functional roles of genes, of the cellular processes in which they participate, and of their regulatory interactions. Thus, gene expression data for many cell types and organisms at multiple time points and experimental conditions are rapidly becoming available (Brazma and Vilo, 2000). In fact, the amounts of data typically gathered in experiments call for computational methods and formal modeling in order to make meaningful interpretations (Huang, 1999). The emerging view is that as biology becomes a more quantitative science, modeling approaches will become more and more usual (Brazma and Vilo, 2000).

One popular computational approach to gene expression analysis is to compare gene expression *profiles*, that is, the dynamic behavior of genes over time points or cell types, and to apply clustering (Eisen *et al.*, 1998; Ben-Dor and Yakhini, 1999) and data reduction and visualization techniques such as the self-organizing map (Tamayo *et al.*, 1999; Kaski *et al.*, 2001) (also see Chapter 5) or principle components analysis (Alter *et al.*, 2000; Holter *et al.*, 2000). An inherent assumption in many such approaches is that if two gene profiles are similar, the respective genes are co-regulated and possibly functionally related (Brazma and Vilo, 2000). Although this assumption does not always hold, such methods can nevertheless be useful in uncovering important underlying mechanisms in gene regulation. Another difficulty with these approaches is that currently, there is no theory on how to choose the best distance or similarity measure (Brazma and Vilo, 2000; Shmulevich and Zhang, 2002a) (e.g. correlation coefficient, rank/ordinal correspondence measures, various norms), and each one may lead to possibly very different results. But perhaps a more fundamental criticism that such approaches have received is that they are essentially "genocentric" to use a term of Huang (1999), in that they focus on functions of individual genes.

In order to understand the nature of cellular function, it is necessary to study the behavior of genes in a holistic rather than in an individual manner. A significant role is played by the development and analysis of mathematical and computational methods in order to construct formal models of genetic interactions. This research direction provides insight and a conceptual framework for an integrative view of genetic function and regulation and paves the way toward understanding the complex relationship between the genome and the cell. Moreover, this direction has provided impetus for experimental work directed toward verification of these models.

There have been a number of attempts to model gene regulatory networks, including linear models (van Someren *et al.*, 2000; D'Haeseleer *et al.*, 1999), Bayesian networks (Murphy and Mian, 1999; Friedman *et al.*, 2000), and neural networks (Weaver *et al.*, 1999). The model system that has received, perhaps, the most attention is the so-called *Random Boolean Network* model originally introduced by Kauffman (Kauffman, 1993), approximately thirty years ago. In this model, gene expression is quantized to only two levels: ON and OFF. The expression level (state) of each gene is functionally related to the expression states of some other genes. These connections are represented by the network 'wiring'.

Recent research seems to indicate that many realistic biological questions may be answered within the seemingly simplistic Boolean formalism, which in essence emphasizes fundamental, generic principles rather than quantitative biochemical details (Huang, 1999; Shmulevich and Zhang, 2002a). Moreover, this is the only model system that has yielded insights into the overall behavior of large genetic networks (Szallasi and Liang, 1998; Wuensche, 1998). For example, the dynamic behavior of such networks corresponds to and can be used to model many biologically meaningful phenomena, such as, for example cellular state dynamics, possessing switch-like behavior, stability, and hysteresis (Huang, 1999).

Besides the conceptual framework afforded by such models, a number of practical uses may be reaped by inferring the structure of the genetic models from experimental data, that is, from gene expression profiles. One such use is the identification of suitable drug targets in cancer therapy. To that end, much recent work has gone into identifying the structure of gene regulatory networks from expression data (Liang *et al.*, 1998; Akutsu *et al.*, 1999; Akutsu *et al.*, 1998; Akutsu *et al.*, 2000; Shmulevich *et al.*, 2002b).

Most of the work, however, has focused on the so-called Consistency Problem, namely, the problem of determining whether there exists a network that is consistent with the examples. While this problem is important in computational learning theory, as it can be used to prove the hardness of learning for various function classes, it may not be applicable in a realistic situation in which noisy observations or errors are contained, as is the case with microarrays. Measurement errors can arise in the data acquisition process or may be due to unknown latent factors. A learning paradigm that can incorporate such inconsistencies is called the *Best-Fit Extension Problem*. Essentially, the goal of this problem is to establish a rule or in our case, network, that would make as few misclassifications as possible.

In order for an inferential algorithm to be useful, it must be computationally tractable. In this chapter, we consider the computational complexity of the Best-Fit Extension Problem for the Random Boolean Network model. We show that for many classes of Boolean functions, the problem is polynomial-time solvable, implying its practical applicability to real data analysis. We first review the necessary background information on Random Boolean Networks and then discuss the Best-Fit Extension Problem for Boolean functions and its complexity for Boolean networks.

2. Boolean Networks

For consistency of notation with other related work, we will be using the same notation as in (Akutsu *et al.*, 1999). A Boolean network $G(V, F)$ is defined by a set of nodes $V = \{v_1, \ldots, v_n\}$ and a list of Boolean functions $F = (f_1, \ldots, f_n)$. A Boolean function $f_i(v_{i_1}, \ldots, v_{i_k})$ with k specified input nodes is assigned to node v_i. In general, k could be varying as a function of i, but we may define it to be a constant without loss of generality as $k = \max_i k(i)$ and allowing the unnecessary variables (nodes) in each function to be *fictitious*. For a function f, the variable x_i is fictitious if

$$f(x_1, \ldots, x_{i-1}, 0, x_{i+1}, \ldots, x_n) =$$
$$f(x_1, \ldots, x_{i-1}, 1, x_{i+1}, \ldots, x_n),$$

for all $x_1, \ldots, x_{i-1}, x_{i+1}, \ldots, x_n$. A variable that is not fictitious is called *essential*. We shall also refer to k as the *indegree* of the network. Each node v_i represents the state (expression) of gene i, where $v_i = 1$ means that gene i is expressed and $v_i = 0$ means it is not expressed. The list of Boolean functions F represents how genes regulate each other. That is, any given gene transforms its inputs (regulatory factors that bind to it) into an output, which is the state or expression of the gene itself. All genes (nodes) are updated synchronously in accordance with the functions assigned to them and this process is then repeated. The artificial synchrony simplifies computation while preserving the qualitative, generic properties of global network dynamics (Huang, 1999; Kauffman, 1993; Wuensche, 1998).

To capture the dynamic nature of these networks, it is useful to consider a 'wiring diagram' $G'(V', F')$ (Akutsu *et al.*, 1999). Let $k(i)$ be the number of essential variables of function f_i in F. We then construct n additional nodes $v'_1, \ldots, v'_n$ and for each $i = 1, \ldots, n$, we draw an edge from v_{i_j} to v'_i, for each $1 \le j \le k(i)$. Then, $V' = \{v_1, \ldots, v_n, v'_1, \ldots, v'_n\}$ and the list F' is actually the same as F, but with the functions being assigned to nodes $v'_1, \ldots, v'_n$ (with inputs from V) while the functions assigned to $v_1, \ldots, v_n$ are just the trivial identity functions, e.g. $f(v_i) = v_i$. In other words, $v'_i = f_i(v_{i_1}, \ldots, v_{i_{k(i)}})$ and thus, the expression pattern $\{v_1, \ldots, v_n\}$ corresponds to the states of the genes at time t (INPUT) and the pattern $\{v'_1, \ldots, v'_n\}$ corresponds to the states of the genes at time $t + 1$ (OUTPUT). Collectively, the states of individual genes in the genome form a *gene activity profile* (GAP) (Huang, 1999).

Consider the state space of a Boolean network with n genes. Then, the number of possible GAPs is equal to 2^n. For every GAP, there is another successor GAP into which the system transitions in accordance with its structural rules as defined by the Boolean functions. Thus, there is a directionality that is intrinsic to the dynamics of such systems. Consequently, the system ultimately transitions into so-called *attractor* states. The states of the system that flow into the same attractor state make up a *basin of attraction* of that attractor (Wuensche, 1998). Sometimes, the system periodically cycles between several *limit-cycle* attractors. It is interesting to note that such behavior even exists for some infinite networks (networks with an infinite number of nodes) (Moran, 1995), such as those in which every Boolean function is the majority function. Moreover, the convergence of a discrete dynamical system to attractors should be well known to many researchers from the area of non-linear signal processing, where convergence to *root signals* has been studied for many classes of digital filters (Gabbouj *et al.*, 1992). Root signals are those signals that are invariant to further processing by the same filter. Some filters are known to reduce any signal to a root signal after a finite number of passes while others possess cyclic behavior.

Although the large number of possible GAPs would seem to preclude computer-based analysis, simulations show that for networks with low connectivity, only a small number of GAPs actually correspond to attractors (Kauffman, 1993). Since other GAPs are unstable, the system is normally not found in those states unless perturbed. In fact, real genetic regulatory networks, at least in bacteria, are known to have very low connectivity (2 or 3) (Thieffry *et al.*, 1998).

3. The Best-Fit Extension Problem

One of the central goals in the development of network models is the inference of their structure from experimental data. In the strictest sense, this task falls under the umbrella of computational learning theory (Kearns and Vazirani, 1994). Essentially, we are interested in establishing "rules" or, in our case, Boolean functions by observing binary INPUT/OUTPUT relationships. Thus, this task can also be viewed as a system identification problem. One approach is to study the so-called Consistency Problem, considered for Boolean networks in (Akutsu *et al.*, 1999).

The Consistency Problem is important in computational learning theory (Valiant, 1984) and can be thought of as a search of a rule from examples. That is, given some sets T and F of "true" and "false" vectors, respectively, we aim to discover a Boolean function f that takes on the value 1 for all vectors in T and the value 0 for all vectors in F. We may also assume that the target function f is chosen from some class of possible target functions. One important reason for studying the complexity of the consistency problem is its relation to the PAC approximate learning model of Valiant (Valiant, 1984). If the consistency problem for a given class is NP-hard, then this class is not PAC-learnable. Moreover, this would also imply that this class cannot be learned with equivalence queries (Angluin, 1987).

Unfortunately, in realistic situations, we usually encounter errors that may lead to inconsistent examples. This is no doubt the case for gene expression profiles as measured from microarrays, regardless of how the binarization is performed. In order to cope with such inconsistencies, we can relax our requirement and attempt to establish a rule that makes the minimum number of misclassifications. This is called The Best-Fit Extension Problem and has been extensively studied in (Boros *et al.*, 1998) for many function classes.

We now briefly define the problem for Boolean functions. The generalization to Boolean networks is straightforward. A *partially defined Boolean function* pdBf is defined by a pair of sets (T, F) such that $T, F \subseteq \{0,1\}^n$, where T is the set of true vectors and F is the set of false vectors. A function f is called an *extension* of pdBf(T, F) if $T \subseteq T(f)$ and $F \subseteq F(f)$, where $T(f) = \{x \in \{0,1\}^n : f(x) = 1\}$ and $F(f) = \{x \in \{0,1\}^n : f(x) = 0\}$. Suppose that we are also given positive weights $w(x)$ for all vectors $x \in T \cup F$ and define $w(S) = \sum_{x \in S} w(x)$ for a subset $S \subseteq T \cup F$ (Boros *et al.*, 1998). Then, the *error size* of function f is defined as

$$\varepsilon(f) = w(T \cap F(f)) + w(F \cap T(f)). \qquad (11.1)$$

If $w(x) = 1$ for all $x \in T \cup F$, then the error size is just the number of misclassifications. The goal is then to output subsets T^* and F^* such that $T^* \cap F^* = \emptyset$ and $T^* \cup F^* = T \cup F$ for which the pdBf(T^*, F^*) has an extension in some class of functions $\mathcal{C}$ (chosen a priori) and so that $w(T^* \cap F) + w(F^* \cap T)$ is minimum. Consequently, any extension $f \in \mathcal{C}$ of pdBf(T^*, F^*) has minimum error size.

It is clear that the Best-Fit Extension Problem is computationally more difficult than the Consistency Problem, since the latter is a special case of the former, that is, when $\varepsilon(f) = 0$. The computational complexity of these problems has been studied for many function classes in (Boros *et al.*, 1998). For example, the Best-Fit Extension Problem was proved to be polynomially solvable for all transitive classes and some others, while for many classes including threshold, Horn, Unate, positive self-dual, it was shown to be NP-hard.

It is important to note here that if the class $\mathcal{C}$ of functions is not restricted (i.e. all Boolean functions), then an extension exists if and only if T and F are disjoint. This can be checked in $O(|T| \cdot |F| \cdot \text{poly}(n))$ time, where poly(n) is the time needed to answer "is $x = y$?" for $x \in T$, $y \in F$. This is precisely why attention has been focused on various subclasses of Boolean functions.

For the case of Boolean networks, we are given n partially defined Boolean functions defined by sets $(T_1, F_1), \cdots, (T_n, F_n)$. Since we are making "genome-wide" observations, it follows that $|T_1 \cup F_1| = \cdots = |T_n \cup F_n| = m$. Given some class of functions $\mathcal{C}$, we say that the network $G(V, F)$ is consistent with the observations if f_i from F is an extension of pdBf(T_i, F_i), for all i. In (Akutsu *et al.*, 1999) it was shown that when $\mathcal{C}$ is the class of Boolean functions containing no more than k essential variables (maximum indegree of the network), the Consistency Problem is polynomially solvable in n and m. In fact, it turns out that if we make no restriction whatsoever on the function class, the Consistency Problem for Boolean networks is still polynomial-time solvable, because for each node v_i, all we need to do is check whether or not $T_i \cap F_i = \emptyset$.

For a restricted class $\mathcal{C}$, we can say that if the Consistency Problem is polynomially solvable for one Boolean function (i.e. one node), then it is also polynomially solvable for the entire Boolean network, in terms of n and m. The reason is that the time required to construct an extension simply has to be multiplied by n - the number of nodes. For example, as shown in (Akutsu *et al.*, 1999), the time needed to construct one extension from the class of functions with k essential variables (k fixed), is $O\left(2^{2^k} \cdot n^k \cdot m\right)$ because there are a total of 2^{2^k} Boolean functions that must be checked for each of the $\binom{n}{k}$ possible combinations of variables and for m observations. Thus, the Consistency Problem for the entire network can be solved in $O\left(2^{2^k} \cdot n^k \cdot m \cdot n\right)$ time, for fixed k.

We now see that the same must hold true for the Best-Fit Extension Problem as well. Consider again the class of functions with k essential variables. Then, all we must do is calculate the error size $\varepsilon(f)$ for every Boolean function f, for each of the $\binom{n}{k}$ possible combinations of variables, over all m observations, and keep track of the minimum error size as well as the corresponding function and its variables. To generalize this for a Boolean network, we must simply repeat the process for every one of the n nodes, essentially multiplying the time needed for obtaining a best-fit extension by n. Consequently, the Best-Fit Extension Problem is polynomial-time solvable for Boolean networks, when all functions are assumed to have no more than k essential variables. Moreover, if $\mathcal{C}$ is the class of all Boolean functions (i.e. no restrictions), then the Best-Fit Extension Problem for Boolean networks can also be solved in polynomial time by virtue of it being polynomially solvable for general Boolean functions (see (Boros *et al.*, 1998)). So, we can say the following:

Proposition 1 *If it is known that the Best-Fit Extension Problem is solvable in polynomial time in n and m for one Boolean function from class $\mathcal{C}$, then the Best-Fit Extension Problem has a polynomial time solution for a Boolean network in which all functions belong to class $\mathcal{C}$.*

For example, it is known that for the class of monotone (positive) Boolean functions, the Boolean function version of the Best-Fit Extension Problem is polynomially solvable (Boros *et al.*, 1998). Then, it immediately follows that the Boolean network version of the Best-Fit Extension Problem is also polynomial-time solvable.

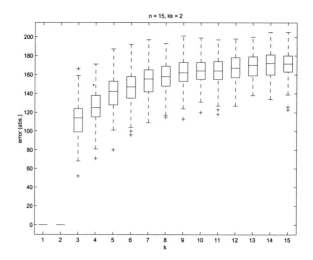

Figure 11.1. Results from the first simulation. The x-axis shows the indegree k of the network, which was used to obtain the original OUTPUT matrix. The y-axis shows the absolute error (i.e. the number of wrong OUTPUTs in the OUTPUT matrix of the network, which was inferred with the Best-Fit method. In this figure, medians, upper and lower quartiles, and extreme values are shown.

4. Simulation Analysis

The idea behind the following simulations was to computationally assess the behavior of the error attained by the Best-Fit method, when presented with given INPUT-OUTPUT pairs, as described in Section 3.

In the first simulation, networks with $n = 15$ (i.e. the number of genes) were used. First, a network with indegree k (in this simulation k varied from 1 to 15) was constructed. Then, 50 INPUTs were created randomly and the constructed network was used to produce 50 OUTPUTs, forming a so-called OUTPUT matrix. For example, for a network with 15 genes, the OUTPUT matrix is of size 15×50. After this, the Best-Fit method was used to infer a network from these 50 INPUT-OUTPUT pairs. For the error in Eq. (11.1), we used $w(x) = 1$ for all vectors, thus simply counting the number of wrong OUTPUTs.

The inference was constrained so that the indegree of the inferred network was set to be $kk = 2$. Another OUTPUT matrix was produced from the inferred network with the same 50 INPUTs. After having these two OUTPUT matrices, the error was calculated by comparing the corresponding elements of these matrices. Every wrong OUTPUT in the matrix that was derived from the inferred network increased the

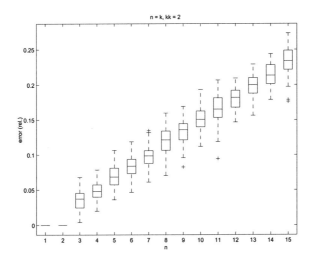

Figure 11.2. Results from the second simulation. The x-axis shows the number of genes in the network. In this simulation, this is also the indegree of that network (i.e. $n = k$). The y-axis shows the relative error; i.e. the number of wrong OUTPUTs was normalized by dividing it by the number of all OUTPUTs. Medians, upper and lower quartiles, and extreme values are presented.

error by one. Thus, because the OUTPUT matrices were of size 15×50, the maximum error was 750.

To produce Figure 11.1, the procedure described above was repeated 250 times for each value of k. Thus, for each k we have 250 different errors that were produced. The x-axis shows the indegree k that was used to create the network that generated the original OUTPUT matrix. The y-axis contains the absolute error, that is, the number of wrong elements in the OUTPUT matrix generated by the inferred network. The figure shows the median of the errors as well as the upper and the lower quartiles. The outliers are also presented and are marked with a '+' symbol. The figure indicates that the behavior of the error appears to be logarithmic.

In the second simulation, we took a different approach. Rather than having a fixed network size n, we set $n = k$ for every indegree k. Then, methods similar to those in the first simulation were used. For every k, a network with $n = k$ was constructed and 50 INPUTs were randomly created. Then an OUTPUT matrix was produced from the constructed network. At this point, the OUTPUT matrix was of size $n \times 50$. As before, the Best-Fit method was used to infer a network from these INPUT-OUTPUT pairs, but under the constraint that the indegree was fixed to be $kk = 2$. The number of wrong OUTPUTs was calculated

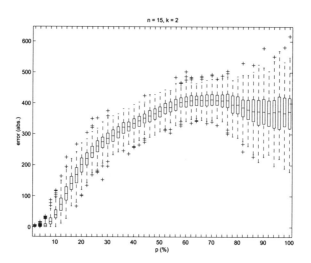

Figure 11.3. Results from the third simulation. The x-axis shows the propability p (in this simulation 2%, 4%, 6%, ... , 98%, 100%) and the y-axis is the absolute error (i.e. the number of wrong OUTPUTs in the OUTPUT-matrix). Medians, upper and lower quartiles, and extreme values are presented.

as before. However, because for each k the OUTPUT matrix was of a different size, the number of wrong OUTPUTs was normalized by dividing by the number of all OUTPUTs (i.e., the percentage of wrong OUTPUTs was calculated). This way the errors in each network could be compared.

To produce Figure 11.2, we repeated the procedure above 250 times. The x-axis shows the network size n and the y-axis shows the relative error. As in Figure 11.1, the medians, upper and lower quartiles, as well as the extreme values are presented. As opposed to the first figure, the behavior of the relative error seems to be linear rather than logarithmic.

Finally, we performed a third simulation designed to observe networks that are inferred from noisy data. Such a situation was considered by Akutsu *et al.* (2000), who proposed so-called noisy Boolean networks together with an identification algorithm, in order to deal with noise present in expression patterns. In that model, they relax the requirement of consistency intrinsically imposed by the Boolean functions. In our simulation, we first constructed a network with $n = 15$ and $k = 2$. Then 50 random INPUTs were created and the corresponding 50 OUT-PUTs were derived from the constructed network. Then each bit (both in the INPUT- and OUTPUT-matrix) was flipped with probability p. After this the Best-Fit method was used to infer a network from these matrices. Having done that, another OUTPUT-matrix was derived from the inferred network with the original INPUT-matrix (i.e., the matrix which has no flipped bits). Finally, the error was calculated by comparing the derived OUTPUT-matrix to the original OUTPUT-matrix. In this simulation $w(x) = 1$ for every vector x so the error is simply the number of wrong OUTPUTs in the OUTPUT-matrix. The results from this simulation are presented in Figure 11.3.

5.   Conclusions

The ability to efficiently infer the structure of Boolean networks has immense potential for understanding the regulatory interactions in real genetic networks. We have considered a learning strategy that is well suited for situations in which inconsistencies in observations are likely to occur. This strategy produces a Boolean network that makes as few misclassifications as possible and is a generalization of the well-known Consistency Problem. We have focused on the computational complexity of this problem. It turns out that for many function classes, the Best-Fit Extension Problem for Boolean networks is polynomial-time solvable, including those networks having bounded indegree and those in which no assumptions whatsoever about the functions are made. This promising result provides motivation for developing efficient algorithms for inferring network structures from gene expression data.

References

Akutsu, T., Kuhara, S., Maruyama, O. and Miyano, S. (1998) Identification of gene regulatory networks by strategic gene disruptions and gene overexpressions. *Proc. the 9th Annual ACM-SIAM Symposium on Discrete Algorithms (SODA '98)*, 695-702.

Akutsu, T., Miyano, S. and Kuhara, S. (1999) Identification of Genetic Networks from a Small Number of Gene Expression Patterns Under the Boolean Network Model. *Pacific Symposium on Biocomputing* 4, 17-28.

Akutsu, T., Miyano, S., and Kuhara, S. (2000) Inferring qualitative relations in genetic networks and metabolic pathways. *Bioinformatics* 16, 727-734.

Alter, O., Brown, P. O. and Botstein, (2000) D. Singular value decomposition for genome-wide expression data processing and modeling. *Proc. Natl. Acad. Sci. USA* 97:18, 10101-10106.

Angluin, D. (1987) Learning regular sets from queries and counterexamples. *Information and Computation*, 75:2, 87-106.

Ben-Dor, A. and Yakhini, Z. (1999) Clustering Gene Expression Patterns *Proc. of the 3rd International Conference on Computational Molecular Biology, 3342.* Lyon, France: ACM Press.

Boros, E., Ibaraki, T., and Makino, K. (1998) Error-Free and Best-Fit Extensions of Partially Defined Boolean Functions. *Information and Computation*, 140, 254-283.

Brazma, A. and Vilo, J. (2000) Gene expression data analysis. *FEBS Letters* 480, 17-24.

Celis, J. E., Kruhøffer, M., Gromova, I., Frederiksen, C., Østergaard, M., Thykjaer, T., Gromov, P., Yu, J., Pálsdóttir, H., Magnusson, N., & Ørntoft, T. F. (2000) Gene expression profiling: monitoring transcription and translation products using DNA microarrays and proteomics. *FEBS Letteers* 480, 2-16.

D'Haeseleer, P., Wen, X., Fuhrman, S. and Somogyi, R. (1999) Linear modeling of mRNA expression levels during CNS development and injury *Pacific Symposium on Biocomputing*, 4, 41–52.

DeRisi, J.L., Iyer, V.R., and Brown, P.O. (1997) Exploring the metabolic and genetic contol of gene expression on a genomic scale. *Science*, 278, 680-686.

Eisen, M. B., Spellman, P. T., Brown, P. O., & Botstein, D. (1998) Cluster analysis and display of genome-wide expression patterns. *Proc. Natl. Acad. Sci. USA*, 95, 14863-14868.

Friedman, N., Linial, M., Nachman, I. and Pe'er, D. (2000) Using Bayesian Network to Analyze Expression Data. *Journal of Computational Biology*, 7, 601-620.

Gabbouj, M., Yu, P-T., and Coyle, E. J. (1992) Convergence behavior and root signal sets of stack filters. *Circuits Systems & Signal Processing*, 11:1, 171-193.

Holter, N. S., Mitra, M., Maritan, A., Cieplak, M., Banavar, J. R., & Fedoroff, N. V. (2000) Fundamental patterns underlying gene expression profiles: Simplicity from complexity. *Proc. Natl. Acad. Sci. USA*, 97, 8409-8414.

Huang, S. (1999) Gene expression profiling, genetic networks, and cellular states: an integrating concept for tumorigenesis and drug discovery. *Journal of Molecular Medicine* 77, 469-480.

Iyer, V. R., Eisen, M. B., Ross, D. T., Schuler, G., Moore, T., Lee, J. C. F., Trent, J. M., Staudt, L. M., Hudson Jr., J., Boguski, M. S., Lashkari, D., Shalon, D., Botstein, D., and Brown, P. O. (1999) The transcriptional program in the response of human fibroblasts to serum. *Science*, 283, 83-87.

Kaski, S., Nikkilä, J., Törrönen, P., Castrén, E., and Wong, G. (2001) Analysis and visualization of gene expression data using self-organizing maps. *IEEE - EURASIP Workshop on Nonlinear Signal and Image Processing (NSIP-01)*, Baltimore, Maryland, June 3-6.

Kauffman, S. A. (1993) *The origins of order: Self-organization and selection in evolution*, Oxford University Press, New York.

Kearns, M. J. and Vazirani, U. V. (1994) *An Introduction to Computational Learning Theory*, MIT Press.

Liang, S., Fuhrman, S. and Somogyi, R. (1998) REVEAL, A General Reverse Engineering Algorithm for Inference of Genetic Network Architectures. *Pacific Symposium on Biocomputing* 3, 18-29.

Moran, G. (1995) On the period-two-property of the majority operator in infinite graphs. *Trans. Amer. Math. Soc.* 347, No. 5, 1649–1667.

Murphy, K. and Mian, S. (1999) Modelling Gene Expression Data using Dynamic Bayesian Networks. *Technical Report, University of California, Berkeley.*

Schena, M., Shalon, D., Davis, R. W., and Brown, P.O. (1995) Quantitative monitoring of gene expression pattern with a complementing DNA microarray. *Science*, 270, 467-470.

Shmulevich, I. and Zhang, W. (in press) Binary Analysis and Optimization-Based Normalization of Gene Expression Data, *Bioinformatics.*

Shmulevich, I., Dougherty, E. R., Kim, S., and Zhang, W. (in press) Probabilistic Boolean Networks: A Rule-based Uncertainty Model for Gene Regulatory Networks. *Bioinformatics.*

Szallasi, Z. and Liang, S. (1998) Modeling the Normal and Neoplastic Cell Cycle With Realistic Boolean Genetic Networks: Their Application for Understanding Carcinogenesis and Assessing Therapeutic Strategies. *Pacific Symposium on Biocomputing 3*, 66-76.

Tamayo, P., Slonim, D., Mesirov, J., Zhu, Q., Kitareewan, S., Dmitrovsky, E., Lander, E. S., & Golub, T. R. (1999) Interpreting patterns of gene expression with self-organizing maps: Methods and application to hematopoietic differentiation. *Proc. Natl. Acad. Sci.* USA 96, 2907-2912.

Thieffry, D., Huerta, A. M., Pérez-Rueda, E., and Collado-Vides, J. (1998) From specific gene regulation to genomic networks: a global analysis of transcriptional regulation in *Escherichia coli. BioEssays*, 20:5, 433-440.

van Someren, E. P., Wessels, L.F.A., and Reinders, M.J.T. (2000) Linear modeling of genetic networks from experimental data. *Intelligent Systems for Molecular Biology (ISMB 2000)*, San Diego, August 19-23.

Valiant, L. G. (1984) A theory of the learnable. *Comm. Assoc. Comput. Mach,* 27, 1134-1142.

Weaver, D.C., Workman, C.T. and Stormo, G.D. (1999) Modeling Regulatory Networks with Weight Matrices. *Pacific Symposium on Biocomputing,* 4, 112-123.

Wen, X., Fuhrman, S., Michaels, G. S., Carr, D. B., Smith, S., Barker, J. L., and Somogyi, R. (1998) Large-Scale Temporal Gene Expression Mapping of Central Nervous System Development. *Proc Natl Acad Sci USA*, 95, 334-339.

Wuensche, A. (1998) Genomic Regulation Modeled as a Network with Basins of Attraction. *Pacific Symp. on Biocomp.* 3, 89-102.

Chapter 12

REGULARIZATION AND NOISE INJECTION FOR IMPROVING GENETIC NETWORK MODELS

Eugene van Someren[1], Lodewyk Wessels[1,2], Marcel Reinders[1], and Eric Backer[1]

[1]*Information and Communication Theory Group*
[2]*Systems and Control Engineering*
Faculty of Information Technology and Systems
Delft University of Technology, Delft, The Netherlands

1. Introduction

Genetic network modeling is the field of research that tries to find the underlying network of gene-gene interactions from the measured set of gene expressions. Up to now, several different modeling approaches have been suggested, such as Boolean networks (Liang et al., 1998), Bayesian networks (Friedman et al., 2000), Linear networks (van Someren et al., 2000a; D'Haeseleer et al., 1999), Neural networks (Weaver et al., 1999; Wahde and Hertz, 1999) and Differential Equations (Chen et al., 1999b). In these approaches, genetic interactions are represented by parameters in a parametric model which need to be inferred from the measured gene expressions over time. Current micro-array technology has caused a significant increase in the number of genes whose expression can be measured simultaneously on a single array. However, the number of measurements that are taken in a time-course experiment has not increased in a similar fashion. As a result, typical gene expression data sets consist of relatively few time-points (generally less than 20) with respect to the number of genes (thousands). This so called *dimensionality problem* and the fact that measurements contain a substantial amount of measurement noise are two of the most fundamental problems in genetic network modeling.

Generally, when learning parameters of genetic network models from ill-conditioned data (many genes, few time samples), the solutions become arbitrary. Comparative studies (van Someren et al., 2001a; Wessels et al., 2001) have recently reported that at least a number of the currently proposed models suffer from poor inferential power when estimated from such data. The performance of these models can be significantly improved by employing additional (biological) knowledge about the properties of real genetic networks such as limited connectivity, dynamic stability and robustness. In this chapter, several methods to impose robustness are considered.

First, we present the fundamental problems of genetic network modeling and review currently known methodologies to overcome these problems. Then we introduce the concept of noise injection, where the original measured data set is artificially expanded with a set of noisy duplicates. In addition, two methods are presented, being ridge regression and lasso regression, that impose robustness by directly minimizing the first derivatives. We show how each of these approaches influence the obtained interaction parameters in the case of linear models and to what extent they are equivalent. Finally, experimental investigations show under which conditions these three regularization methods improve the performance in comparison with other models that are currently proposed in literature.

2. Current Approaches to Tackling the Dimensionality Problem

In order to capture the *combinatorial* nature of genetic relations (Holstege et al., 1998), gene regulation should be modelled as a *network* of genetic interactions rather than by methods that are based solely on pair-wise comparisons (Arkin et al., 1997; Chen et al., 1999a). The inference of genetic network models is, however, *especially* hampered by the dimensionality problem. Fortunately, (biological) knowledge about the general properties of genetic networks can be employed to alleviate some of the data requirements. In particular, true genetic networks are assumed to be 1) *sparsely connected*, because genes are only influenced by a limited number of other genes (Arnone and Davidson, 1997), 2) *robust against noise*, because small fluctuations in the expression-state may not lead to large differences in the consecutive expression-states, in other words, the general behavior of the system is not influenced by the small stochastic disturbances which are inevitable in biological systems (D'Haeseleer et al., 1999), 3) *redundant*, because genes are known to share functionality (Wen et al., 1998), and 4) *dynamically stable*, because the expression of a gene cannot grow without bound and 5) the activity level of genes are assumed to behave *smoothly* over time (D'Haeseleer et al., 1999). This knowledge can be used to alleviate the dimensionality problem in the modeling process either by a *pre-processing* step that modifies the data set prior to the inference process or by directly controlling the inference process through *direct regularization*.

Pre-processing: With a pre-processing step the dimensionality problem can be reduced by either artificially reducing the number of genes or by artificially increasing the number of time-points. Reduction of the number of genes can be achieved by thresholding and/or clustering the data (Wahde and Hertz, 1999; van Someren et al., 2001a; Chen et al., 1999a). Thresholding is based on the fact that small signals should be rejected because of their strong corruption with measurement noise. Clustering solves the ambiguity problem by considering co-expressed genes as a single 'meta-gene' and is biologically inspired by the redundancy assumption. Unfortunately, the use of thresholding and clustering is limited to what is biologically plausible and can therefore only partly reduce the dimensionality problem. By exploiting the smoothness of gene expression as a function of time, the number of time-points can be increased by employing interpolation (D'Haeseleer et al., 1999). However, our experience shows that interpolation reduces the dimensionality problem only marginally regardless of the number of time-points added.

Direct Regularization: The use of biological constraints to regulate the inference process directly has only been applied in a few earlier papers. In previous work (van Someren et al., 2000a; van Someren et al., 2001b), we exploited the fact that networks have a limited connectivity to search directly for sparse linear networks. Weaver (Weaver et al., 1999) also enforce the sparsity, but only briefly describe an approach that iteratively sets small weights in the gene regulation matrix to zero. D'Haeseleer employs weight decay and weight elimination when learning recurrent neural networks to tackle the dimensionality problem (D'Haeseleer et al., 1999).

Thus far, none of the reported modeling approaches have incorporated methodologies that explicitly increase robustness or dynamic stability of the obtained gene regulation matrices. In this chapter, we focus on methods that impose robustness.

3. Learning Genetic Network Models

First, in this section, the representation of gene expression data and genetic network models is defined and we describe how genetic interactions can be learned from data. Throughout this chapter, square brackets are used to denote concatenation of vectors or matrices and

the following notation for positive integers is used: $N_K^+ = 1, 2, \ldots, K$.
Data: Let $g_i(t)$ represent the gene expression level of gene i at time-point t and let N genes be measured. Then an expression-state is represented by a column vector of all gene expression values measured at a certain point in time, i.e. $\mathbf{g}(t) = [g_1(t), \ldots, g_N(t)]^T$. A typical time-course gene expression data set reflects how one state of gene expression levels is followed by a consecutive state of gene expression. For learning a genetic network model, the data set is best represented as a training set of Q state-transition (input-output) pairs, represented by $\mathcal{S} = \langle \mathbf{X}; \mathbf{T} \rangle$. Here, input matrix $\mathbf{X} = [\mathbf{x}^1, \mathbf{x}^2, \ldots \mathbf{x}^Q]$ and target (output) matrix $\mathbf{T} = [\mathbf{t}^1, \mathbf{t}^2, \ldots \mathbf{t}^Q]$ are matrices containing the concatenated input-states and target output-states respectively. The pairs are constructed such that each input state $\mathbf{x}^q$ directly precedes the target (output) state $\mathbf{t}^q$, i.e. if $\mathbf{x}^q = \mathbf{g}(t)$ then $\mathbf{t}^q = \mathbf{g}(t+1)$.
Genetic Network Model: A genetic network model is a representation of the genetic interactions such that for a given state of gene expression it can predict (output) the consecutive state(s). In general, a genetic network model represents a parameterized non-linear mapping, $\mathbf{y} = f(\mathbf{x})$, from an N-dimensional input vector into an N-dimensional output vector. One of the simplest continuous models is a linear genetic network model. This model is studied in this chapter because it allows insight in the consequences of the dimensionality problem and the effects of regularization, while its relatively limited number of parameters makes it less sensitive to the dimensionality problem. In addition, several other continuous genetic network models are just extensions of the linear model (Wessels et al., 2001). The linear model assumes that the gene expression level of each gene is the result of a weighted sum of all other gene expression levels at the previous time-point, or in vector notation:

$$\mathbf{y}^q = \mathbf{W} \cdot \mathbf{x}^q \qquad q \in N_Q^+ \tag{12.0}$$

The interaction parameter, w_{ij}, represents the existence ($w_{ij} \neq 0$) or absence ($w_{ij} = 0$) of a controlling action of gene j on gene i, whether it is activating ($w_{ij} > 0$) or inhibiting ($w_{ij} < 0$), as well as the strength ($|w_{ij}|$) of the relation. The complete matrix of interactions, $\mathbf{W}$, is called the gene regulation matrix (GRM). To learn the gene regulation matrix, we simply require that the predicted states, $\mathbf{y}^q$, are as close as possible to the target states, $\mathbf{t}^q$. For example, by requiring that the mean square error, E, should be minimal, i.e.:

$$\hat{\mathbf{W}} = \arg \min_{\mathbf{W}} E \quad , with \quad E = \frac{1}{2Q} \sum_{q=1}^{Q} ||\mathbf{y}^q - \mathbf{t}^q||^2 \tag{12.1}$$

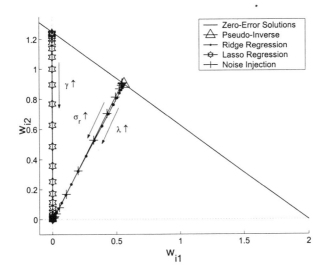

Figure 12.1. Weight space of an example genetic network with two inputs, one output and one state-transition pair given. Indicated are the set of zero-error solutions, the pseudo-inverse and a range of solutions for different parameter settings of ridge regression and lasso regression.

For problems affected by the dimensionality problem, minimizing E results in infinitely many GRM's, which all give a zero mean square error. In the linear case this is obvious, since each of the Q elements in the training set constitute one linear equation of N unknowns and typically $N >> Q$, i.e. many unknowns and only a few equations to solve them. For example, in the case of a genetic network with two inputs, one output and one state-transition pair, this set of zero-error solutions corresponds to a line in the space of all solutions as is illustrated in Figure 12.1.

Pseudoinverse: A well-known solution to this problem is to select from the set of zero-error solutions, the one with minimal sum of squared weights. This solution is known as the Moore-Penrose pseudoinverse, for which the following analytical solution exists (See also Figure 12.1):

$$\hat{\mathbf{W}}_{PSEUDO} = \mathbf{T} \cdot (\mathbf{X}^T \cdot \mathbf{X})^{-1} \cdot \mathbf{X}^T \qquad (12.2)$$

This is one possible way to select a weight matrix from a whole array of possibilities. However, many other ways of regularizing the problem exist. In this chapter, we will focus on choosing the GRM such that the resulting genetic network model is robust.

4.   Robust Methods

A genetic network model is robust against noise in the input, when small fluctuations in the input-state do not lead to large differences in the output-state.

Noise Injection: The basic idea of noise injection is that robustness can be imposed by augmenting the original data set with state-transition pairs onto which small distortions are added, such that the model automatically learns to be robust. To achieve this, an extended training set, $\mathcal{S}' = \langle[\mathbf{X}, \mathbf{X}']; [\mathbf{T}, \mathbf{T}']\rangle$, is constructed that contains for each state-transition pair in the original training set, the original pair *and* N_r repetitions of the original pair corrupted with zero-mean additive Gaussian noise of given standard deviation, σ_r, i.e.:

$$
\begin{aligned}
\mathbf{X}' &= [\mathbf{x}^{11}, \mathbf{x}^{12}, \dots, \mathbf{x}^{qm}, \dots, \mathbf{x}^{QN_r}], \; where \; \mathbf{x}^{qm} = \mathbf{x}^q + \tilde{\mathbf{x}}^{qm}, \\
\mathbf{T}' &= [\mathbf{t}^{11}, \mathbf{t}^{12}, \dots, \mathbf{t}^{qm}, \dots, \mathbf{t}^{QN_r}], \; where \; \mathbf{t}^{qm} = \mathbf{t}^q + \tilde{\mathbf{t}}^{qm}, \\
& q \in N_Q^+, \; m \in N_{N_r}^+, \; \tilde{\mathbf{x}}^{qm}, \tilde{\mathbf{t}}^{qm} \sim \mathcal{N}(\mathbf{0}, \sigma_r \mathbf{I})
\end{aligned} \tag{12.3}
$$

Here, the double superscript qm indicates the m-th noise corrupted copy of the q-th original. A pleasant side-effect of using the augmented training set, $\mathcal{S}'$, is that, already for relatively small N_r, the (linear) model becomes over-determined. Consequently, the least squares estimation can be employed to infer the gene regulation matrix from the data set $\mathcal{S}'$. We denote this new approach as noise injection, for which a solution is obtained as follows:

$$
\hat{\mathbf{W}}_{NOISE} = \mathbf{T}' \cdot \mathbf{X}'^T \cdot (\mathbf{X}' \cdot \mathbf{X}'^T)^{-1} \tag{12.4}
$$

The number of repetitions should always be chosen as high as is computationally feasible to avoid random effects. For the experiments in this chapter, we let it grow proportional to the number of genes, i.e. $N_r = 5N$.

Ridge Regression: An alternative approach to make a model less sensitive to deviations in the input is by directly minimizing the model's first derivatives of the outputs with respect to the inputs, $\partial \mathbf{y}/\partial \mathbf{x}$. This can be achieved by augmenting the original mean squared error (Eq 12.2) with, amongst other, a *quadratic* term (Hastie et al., 2001). In the case of a linear model, with first derivatives $\partial y_i/\partial x_j = w_{ij}$, this results in:

$$
\hat{\mathbf{W}}_{RIDGE} = \arg\min_{\mathbf{W}} \frac{1}{2Q} \sum_{q=1}^{Q} ||\mathbf{W} \cdot \mathbf{x}^q - \mathbf{t}^q||^2 + \lambda \sum_{i=1}^{N} \sum_{j=1}^{N} w_{ij}^2 \tag{12.5}
$$

For this problem, an analytic solution exists, known as ridge regression (Hoerl and Kennard, 1970):

$$\hat{\mathbf{W}}_{RIDGE} = \mathbf{T} \cdot \mathbf{X}^T \cdot (\mathbf{X} \cdot \mathbf{X}^T + \lambda \mathbf{I})^{-1}, \qquad \lambda > 0 \qquad (12.6)$$

This method utilizes λ as a parameter that provides a tunable trade-off between data-fit and shrinkage of the weights. In general, ridge regression tends to produce solutions with many small, but non-zero weights.
Lasso Regression: Alternatively, the original mean squared error (Eq. 12.2) can be augmented with a slightly different penalty term, namely one that shrinks the sum of the absolute values of the weights (Tibshirani, 1994):

$$\hat{\mathbf{W}}_{LASSO} = \arg\min_{\mathbf{W}} \frac{1}{2Q} \sum_{q=1}^{Q} ||\mathbf{W} \cdot \mathbf{x}^q - \mathbf{t}^q||^2 + \gamma \sum_{i=1}^{N} \sum_{j=1}^{N} |w_{ij}| \quad (12.7)$$

A solution to this equation can, generally, be found using only a few iterations of the EM-algorithm (Grandvalet, 1998). This method is called Least Absolute Shrinkage and Selection Operator, because it tends to shrink the weights such that only a few weights remain non-zero. Lasso regression is especially suited for genetic network modeling as it tries to produce models that exhibit robustness *as well as* limited connectivity.
Parameter Setting: The performance of all three methods depends heavily on the right setting of the free parameter, i.e. the noise strength, lambda and/or gamma. Figure 12.1 depicts the different solutions of ridge and lasso regression for a range of lambda and/or gamma settings. The right setting of this parameter depends on the number of genes, the original connectivity, the underlying type of model, the number of measurements, but most heavily on the expected amount of measurement noise. At the expense of increased computational costs we propose to employ a leave-one-out cross-validation procedure to automatically determine the right parameter for each new data set. Given a set of k possible parameter settings, at each step, this procedure 1) sets aside one sample out of the training set, 2) learns, using the remaining samples, k different model solutions for each of the k different parameter settings, 3) predicts, using the input of the left-out sample, k different outputs for each of the found k solutions, 4) determines, using the output of the left-out sample and the k predictions, the corresponding prediction errors. This procedure is repeated for each sample in the training-set and the resulting prediction-errors, corresponding to one parameter setting, are averaged. The final model solution is then learned on the complete training-set using the parameter setting that showed the lowest average prediction-error on the unseen data.

Example: To further illustrate the difference in shrinkage, the solutions and corresponding errors of noise injection, ridge regression and lasso regression for different parameter settings are depicted in Figure 12.2 on an example of a 27-gene network. The three plots on the left show, how, for increasing parameter settings, the weights of each approach are shrunk. For further illustration, in all plots, the solution marked with stars on the y-axis does not corresponds to the lowest parameter setting, but to the pseudo-inverse solution. All three methods start for low parameter settings $(\sigma_r, \lambda, \gamma \downarrow 0)$ with a different solution, i.e. while noise injection is only approximately similar to the pseudo-inverse, it is clear that ridge regression is exactly equivalent to the pseudo-inverse, whereas lasso regression directly sets a substantial number of weights to zero. For very large parameter settings $(\lambda, \gamma \to \infty)$ noise injection approaches some fixed solution, while ridge and lasso regression solutions approach the zero matrix (all weights to zero). The right-most plots depict the corresponding training error and the average leave-one-out error. The parameter value that will be chosen based on the minimum of the leave-one-out error is indicated in all plots with a vertical dashed line.

5. Noise Injection is Equivalent to Regularization

Bishop studied the relation between adding noise to a training set and regularization for a set of general non-linear models (Bishop, 1994). He has shown that, for small noise strength, adding noise to the data is equivalent to minimization of the following error function:

$$\tilde{E} \;=\; E + \eta^2 E^R \tag{12.8}$$

Here, E^R is controlled by the strength of the noise, η^2, and can be further approximated by (Eq. 19 in (Bishop, 1994):

$$\hat{E}^R \;=\; \frac{1}{2Q}\sum_{q=1}^{Q}\sum_{i=1}^{N}\sum_{j=1}^{N}\left(\frac{\partial y_i^q}{\partial x_j^q}\right)^2 \tag{12.9}$$

Index i labels the N predicted genes and index j labels the N genes used as input. In the case of genetic network models the parameters remain constant for each training sample, so we can sum over all q and in the case of a linear model, the partial derivatives simply are the corresponding weights in the gene regulation matrix. Thus, Eq. (12.10) simplifies to:

$$\hat{E}^R \;=\; \frac{1}{2}\sum_{i=1}^{N}\sum_{j=1}^{N} w_{ij}^2 \tag{12.10}$$

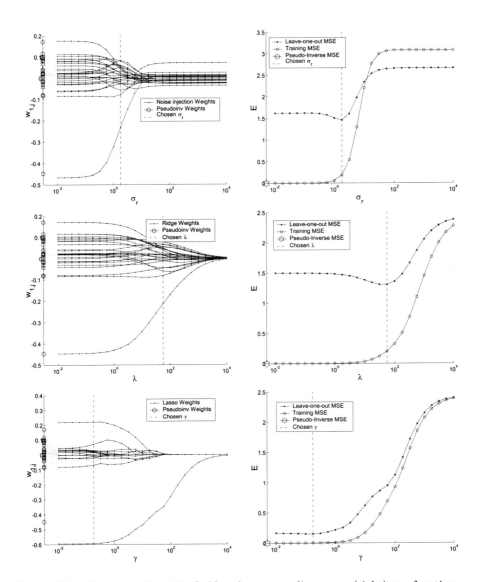

Figure 12.2. Shrinkage of weights (left) and corresponding errors (right) as a function of the free parameter for noise injection (top), ridge regression (middle) and lasso regression (bottom). Weights corresponding to the pseudo-inverse are marked with stars

Consequently, for small noise strength and a large number of repetitions, noise injection is equivalent to ridge regression. Furthermore, in the under-determined case and in the limit of very small λ and/or σ_r^2 respectively, both are equivalent to the application of the Moore-Penrose pseudo-inverse to the original data set. Obviously, the mean square error, E, then becomes zero since a perfect solution exists.

6.   Comparison with other models

To show that imposing robustness improves the performance of genetic network models, the four types of "robust" models (*pseudo-inverse, noise injection, ridge regression* and *lasso regression*) are compared with a *reference* model that always returns a random stable GRM of connectivity four and a model that employs hierarchical clustering (van Someren et al., 2000b), denoted as the cluster model.

Experimental Setup: For a given number of genes, N, and a given connectivity (number of non-zero weights in each row of the GRM), C, a uniformly random gene regulation matrix is generated. To ensure that the random GRM is stable, all its elements are divided by the magnitude of the largest eigenvalue. Then, using a linear model and a given number of time-points, T, a subset is generated from a randomly chosen initial state. Measurement noise is added by corrupting all elements of this data set with zero-mean additive Gaussian noise of given standard deviation, σ_m.

The performance was evaluated by monitoring the inferential power, the predictive power, the number of connections and the computational costs. The principle goal of genetic network modeling is to maximize the inferential power. This criterium expresses the correlation between the true genetic interactions and the relationships inferred by the model and is defined by:

$$IP = \frac{1}{2}(PC(\mathbf{W}, \hat{\mathbf{W}}) + 1) \qquad (12.12)$$

Here, PC, returns the Pearson correlation coefficient between the two vectors that are obtained by transforming the entries of each matrix into one vector. The inferred model-parameters are, however, determined by maximizing the predictive power, i.e. by minimizing the mean squared error (Eq. 12.2). The predictive power is given by:

$$PP = \frac{1}{1 + E} \qquad (12.13)$$

An important cause of decreased inferential power (which easily arises with insufficient regularization) is the variability of the obtained solutions when small modifications are made to the data set. This inferential stability is determined by measuring the average inferential power between all solutions, $\hat{\mathbf{W}}^k$, obtained by the inference on K corresponding noise corrupted data sets, $\mathbf{X}^k = \mathbf{X} + \tilde{\mathbf{X}}$, where the elements of $\tilde{\mathbf{X}}$ are given by $\tilde{x}_{nq} \sim \mathcal{N}(0, .01)$. Formally, the inferential stability is defined as:

$$IS = \frac{2}{K(K-1)} \sum_{k=1}^{K-1} \sum_{m=k+1}^{K} IP(\hat{\mathbf{W}}^k, \hat{\mathbf{W}}^m) \qquad (12.14)$$

Experimental settings: In subsequent experiments, either the number of genes, N, was varied and the connectivity, C, was fixed at four or the connectivity was varied while the number of genes was fixed at thirty. Available micro-array data, generally, consist of multiple small (time-course) experiments. Therefore, data sets were generated by merging four differently initialized subsets of $T = 5$ time-points each. Note that one such data set, consists of $Q = 4 \cdot (T - 1) = 16$ state-transition pairs. For each data set, σ_m was chosen such that the peak-signal-to-noise-ratio (Psnr) was 40 dB. Parameter values of σ_r, λ and γ were set to their optimal value (determined by the leave-one-out procedure). To filter out random effects, the performance of the models at each fixed condition (data-point) was determined by the average over 40 data sets. Experiments were done using GENLAB, a gene expression analysis toolbox developed by the authors and available at our website http://www.genlab.tudelft.nl.

Results for small networks: Figure 12.3 depicts the inferential power, predictive power on a test-set, connectivity and computational costs of the different models for small networks up to thirty genes. The gray patch denotes the average performance of the reference model plus and minus two standard deviations. The cluster model performs well for $N < 16$, however, it tackles the dimensionality problem only marginally and performs poorly for $N > 16$. A remarkable observation is that all robust models outperform the reference and cluster model in terms of inferential power, especially in the under-determined case ($N > 16$). This illustrates the suggestion that different ways of imposing robustness all improve the inferential power. Around $N \approx Q$ the pseudo-inverse shows a dip in performance with respect to the other robust models. This phenomenon is also observed in classification tasks and is probably caused by the fact that the covariance matrix becomes ill-conditioned when the number of samples is close to the number of dimensions and, effectively, no regulation takes place. Both ridge regression and noise injection exhibit good and similar inferential power and do not show the same dip as the pseudo-inverse. Most striking is the superior performance of lasso regression which tries to find solutions that are not only robust, but

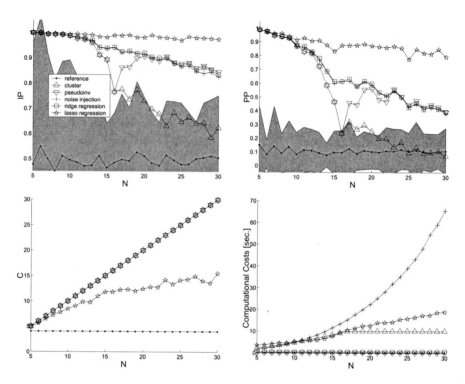

Figure 12.3. Inferential power, predictive power on test-set, connectivity and computational costs of six genetic network models (see legend) for small networks as a function of no. genes (x-axis). See text for more detailed description.

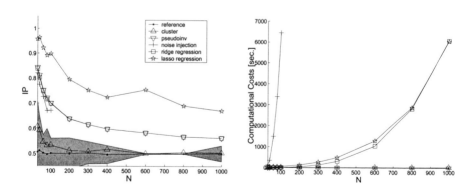

Figure 12.4. Inferential power and computational costs of six genetic network models (see legend) for large networks as a function of no. genes (x-axis).

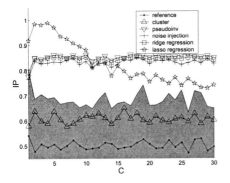
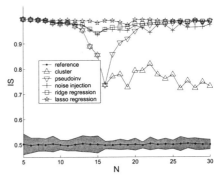

Figure 12.5. Inferential power of six genetic network models (see legend) for a 30-gene network as a function of connectivity (x-axis).

Figure 12.6. Inferential stability of six genetic network models (see legend) as a function of no. genes (x-axis).

also exhibit limited connectivity (since weights are set to zero). The inferential power shows strong correlation with the predictive power on a separate test-set, supporting the use of cross-validation to determine the right parameter setting. The connectivity of the obtained solutions shows that the pseudo-inverse, ridge regression and noise injection, return fully connected solutions whereas lasso regression limits the connectivity. However, lasso regression still over-estimates the connectivity (which should be four) under all conditions. Finally, it is shown that, as expected, the analytical solutions (pseudo-inverse and ridge regression) are considerably faster and thus more likely to scale up. Noise injection is the slowest model, because it needs to solve more equations due to the extension of the data set.

Results for large networks: The inferential power and computational costs for larger networks are depicted in Figure 12.4. Although decreasing performance for larger networks is to be expected, it becomes apparent that the performance of all models drop sharply when the network size increases from 30 to 200 genes, whereas the performance decreases more asymptotically for larger networks. At the same time, the exponential increase in the computational costs for noise injection caused us to refrain from applying noise injection for networks larger than 100 genes. For a hundred gene network, it took already almost two hours to determine the connections going into a single node on a 450 Mhz Pentium II.

Results for varying connectivity: The difference in performance of the models is expected to depend strongly on the actual complexity of the original GRM. Therefore, the models were also tested for a 30-gene GRM with connectivity varying from 1 to 30. The results of this experiment are depicted in Figure 12.5. The results indicate that the pseudo-inverse, noise injection and ridge regression perform equivalently and consistent for different connectivity conditions. The remarkably constant performance seems to indicate that the models are not affected by the addition of more connections, suggesting that the individual contributions of each gene can still be recognized. Furthermore, it becomes apparent that the lasso regression favors data sets created by GRM's with low connectivity and performs worse than the pseudo-inverse, noise injection and ridge regression when the connectivity becomes larger than half the number of genes.

Results of inferential stability: The inferential stability as a function of the number of genes for small networks is depicted in Figure 12.6. Apart from a downward trend, the inferential stability shows strong correlation with the inferential power, depicted in Figure 12.3. These figures illustrate that low inferential stability implies low inferential power. Thus, it becomes apparent that the dip in the inferential power of the pseudo-inverse and the marginal performance of the cluster model are both caused by inferential instability. By imposing robustness (noise injection, ridge regression and lasso regression) the models become inferentially stable and the general performance is improved. However, the inferential power can be further improved by imposing other constraints that do not necessarily improve inferential stability. For example, the superior inferential power of lasso regression, that imposes limited connectivity, cannot fully be explained by significantly improved inferential stability.

7. Discussion

In this chapter, we've shown that imposing robustness, generally, improves the performance of genetic network models. In our view, direct regularization is to be preferred to adding noise and we believe it is an essential tool for successful genetic network modeling. However, when a good regularization term is hard to define or the inference process does not allow inclusion of such a term, adding noise will prove to be a very simple alternative to regularization. As a rule-of-thumb, the number of added samples should be chosen as high as is computationally feasible and the added noise strength should be chosen based on cross-validation. Whether these observations hold for non-linear models or even realistic data sets needs to be further investigated. Nevertheless, in addition to thresholding, clustering and interpolation, adding noise provides a simple and general tool to tackle the dimensionality problem and is definitely the most effective method thus far.

Acknowledgments

This work was funded by the Dioc-IMDS program of the Delft University of Technology.

References

Arkin, A., Shen, P., and Ross, J. (1997). A test case of correlation metric construction of a reaction pathway from measurements. *Science*, 277.

Arnone, A. and Davidson, B. (1997). The hardwiring of development: Organization and function of genomic regulatory systems. *Development*, 124:1851–1864.

Bishop, C. (1994). Training with noise is equivalent to tikhonov regularization. *Neural Computation*, 7(1):108–116.

Chen, T., Filkov, V., and Skiena, S. (1999a). Identifying gene regulatory networks from experimental data. In *Proceedings of the third annual international conference on Computational molecular biology (RECOMB99)*, pages 94–103. Association for Computing Machinery.

Chen, T., He, H., and Church, G. (1999b). Modeling gene expression with differential equations. In *Pacific Symposium on Biocomputing '99*, volume 4, pages 29–40. World Scientific Publishing Co.

D'Haeseleer, P., Wen, X., Fuhrman, S., and Somogyi, R. (1999). Linear modeling of mrna expression levels during cns development and injury. In *Pacific Symposium on Biocomputing '99*, volume 4, pages 41–52. World Scientific Publishing Co.

Friedman, N., Linial, M., Nachman, I., and Pe'er, D. (2000). Using bayesian networks to analyze expression data. *Journal of Computational Biology*, 7:601–620.

Grandvalet, Y. (1998). Least absolute shrinkage is equivalent to quadratic penalization. In *International Conference on Artificial Neural Networks (ICANN'98)*, Skövde, Sweden.

Hastie, T., Tibshirani, R., and Friedman, J. (2001). *Elements of Statistical Learning: Data Mining, Inference and Prediction*. Springer-Verlag, New York.

Hoerl, A. and Kennard, R. (1970). Ridge regression: Applications to nonorthogonal problems. In *Technometrics*, volume 17, pages 69–82.

Holstege, F., Jennings, E., Wyrick, J., Lee, T., Hengartner, C., Green, M., Golub, T., Lander, E., and Young, R. (1998). Dissecting the regulatory circuitry of a eukaryotic genome. *Cell*, 95(5):717–728.

Liang, S., Fuhrman, S., and Somogyi, R. (1998). Reveal, a general reverse engineering algorithm for inference of genetic network architectures. In *Pacific Symposium on Biocomputing '98*, volume 3, pages 18–29. World Scientific Publishing Co.

Tibshirani, R. (1994). Regression selection and shrinkage via the lasso. Technical report.

van Someren, E., Wessels, L., and Reinders, M. (2000a). Information extraction for modeling gene expressions. In *Proceedings of the 21st Symposium on Information Theory in the Benelux*, pages 215–222, Wassenaar, The Netherlands. Werkgemeenschap voor Information Theory in the Benelux.

van Someren, E., Wessels, L., and Reinders, M. (2000b). Linear modeling of genetic networks from experimental data. In Altman, R., Bailey, T., Bourne, P., Gribskov, M., Lengauer, T., Shindyalov, I., Eyck, L. T., and Weissig, H., editors, *Proceedings of the Eighth International Conference on Intelligent Systems for Molecular Biology*, pages 355–366, La Jolla, California. AAAI.

van Someren, E., Wessels, L., and Reinders, M. (2001a). Genetic network models: A comparative study. In *Proceedings of SPIE, Micro-arrays: Optical Technologies and Informatics*, volume 4266, San Jose, California.

van Someren, E., Wessels, L., Reinders, M., and Backer, E. (2001b). Searching for limited connectivity in genetic network models. In *Proceedings of the Second International Conference on Systems Biology (ICSB'01)*, Pasadena, California.

Wahde, M. and Hertz, J. (1999). Coarse-grained reverse engineering of genetic regulatory networks. In *IPCAT 99*.

Weaver, D., Workman, C., and Stormo, G. (1999). Modeling regulatory networks with weight matrices. In *Pacific Symposium on Biocomputing '99*, volume 4, pages 112–123, Hawai. World Scientific Publishing Co.

Wen, X., Fuhrman, S., Michaels, G., Carr, D., Smith, S., Barker, J., and Somogyi, R. (1998). Large-scale temporal gene expression mapping of central nervous system development. In *Proceedings of the National Acadamy of Sciences of the USA*, volume 95 of *1*, pages 334–339. National Academy of Sciences.

Wessels, L., van Someren, E., and Reinders, M. (2001). A comparison of genetic network models. In *Pacific Symposium on Biocomputing 2001*, volume 6, pages 508–519, Hawai. World Scientific Publishing Co.

Chapter 13

PARALLEL COMPUTATION AND VISUALIZATION TOOLS FOR CODETERMINATION ANALYSIS OF MULTIVARIATE GENE EXPRESSION RELATIONS

Edward B. Suh[1], Edward R. Dougherty[2], Seungchan Kim[2], Michael L. Bittner[3], Yidong Chen[3], Daniel E. Russ[1], and Robert L. Martino[1]

[1] *Division of Computational Bioscience, Center for Information Technology, National Institutes of Health, Bethesda, Maryland, USA*

[2] *Department of Electrical Engineering, Texas A&M University, College Station, Texas, USA*

[3] *Cancer Genetic Branch, National Human Genome Research Institute, National Institutes of Health, Bethesda, Maryland, USA*

1. Introduction

A key goal of functional genomics is to develop methods for determining ways in which individual actions of genes are integrated in the cell. Kim *et al.* have proposed using the nonlinear coefficient of determination for finding associations between genes (Kim *et al.*, 2000a; Kim *et al.*, 2000b; Dougherty *et al.*, 2000). The method assesses the codetermination of gene transcriptional states based on statistical evaluation of reliably informative subsets of data derived from large-scale gene-expression measurements. It measures the degree to which the transcriptional levels of a small gene set can be used to predict the transcriptional state of a target gene in excess of the predictive capability of the mean level of the target. Conditions besides transcriptional features can be used as predictive elements, thus broadening the information that can be evaluated in modelling biological regulation.

Application of the statistical framework to a large set of genes requires a prohibitive amount of computer time on a classical single-CPU computing machine. To meet the computational requirement, we have developed a parallel implementation of the codetermination method, named the *Parallel Analysis of Gene Expression* (PAGE) program. PAGE has performed a codetermination analysis of melanoma skin cancer expression data in *two weeks* of computer time using 64 processors in a Beowulf parallel computing machine (Sterling *et al.*, 1999) at the Center for Information Technology, National Institutes of Health (NIH). This analysis would have required about *one and half years* on a sequential computing machine.

PAGE consists mainly of three modules. The first module generates predictor gene sets out of a large gene pool. It has an exclusionary feature that eliminates any set that contains a subset in an exclusion list. The second computes the *coefficient of determination* (CoD) for the gene sets generated by the first module. The third module checks to see if the CoD calculated from the data satisfies a certain theoretical requirement. If it does not, then the module adjusts it accordingly. This chapter analyzes the performance of the parallel implementation.

PAGE provides a massive amount of computationally derived information that is too large for a user to directly review. In order to facilitate the intelligent navigation and further analyses of the computationally derived information, computer-aided data visualization and analysis tools have been developed. The data visualization software provides tools within a *graphical user interface* (GUI) environment for analysis, interpretation, and application by geneticists. Aspects of this data visualization program, *Visualization of Gene Expression* (VOGE), are illustrated.

2.   Codetermination algorithm

When using multiple microarrays, raw signal intensities vary extensively due to the processes of preparing and printing the *expressed sequence tag* (EST) elements, and of preparing and labelling the cDNA representations of the RNA pools. This problem is solved via internal standardization. An algorithm calibrates the data internally to each microarray and statistically determines whether the data justify the conclusion that an expression is up- or down-regulated (Chen *et al.*, 1997). Prior to analysis, data from a microarray is reduced by thresholding changes in transcript level into ternary form: -1 (down-regulated), $+1$ (up-regulated), or 0 (invariant).

The determination paradigm involves nonlinear multivariate predictors acting on the ternary input data. A nonlinear predictor is a logical system L that estimates a target expression level Y based on a set of input expression levels $X_1, X_2, ..., X_k$. The target Y is the output of a biological system S that is observed when $X_1, X_2, ..., X_k$ are input to S. The optimal predictor yields a logical value Y_{pred} that best predicts Y. The target value is random owing to the various possible values of $X_1, X_2, ..., X_k$ that might be input to S, as well the fact that our knowledge of S is necessarily incomplete. Statistical training uses only that $X_1, X_2, ..., X_k$ are among the inputs to S, the output Y of S can be measured, and a logical system L can be constructed whose output Y_{pred} statistically approximates Y. The logic of L represents an operational model of our understanding.

The theoretically optimal predictor, which has minimum error across the population, is unknown and must be statistically designed (estimated) from a sample. How well a designed predictor performs depends on the training procedure and sample size n. The error ε_n of a designed predictor exceeds the error ε_{opt} of the optimal predictor. For a large number of microarrays, $\varepsilon_n \approx \varepsilon_{opt}$, but for the small samples typically used, ε_n exceeds ε_{opt}.

The *coefficient of determination* (CoD), θ_{opt}, of the optimal predictor is the relative decrease in error owing to the observed variables:

$$\theta_{opt} = \frac{\varepsilon_b - \varepsilon_{opt}}{\varepsilon_b} \tag{13.1}$$

where ε_b is the error for the best predictor in the absence of observations. Since the error, ε_{opt}, of the optimal predictor cannot exceed ε_b, $0 \le \theta_{opt} \le 1$. If optimization is relative to *mean-square-error* (MSE), which is the expectation $E[|Y_{pred} - Y|^2]$, then the best predictor based on the predictor variables $X_1, X_2, ..., X_k$ is the conditional expectation of Y given $X_1, X_2, ..., X_k$. The best predictor of Y in the absence of observations is its mean, μ_Y, and the corresponding error is the variance of Y. Predicting Y by its mean might yield small or large MSE depending on the variance of Y. Thus, there is normalization by ε_b in Equation 13.1 to measure the effect of the observations. Here, the conditional expectation is quantized to ternary values, and ε_b is the error from predicting Y by applying the ternary threshold of μ_Y. For designed predictors, ε_b and ε_{opt} are replaced by sample estimates, and errors are estimated by cross-validation. The coefficient of determination measures normalized predictive capability in a probabilistic, not causal, sense. A high CoD may indicate causality of the predictors on the target or vice versa. It may also indicate diffuse control along network pathways, or co-regulation of both predictors and target by another mechanism. Hence, the CoD is being called the *codetermination* in biology.

3.   Prediction system design

In this and the next two sections, system design, parallel-program modules, and visualization of algorithm output are discussed. Although we have previously implemented a perceptron-based prediction system (Kim *et al.*, 2000a; Kim *et al.*, 2000b; Dougherty *et al.*, 2000), we focus on the design of a logic-filter-based prediction system in this chapter.

Rather than use the ternary-quantized conditional expectation for the predictor, we use the conditional mode. This requires less computation, the differences between it and the conditional expectation are small, and it avoids predicting values that have not occurred in the samples, a desirable property for such coarse quantization. The ternary logic filter is constructed via the conditional mode. For any observation vector $\mathbf{x}$, $\Psi(\mathbf{x})$ is the value of Y seen most often with $\mathbf{x}$ in the sample data. The size of table defining the predictor grows exponentially with the number of predictor variables, and the number of conditional probabilities to estimate increases accordingly.

For gene expression ratio data, the number of pairs to use in filter design is very limited; in fact, we often do not observe all vectors. When applying the filter to test data, there may be inputs not observed during design. A value needs to be assigned to such inputs. The algorithm provides several choices.

For a given target gene and predictor-gene combination, it first designs a nonlinear prediction model, *ternary logic filter*, with a training data set. It then uses a test data set to evaluate predictor performance based on the coefficient of determination. A user may pre-select t target genes of interest. To find gene sets with high CoDs for a target-gene requires searching through combinations of one-, two-, three-, or up to k-predictor-gene combinations. There may exist prior knowledge to guide this combinatorial search, or we may search through the full space that grows exponentially as k and the number of genes in a gene pool increase.

For estimating the CoD, a user can choose a number, N, of repetitions for data splitting. The higher N, the better estimated the expectation. The system has nested loops to go through N repetitions of data splitting, and has $^{m}C_{k} = m!/(k!(m-k)!)$ combinations to compute for each of t target genes. Since incremental relations between smaller and larger predictor sets are important, it is necessary to calculate the CoD for k predictor gene combinations, for each k of 1, 2, 3, to some stopping point. The overall runtime, computing complexity, of these operations is $O(N \cdot {}^{m}C_{k} \cdot t)$. While design and validation (error estimation) of a prediction model for a single predictor combination for a target gene may not require significant computational resources, the prediction system does require significant parallel computing performance as k and m are increased.

4. Parallel analysis of gene expression (PAGE)

As more and newer types of biological information are obtained from the biomedical research community, the computational performance of parallel computing is required to analyze this information with new methods in reasonable time (Martino *et al.*, 1997; Suh *et al.*, 1998). This section presents PAGE, a parallel program for analyzing the code-termination of gene transcriptional states from large-scale simultaneous gene expression measurements with cDNA microarrays.

4.1 The three sequential algorithms and motivation for parallel implementation

The program module *gcombi* generates k-predictor gene combinations from a set of m genes in lexicographic order. A k-predictor gene combination of an m gene pool set is obtained by selecting k distinct integers out of m and arranging them in increasing order. The total number of distinct k-combinations of an m-set is ^mC_k. The computational complexity of the sequential *gcombi* program is $gcombi(n, k) = O(^mC_k \cdot k)$. The run-time is proportional to this computational complexity. Parallelization of *gcombi* is necessary for large values of k and m. Optionally, a k-predictor combination can be excluded it if contains a subset of $(k-1)$ genes that appears unimportant in predicting the target gene.

glogic uses a logical-filter-based design to find the CoD between expression values of different k-predictor gene combinations for a given target gene. Design and validation are performed independently on each k-predictor combination. Each k-predictor gene combination also takes about the same amount of computing time for the design and validation operation.

Theoretically, adjoining predictors to a predictor set gives no worse performance; however, owing to estimation error, adding more predictors may yield a lower estimated CoD (Kim *et al.*, 2000a; Kim *et al.*, 2000b; Dougherty *et al.*, 2000). The *gadjust* module adjusts the CoD value of a k-predictor gene combination by comparing it to the CoD values of its $(k-1)$-predictor subsets and changing it to the maximum of these CoD values if it does not exceed them all. The computational complexity of this adjustment operation has an upper bound of $O(^mC_k \cdot m)$.

4.2 Parallel implementation

As each of the three modules is computationally intensive, we implemented parallel versions of each of these modules to create three phases of the parallel program, PAGE. PAGE was implemented on a Beowulf cluster parallel architecture (Sterling *et al.*, 1999), running the Redhat Linux operating system.

The Beowulf cluster is a distributed memory parallel machine that facilitates *single-program-multiple-data* (SPMD) and *multiple-program-multiple-data* (MPMD) parallel computing environments (Hwang, 1993). We use both the SPMD and the MPMD approaches with the industry standard *message-passing interface* (MPI) for handling communication between the processing nodes in our parallel program, PAGE.

The measures we use to evaluate the parallel performance of PAGE are speedup and efficiency. Speedup is defined as $\chi_p = T_{\text{seq}}/T_p$, where T_{seq} is the sequential runtime on a single processor and T_p is the parallel runtime on P processors. T_p is the runtime of the processor that completes its assigned tasks last among the P processors. Ideal speedup is P. Efficiency, defined as $E_p = \chi_p/P$, is a measure of how effectively a parallel program is using the processors. Ideal efficiency is 100%.

4.3 Parallelization methods

There exist several ways of distributing the computational work when implementing a parallel version of an algorithm including, uniform load assignment and dynamic load assignment.

Uniform Load Assignment. The *uniform load assignment* strategy was used for the first and second module of PAGE. This strategy decomposes the data, $D = \{d_1, d_2, \ldots, d_q\}$, where $q = {}^mC_k$, into P contiguous partitions, $S_1 = \{d_1, d_2, \ldots, d_i\}$, $S_2 = \{d_{i+1}, d_{i+2}, \ldots, d_j\}$, , $S_p = \{d_{t+1}, d_{t+2}, \ldots, d_q\}$, with an equal number of data elements in each partition, where $(i-1) = (j-i+1) = (q-t+1)$. It then uniquely maps P partitions onto P processors. Processors work in parallel on the data elements in their assigned partitions. Uniform load assignment works well if the data elements require equal computation times (workloads). If data elements require different computation times, workload imbalance among the P processors will cause the processors to complete computations at different times, thereby idle processors wait for the final processor to complete its job. Therefore, the parallel execution time T_p is determined by the maximum execution time T_{max} among the processors, which can severely lower the efficiency.

Dynamic load assignment. To improve the parallel performance by reducing the load imbalance, we have employed a *dynamic load partition and assignment* scheme, for the third phase of PAGE. Load mapping (load balancing) consists of two operations: (1) partitioning a data domain into data partitions, and (2) assigning each partition to a certain processor in a parallel machine. Given a data domain $D = \{d_1, d_2, \dots, d_q\}$ with q data elements, the partitioning operation computes the total load W_D of D to determine an ideal partition size Δ. Assuming e_j is the computational weight of data element d_j, total load is obtained by $W_D = \sum_{j=1}^{q} e_j$. An ideal partition size, Δ, is then obtained for P partitions $(q \geq P)$ by

$$\Delta = \frac{1}{p} W_D = \frac{1}{p} \sum_{j=1}^{q} e_j \approx \lambda_{avg}. \qquad (13.2)$$

Using Δ (Equation 13.2), the partitioning operation must then partition D into an evenly weighted data partition, $D = S_0 \cup S_1 \cup \cdots \cup S_{p-1}$. The size of partition S_i is determined by

$$S_i = \left\{ d_{k_i} \mid 1 \leq k_i \leq q, \sum_{i=0}^{p-1} k_i = q \right\}, \qquad (13.3)$$

$$\Delta \approx \lambda_{s_0} \approx \lambda_{s_1} \cdots \approx \lambda_{s_{p-1}} = \sum_{k_p} e_{k_p}, \qquad (13.4)$$

where k_i is a number of data elements in data partition S_i, λ_{S_i} is the total load of S_i, and e_{k_i} is the computational weight of data element d_{k_i} in S_i. The S_i may not contain equal numbers of data elements because each data element has a different computational weight. The partitioning operation is highly problem dependent, and is known to be NP-hard (Garey *et al.*, 1979).

The assignment operation assigns each data partition S_i to a processor P_j in a way that minimizes the interconnection distance between communicating processors (*dilation*). Assuming a cut-through wormhole routing, dilation is considered one if processors P_j and P_k are directly connected. Otherwise, dilation is the hop count (the number of intermediate processors to connect on the shortest interconnection path between processors P_j and P_k). The assignment operation is highly dependent on parallel machine architecture, and is known to be computationally intractable (Bokhari, 1981; Garey *et al.*, 1979).

Number of Processors	Runtime (hours)	Speedup	Efficiency (%)
1	24.51	1.0	100.0
2	12.25	2.0	100.0
4	6.12	4.0	100.0
8	3.02	8.1	101.3
16	1.52	16.1	100.6
32	0.77	31.8	99.4
64	0.39	62.8	98.1

Table 13.1. Parallel performance of gcombi for $^{587}C_3$ with an exclusion of 43,000 two-gene predictors.

To reduce processor idle time, it is necessary to perform the partitioning and the assignment operations for an equitable load distribution over the processors. A dynamic load assignment scheme determines the partition size for the processors during run-time (Choudhary *et al.*, 1993; Suh *et al.*, 1998). In applying the dynamic load assignment scheme to *gadjust*, we group the k-predictor gene combinations into a linear chain of N disjoint partitions with equal computational costs, not equal numbers of combinations. A mapping $\pi : \{1, \ldots, {}^m C_k\} \mapsto \{1, \ldots, p\}$, from the combinations to the processors, generates for each processor, $1 \leq i \leq p$, a pair $(start[i], end[i])$. Processor i gets all combinations j where $start[i] \leq j \leq end[i]$ (Choudhary *et al.*, 1993; Martino *et al.*, 1997; Suh *et al.*, 1998). The load λ_l on a processor l, under mapping π, is the sum of the loads of the combinations assigned to the processor: $\lambda_l = \sum_{\{i|\pi(i)=l\}} load[i]$. We now assign combinations to a processor l starting from the combination $start[i]$ to some combination $end[i]$ until λ_l becomes $(\lambda_{avg} - \delta) \leq \lambda_l \leq (\lambda_{avg} + \delta)$. This approach will yield a good load balancing, however, the overhead associated with the computation of λ_{avg} and λ_l might be high.

4.4 Parallel versions of algorithms

We implemented a parallel version of *gcombi* with exclusion by first having each processor determine all k predictor gene combinations of an m gene pool set. Next, the combinations are linearly grouped into P partitions of equal size. The P partitions are distributed to P processors in the third step, each processor being assigned an equal number of combinations. Finally, each processor compares and excludes any k-predictor combination containing an excluded $(k - 1)$-predictor subset. Table 13.1 shows the parallel performance of *gcombi* for the generation of 3-predictor gene combinations of a 587 gene pool with an exclusion of 43,000 2-predictor gene combinations.

Glogic performs the same operation of design and validation on each k-predictor combination, independently. Each k-predictor combination takes about the same computing time for design and validation. Hence, parallelization can be achieved via a uniform load assignment scheme. Table 13.2 shows the parallel performance of *glogic* for the 3-predictor gene combination analysis of a single target gene out of 58 target genes in a 587 gene pool set. The parallel *glogic* achieves almost linearly increasing parallel performance with increasing numbers of processors.

Number of Processors	Runtime (hours)	Speedup	Efficiency (%)
1	72.71	1.0	100.0
2	36.35	2.0	100.0
4	18.36	4.0	99.0
8	9.29	7.8	97.8
16	4.69	15.5	96.9
32	2.37	30.7	95.9
64	1.21	60.1	93.4

Table 13.2. Parallel performance of glogic for a single target gene with $k = 3$ and $m = 586$. (Note: When a gene is selected as a target gene, the selected target gene is not used as a predictor. For each selected target gene, it is only necessary to compute CoD for $^{586}C_3 = 586!/(3!(586 - 3)!) = 33,366,840$ of 3-predictor gene combinations, where $m = 587 - 1 = 586$.)

To analyze the data set for a *melanoma* study with 587 genes in the gene pool and 58 target genes, we designed and tested the model of a *ternary logic filter* with a full combination of up to three predictor genes on the NIH Beowulf machine. It took less than 3 days of computer time to analyze 58 target genes, using 32 nodes (64 processors). Without the parallelization, it would have taken about 178 days of computer time.

For the *gadjust* module, parallelization based on uniform partitioning did not perform well owing to load imbalance. Therefore, we implemented a dynamic load assignment scheme to minimize processor idle time and improve parallel performance. Using the dynamic load assignment scheme on the Beowulf system, we achieved the parallel performance shown in Table 13.3, which shows a detailed timing breakdown of *gadjust* for 58 target genes.

Adjustment Runtime is the computer time taken to adjust the CoD values. As seen in the table, it scales well and achieves high parallel efficiency with an increasing number of processors. It uses the load partition and assignment information calculated by the dynamic load partition and assignment. Total runtime is the time taken to do both the CoD adjustment and the file transfer. As the number of processors

Number of Processors	Adjustment Runtime (hours)	Efficiency (%)	I/O Time (hours)	Load Assignment (hours)	Total Runtime (hours)	Efficiency (%)
1	35.43	100.0	9.19	0	44.62	100.0
2	20.63	85.9	9.08	0.19	29.90	74.6
4	10.75	82.4	5.62	0.09	16.46	67.7
8	5.41	81.8	2.37	0.05	7.83	71.2
16	2.63	84.2	1.62	0.05	4.30	64.8

Table 13.3. Parallel performance of dynamic load assignment scheme on gadjust for 58 target genes. (Note: Total Runtime = Adjustment Runtime + I/O Time + Load Assignment)

increases, efficiency drops because the file I/O time becomes dominant as compared to the CoD adjustment time.

5. Visualization of gene expression (VOGE)

PAGE produces a massive amount of computationally derived information (approximately 200Gbytes) that is too large to study and interpret directly. Without the aid of an appropriate data visualizer, it would be almost impossible to interpret the derived information to determine associations between genes based on their expression patterns. To aid geneticists in their interpretation of the derived information, we have developed an interactive graphical visualization program, VOGE, that provides tools to display, manipulate, and interpret computationally derived information using the following integrated GUIs: decision tree graphs, tables, filters, histograms, and links to other genetic databanks.

Figure 13.1 is a screen snap shot of VOGE that shows data manipulation, filtering, and raw data browsing facilities. The right window lists the target genes in a gene pool. Upon selection of a target gene, VOGE displays the CoD values in a spreadsheet-like environment with some filtering functions that have been suggested, by biologists at NHGRI/NIH, to be useful in extracting relevant relationships for a given study. If a user wants to see the primary data, microarray ratios, then he can pull up a window displaying them.

Once a group of predictor genes is chosen, a user can visualize their relationships to target genes in an arrow-diagram as shown in Figure 13.2. The highlighted blue circle in the center is the target gene of interest and it is expanded to display those gene combinations that have significant CoDs with respect to that target gene. The predictor genes in the selected combinations are displayed as red circles. If any predictor

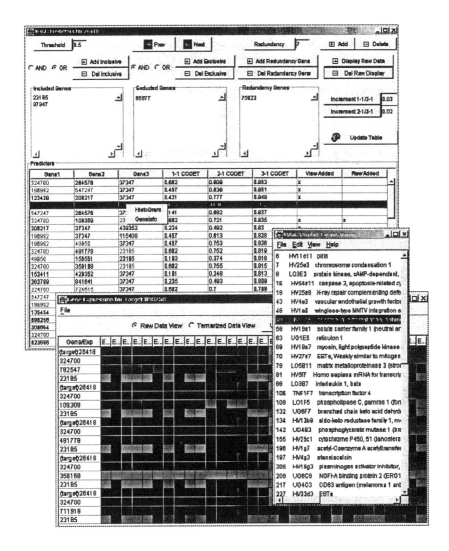

Figure 13.1. VOGE: spreadsheet-like environment. (see color plates)

gene for a given target gene is also a target gene itself, it is displayed as
a blue circle. The predictor gene that is also a target gene is considered
a sub-target gene for this display. Any sub-target gene can be expanded
to display its predictor combinations. This GUI provides a *tree-like* view
or a *circular tree* view with the capability to rotate or change viewing
angles. It also provides an *auto-plot* feature that constructs a tree of gene
combinations for a given target gene with a specified depth and number
of children. It has a zoom function to focus or defocus on the region of

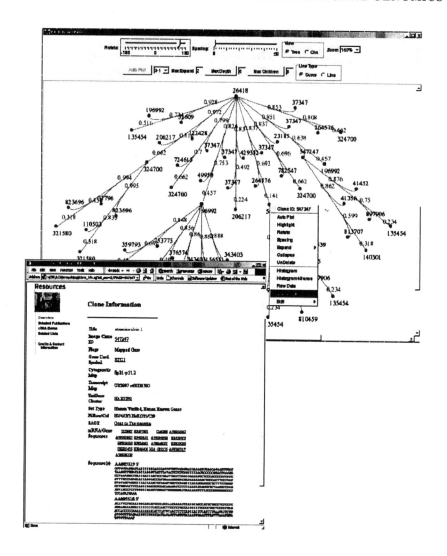

Figure 13.2.   VOGE: arrow-diagram visualizer with web-based UniGene connectivity. (see color plates)

interest. Those genes that are displayed more than once can be cut or copied onto each other. Upon a click on a gene, a popup menu appears with various options, such as a raw data visualizer and links to various genetic databanks to retrieve information on the chosen gene. The goal of this GUI is to provide access to the broad types of information that could be used to construct a hypothetical genetic regulatory network for a given target gene.

While using the visualization environment, a user can connect via a Web browser interface to the UNIGENE database as well as other genetic databases to obtain relevant information, as shown in the left window of Figure 13.2. Access to all sources of prior knowledge concerning genes that a given gene has been shown to interact with, as well as processes a given gene has been shown to influence are of tremendous value in pruning the large number of possible connections down to the most likely few. Currently, many tools are being constructed to carry out experimental searches for gene-to-gene interactions and to catalogue those interactions that have already been reported in the literature. VOGE is designed to allow rapid connection to online representations of these data sets as soon as they are available. New features are continually being added to VOGE as NHGRI biologists gain more experience with this tool.

6.   Summary and conclusions

This chapter presents PAGE, a parallel program for analyzing the codetermination of gene transcriptional states from large-scale simultaneous gene expression measurements with cDNA microarrays, and its application to a large set of genes. Using PAGE, it was possible to compute coefficients of determination for all possible three-predictor sets from 587 genes for 58 targets in a reasonable amount of time. Given the limited sample sizes currently being used for microarray analysis, it is not necessary to go beyond three predictors at this time since the data are insufficient for four-predictor CoD estimation. As shown in Tables 13.1, 13.2, and 13.3, significant speedups are achieved by the parallelization when compared to the sequential version of the program modules.

A related data visualization program, VOGE, helps geneticists navigate, manipulate, and interpret the massive amount of computationally derived information produced by PAGE. Tools provided in VOGE enhance the ability of researchers to interpret and apply the computationally derived information for understanding the functional roles of genes.

Acknowledgments

All biological data used in our experiments came from the Cancer Genetics Branch at the National Human Genome Research Institute. We are indebted to the Division of Computer System Services at the Center for Information Technology (CIT) of the National Institutes of Health (NIH) for providing computational resources and expertise including the use of the CIT Beowulf system. The authors from the CIT, NIH, would also like to thank Thomas Pohida of the CIT, NIH, for introducing us to the exciting area of gene expression analysis.

References

Bokhari, S. H. (1981) On the Mapping Problem. *IEEE Transaction on Computers, C30*, 207-214.

Chen, Y., Dougherty, E. R., and Bittner, M. L. (1997) Ratio-based decisions and the quantitative analysis of cDNA microarray images. *J. Biomed. Optics, 2*, 364-374.

Choudhary, A. N., Narahari, B., and Krishnamurti, R. (1993) An Efficient Heuristic Scheme for Dynamic Remapping of Parallel Computation. *Parallel Computing, 19*, 633-649.

Dougherty, E. R., Kim, S., and Chen, Y. (2000) Coefficient of determination in nonlinear signal processing. *Signal Processing, 80*, 2219-2235.

Garey, M. R., and Johnson, D. S. (1979) *Computer and Intractability: A Guide to the Theory of NP-Completeness*, Freeman.

Hwang, K. (1993) *Advanced Computer Architecture: Parallelism, Scalability, Programmability*, McGraw-Hill, Inc.

Kim, S., Dougherty, E. R., Bittner, M. L., Chen, Y., Sivakumar, K., Meltzer, P., and Trent, J. M. (2000) A general nonlinear framework for the analysis of gene interaction via multivariate expression arrays. *J. Biomed. Optics, 5*, 411-424.

Kim, S., Dougherty, E. R., Chen, Y., Sivakumar, K., Meltzer, P., Trent, J. M., and Bittner, M. (2000) Multivariate measurement of gene expression relationships. *Genomics, 67*, 201-209.

Martino, R. L., Yap, T. K., and Suh, E. B. (1997) Parallel algorithms in molecular biology. *High-Performance Computing and Networking*, Springer-Verlag, 232-240.

Sterling, T. L., Salmon, J., Becker, D. J., and Savarese, D. F. (1999) *How to Build a Beowulf: A guide to the Implementation and Application of PC Clusters*, The MIT Press.

Suh, E., Narahari, B., and Simha, R. (1998) Dynamic load balancing schemes for computing accessible surface area of Protein molecules. *Proceedings of International Conference on High-Performance Computing (HiPC'98)*, 326-333.

Chapter 14

HUMAN GLIOMA DIAGNOSIS FROM GENE EXPRESSION DATA

Gregory N. Fuller[1], Kenneth R. Hess[2], Cristian Mircean[4], Ioan Tabus[4],
Ilya Shmulevich[1], Chang Hun Rhee[5], Kenneth D. Aldape[1], Janet M.
Bruner[1], Raymond A. Sawaya[3], and Wei Zhang[1]

Departments of [1]Pathology, [2]Biostatistics, and [3]Neurosurgery
The University of Texas M. D. Anderson Cancer Center, Houston, Texas, USA
[4]*Institute of Signal Processing, Tampere University of Technology, Tampere, Finland*
[5]*Department of Neurosurgery, Korea Cancer Center Hospital, Seoul, Korea*

1. Introduction

One goal for gene expression profiling in cancer research is to develop a new paradigm for the molecular classification and objective diagnosis of tumors. Motivation for such a new diagnostic tool is the subjectivity and frequent ambiguity associated with classical histological diagnosis based on morphological features. Current molecular approaches suffer from several drawbacks. For example, molecular classification based on single gene molecular markers is vulnerable to the vicissitudes of non-representative tissue sampling in the face of well known tumor heterogeneity. At the other extreme, current gene expression array platforms typically display hundreds to thousands of genes on a single microarray. Diagnosis using such a large number of genes is not practical and is likely to yield suboptimal results. We suggest, alternatively, that the ideal diagnostic tool for clinical use would be a customized expression array comprised of less than 100 genes, with each of these possessing robust differentiating power with respect to the particular cancer under investigation.

In support of this hypothesis, we have carried out a gene expression pilot study using 25 glioma tissue samples that encompass four different glioma subtypes: oligodendroglioma, anaplastic oligodendroglioma, anaplastic astrocytoma, and glioblastoma. The complete microarray platform employed contains 597 genes. To develop a cDNA microarray suitable for reliable diagnosis of the four glioma subtypes, a robust algorithm is required to select a subset of specific genes that have the requisite diagnostic power.

In this chapter, we provide an introduction and overview of current glioma diagnosis and classification, which is largely based on a classical histopathological approach, followed by a discussion of the possibilities for molecular classification based on a gene expression approach. Finally, we describe a novel method for the selection of subsets of genes required for diagnostic microarrays.

2.   Glioma diagnosis: from histology to microarray

2.1   Histopathological glioma classification and diagnosis

The current paradigm for brain tumor diagnosis and classification, as exemplified by the recently revised World Health Organization Classification of Tumours of the Nervous System (Kleihues and Cavenee, 2000), is based primarily on morphologic pattern recognition: the identification of similarities between the phenotypic characteristics expressed by tumor cells compared to those of normal central nervous system constituents as assessed by light microscopic examination of H&E-stained tissue sections, immunohistochemistry, and electron microscopy. Although the morphologic approach has unquestionably been of considerable utility, there are nevertheless a number of shortcomings. Using traditional phenotypic criteria, for example, the identity and classification of some tumor types, such as mixed oligoastrocytomas, is highly subjective and overly dependent upon the individual pathologist's relative weighting of various morphologic characteristics. Currently prevailing histology-based classification methods also do not permit accurate prediction of clinical behavior or response to specific therapeutic agents and regimens for individual patients within a given histologic rubric, as, for example, is the case for anaplastic astrocytoma, in which individual patient response to treatment and survival varies significantly. Another problem is the failure of morphology-based classifications to accurately predict individual patient sensitivity to the toxic effects of various therapies, such as radiation necrosis.

There is thus a need on many levels for a more precise, effective and objective approach to brain tumor diagnosis, classification, grading, and prognosis. The first significant development in the molecular classification of brain tumors was the recently recognized association between the combined loss of heterozygosity (LOH) for chromosomes 1p and 19q and therapeutic responsiveness for a subset of gliomas that display oligodendroglial differentiation features (Cairncross *et al.*, 1998; Caskey *et al.*, 2000; Sasaki *et al.*, 2002). This significant discovery is undoubtedly only the beginning of meaningful stratification of the diffuse gliomas based on molecular characterization. A variety of methods currently exist for the detection of allelic loss in these tumors. LOH analysis requires matched tumor and normal DNA from the same patient, and the presence of matched pairs can be a limiting factor, especially for retrospective studies of rare tumor types. Fluorescent in situ hybridization (FISH) is a method that does not require matched normal tissue and has been used successfully for the detection of 1p/19q loss in gliomas (Fuller and Perry, 2002; Smith *et al.*, 2000). A high-throughput, real-time PCR method using TaqMan chemistry has also been described for the detection of copy number in these tumors. By utilizing the frequent CA microsatellite repeats that exist in the human genome, a single TaqMan probe can be used to evaluate copy number at multiple loci for the molecular classification of oligodendrogliomas (Aldape *et al.*, 2002; Nigro *et al.*, 2001). It is likely that the need for molecular subclassification of human tumors will increase in the future, and such high-throughput methods offer the ability to perform such assays in a time frame consistent with that needed for clinical practice.

2.2 The promise of molecular classification of gliomas by cDNA microarray

To aid molecular classification of tumors, a number of contemporary technologies are available for detailed characterization of the tumor genome, transcriptome and proteome. Of these, transcriptome profiling is currently the most widely employed in molecular classification. Many different statistical methods, both supervised and unsupervised, have been used in attempts to construct molecular classifications based on gene expression profiles. One commonly used approach is multidimensional scaling (MDS) (Borg and Groenen, 1997). MDS, also called principal coordinate analysis, is similar to principal components analysis; it shares the same goal of constructing a low-dimensional representation of high-dimensional data. Unlike MDS, however, principal components analysis cannot take into account nonlinear structures because it describes the data in terms of a linear subspace. MDS attempts to find configurations of the data in the lower dimensional representation such that the relative distance between objects is similar to that in the original, higher dimensional representation. It has an advantage over hierarchical clustering in that multiple dimensions are used to relate objects

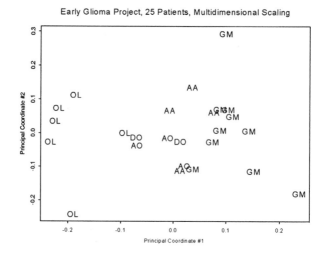

Figure 14.1. Classification of diffuse gliomas by multidimensional scaling (MDS) analysis of gene expression profiles. A scatter plot is shown of the first two MDS coordinates, with each sample labeled with the tumors' histologic subtype.

to one another (in an unconstrained manner). Clusters of objects can be identified on an MDS plot by their relative proximity (i.e., relatively tightly bunched points that are separated from other points on the plot).

We applied MDS to our pilot gene expression profiling study of 25 gliomas. The data contain expression levels for 597 genes obtained using the Clontech Atlas I array (gene expression profiles). Twenty-three of the 25 samples belong to one of four glioma classes: oligodendroglioma (OL), anaplastic oligodendroglioma (AO), anaplastic astrocytoma (AA), or glioblastoma multiforme (GM). Two samples share pathological features of both AO and GM. Gliomas that exhibit these features are sometimes classified as grade D oligodendrogliomas, and the two cases are thus labeled DO in the MDS plot.

As shown in the MDS plot of Figure 14.1, gene expression profiles alone do exhibit some power of separation. Consistent with the heterogeneous (multiform) nature of GM, the GMs are more diverse with greater dispersion. In contrast, most OLs are segregated closely together. There is no separation between AA and AO cases. Although the precise meaning of the dimensions on the MDS plots is not clearly defined, it is interesting that the first principal coordinate (x-axis in Figure 14.1) appears to correlate with grade, with lower grades distributed towards the left and higher grades on the right.

Although MDS may be useful for coarse grain classification, it is clear that it is quite noisy even for histopathologically typical cases. A possible explanation is that when all genes are included in the analysis, the data are inevitably noisy, i.e., many genes do not contribute to class separation but rather make the separation more obscure. With unsupervised statistical methods, more data does not always equate with better results. It is thus evident that although high-density arrays provide far more raw expression data than do low density arrays, this is not a requirement for meaningful molecular classification and tumor stratification and may actually contribute to obfuscation of meaningful group associations. In contrast, the use of a relatively small set of genes that exhibit highly differential expression patterns may provide robust separation for class distinction and class discovery studies. In this regard, glioma genomics research employing expression array profiling followed by immunohistochemical tissue microarray confirmation has already identified one progression- and survival-associated marker, insulin-like growth factor binding protein 2 (IGFBP2), which is uniformly and differentially overexpressed only in glioblastoma (Fuller *et al.*, 1999; Sallinen *et al.*, 2000; Wang *et al.*, 2002; Zhang *et al.*, 2002). We anticipate that a group of robust separator genes such as IGFBP2, identified from microarray experiments and confirmed by other methodologies such as the tissue array, will be the ultimate classifiers used in clinical settings.

Linear Discriminant Analysis.   The above discussion illustrates the need for a focused small set of genes that can be used for clinical diagnosis. This gene set should have high discriminative power for robust disease classification. There are many different approaches that can be used for the selection of genes that show differential expression among tumors. Some of these approaches identify any gene that shows differential expression for one particular tumor subtype. Ideally, however, the set of genes selected should be able to separate several known subtypes simultaneously, as proposed in (Kim *et al.*, in press).

Let us illustrate the need for a small gene set via an example using linear discriminant analysis (LDA). While MDS falls within the general category of unsupervised classification methods, LDA is a supervised statistical method that computes new dimensions that maximally separate the classes. We first ranked the genes using the ratio of between-class sums of squares to within-class sums of squares, which is a criterion of class separability, and then performed LDA analysis using only a selected set of top ranked genes. As Figures 14.2 and 14.3 show, LDA

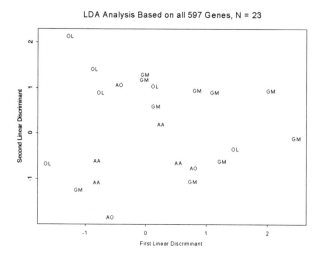

Figure 14.2. Classification of the four glioma subtypes by the LDA method using all 597 genes.

using only the top ranked 20 genes achieves much better separation of the four glioma classes than using all 597 genes.

k-Nearest Neighbor classification with quantized data. In this section, we discuss another supervised classification approach – the k-Nearest Neighbor (k-NN) method. The k-NN algorithm is conceptually very simple and intuitive: given a data set containing labelled vectors (vectors with known classifications), for each new vector to be classified, locate first its k nearest vectors and then select, out of the k nearest vectors' labels, that label which is in majority. However, there may be different variants of it, differing in the distance used when selecting the nearest neighbors or in the policy of choosing the winning label. The k-NN algorithm was introduced by Fix and Hodges (1951) based on a variant of nearest neighbor density estimation of the group-conditional densities and subsequently improved by Stone in his paper on consistent nonparametric regression (Stone, 1977).

Mircean *et al.* (2002) proposed a modified k-NN classification algorithm that achieved a reduction of the error rate by at least a half, compared to all other tested algorithms (cf. Dudoit *et al.*, 2000), when applied to the data set published in (Golub *et al.*, 1999). We now briefly describe the main features of the k-NN method focusing on the modifications introduced in (Mircean *et al.*, 2002).

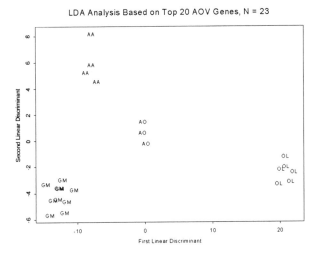

Figure 14.3. Classification of the four glioma subtypes by the LDA method using only a subset consisting of 20 genes. The groups are well separated, as compared with Figure 14.2 in which all 597 genes were used.

It is often the case that before a classification method (e.g., learning vector quantization (LVQ)) is used, a non-supervised clustering method is first applied to the data (e.g., self-organizing map (SOM)). Although the clustering-classification paradigm has been known for a long time, such as in the above mentioned instance of SOM-LVQ, for the simple k-NN classification algorithm such a combination does not appear to exist (at least to our knowledge). There are several reasons for quantizing the data prior to applying a classification method. Mistakenly, quantization is sometimes presented as being merely a loss of data precision. This is a true in the case of perfect data – an assumption that is rarely realistic. The struggle in data processing can be summarized as trying to find the right balance between properly capturing the information content in the data and filtering out the non-informative and detrimental noise. In methods of classification, this principled struggle goes as far as discarding all unwanted information, yielding at the end of the day only a simple decision: Class A, or Class B, etc. Our reasons for quantization are therefore two-fold: firstly, we want to capture the meaningful clusters in the data, represented by their centers, and secondly, we want to make our subsequent decisions more robust with respect to noise present in the data. Gene expression analysis based on quantized data (binary) has already been explored by Shmulevich and Zhang (in press), by Akutsu and Miyano in Chapter 6, and by Tabus *et al.* in Chapter 10 in this book.

We propose to quantize the gene expressions in a number of relevant quantization levels. Based on the previous results reported in (Mircean *et al.*, 2002) and on the current experiments, an empirical rule can be stated: use a number of quantization levels at least equal to the number of classes in the classification problem at hand. That is heuristically plausible, since we like to give to each gene the chance to express its "vote" for each class type. On the other extreme, if we allow an infinite number of quantization levels, the data remain un-quantized; in this case it is certain that no useful information is lost, but on the other hand we do not advance at all towards our goal of giving up the unwanted information. Thus, for the present experiment, we chose to use quaternary quantized data.

Once the number of quantization levels is defined, we focus on seeing the data quantization as a means of removing noise from data. One of the most central concepts in information theory concerning the "controlled" loss of information from data is the rate-distortion theory (Cover and Thomas, 1991), which has proved to be productive in recent advances in communication technology. The rate of the quantizer is fixed once the number of quantization levels is chosen. The representatives of each quantization level (i.e., cluster center) should then be selected such that the distortion (the expected square distance between the gene expression value and the closest cluster center) is minimized. This is accomplished by an iterative algorithm, which in information theory is known as the Lloyd algorithm (Gersho and Gray, 1992) and in pattern recognition is referred to as the k-means algorithm.

Let n denote the selected number of quantization levels for each gene expression, and denote an entry in the data array as $M(i,j)$, where $i \in \{1, \ldots, N\}$ is the index of the gene and $j \in \{1, \ldots, m\}$ is the index of the sample/tumor. Our goal is to replace the value $M(i,j)$ by a quantized value $Q(M(i,j))$, where $Q(\cdot)$ is the quantizer function, associating to each real scalar x a quantization value $c_q = Q(x) \in \mathcal{C}$ ($\mathcal{C}$ is the codebook of the quantizer and $V_q = \{x | Q(x) = c_q\}$ is the Voronoi cell of the quantization value c_q). The association $x \mapsto c_q$ is done by the nearest neighbor function: for a given x we select that $c_q \in \mathcal{C}$ that is closest to x in Euclidean norm.

Lloyd quantization starts with arbitrary positions of the n quantization values and then iterates two steps until convergence: 1) find the nearest quantization value for each data point (re-distribution of data point to Voronoi cells); 2) for each Voronoi cell, find the average of the data points associated to it, and assign this average to be the new center of the Voronoi cell (new quantization value). Each step of the algorithm enforces one necessary condition for optimality, and this iterative algorithm is expected to reach a local optimum of the sample distortion

$$D = \sum_{i=1}^{N} \sum_{j=1}^{m} (M(i,j) - Q(M(i,j)))^2.$$

We can design one codebook for each gene, but we prefer to design a common codebook for all genes. The quantized data will tell which of the Voronoi cells the original data belong to, and we prefer that the Voronoi cells be identical for all genes.

The next task is to select those genes that are the most informative in our classification problem. This is usually realized by using a filter approach (Dudoit *et al.*, 2000), as opposed to wrapper approaches (Kohavi and John, 1997). For each gene i, we compute the mean conditional on the class $\ell \in \mathcal{L}$ as $\overline{M}_{i,\ell} = \frac{1}{n_\ell} \sum_{j|\ell(j)=\ell} M(i,j)$, where n_ℓ is the number of samples in class ℓ. We also define the number of classes by L, the within class variance as

$$W_i = \sum_{\ell \in \mathcal{L}} \sum_{j|\ell(j)=\ell} (M(i,j) - \overline{M}_{i,\ell})^2 / (m - L)$$

and between class variance as

$$B_i = \sum_{\ell \in \mathcal{L}} n_\ell (\overline{M}_{i,\ell} - \frac{1}{m} \sum_{j} M(i,j))^2 / (L - 1).$$

We compute for each gene the ratio between-over-within-variance

$$R_{BW}(i) = B_i / W_i,$$

rank all genes according to their ratios and select as feature genes those genes having the largest ratios, up to a predefined number of feature genes, p.

In order to report good estimates of the classification errors, we use the traditional cross-validation experiments: we split the patients randomly in two groups: the training group and the test group. The cross-validation setting for selection of the parameters (value of k and number of feature genes, p) and evaluation of the performance of the classification algorithm is a variation of 2:1 sampling, where special care is paid to always have representatives from all classes in the k-NN training set. To this end, we considered as cross-validation split policies the ones compatible with the (training set, test set) split shown in Table 14.1.

Number	Class	Total Number of samples	Number of samples in test set	Number of samples in training set
1	AA	4	1	3
2	AO	3	1	2
3	OL	6	2	4
4	GM	10	3	7
5	AO / GM	2	0	0

Table 14.1.   Cross-validation split policies

We choose from the 21,600 (4x3x15x120) total number of possible combinations (compatible with the Table 14.1) at random 150 different splits to perform 150 cross-validation experiments. The k-NN algorithm assigns to the tested profile the dominating class of the k-nearest samples in the training set. If the maximum number of votes is received by two classes, we resort to $(k - 1)$-nearest profile voting, and recursively do so until the class with majority vote is uniquely defined.

The minimum estimated classification error rate is 2%, achieved when $p = 40$ genes are selected as feature genes (the best in the series of experiments with different numbers of feature genes, $p = 20, 40, 80, 160,$ and 320) and $k = 1$ (see Figure 14.4). The distribution of the classification error rate over the samples is not uniform, and changes with the experimental parameters (p and k). The most erroneous samples are from the AA and AO classes due to the small number of samples in the training set (graphs not shown).

Finally, as before, we applied MDS in order to visualize the class partitioning of the data. It should be noted that MDS here is used merely as a *post hoc* visualization of the ability of the top quantized 40 feature genes to separate the classes and is in no way related to the k-NN classification algorithm. In Figure 14.5, the original data (597×25) is presented in a three dimensional representation, with classes rendered by various symbols. There is no grouping of classes in this 3-dimensional representation, with the use of all genes leading to class overlap. If we apply the quantization and perform the selection of the best 40 feature genes according to the between-over-within-variance ranking, the 3-dimensional representation is able to show the interesting topological connections of profiles in the same class, as shown in Figure 14.6. In the figures, we used the following symbols: circles for AA class; diamonds for AO class; triangles for OL class; stars for GM class and, separately for the two profiles with ambiguous AO/GM classes, we used squares.

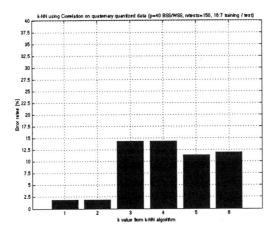

Figure 14.4. Estimated error rate of k-NN algorithm using correlation coefficient on the quaternary quantized data; $p = 40$ genes; 150 runs

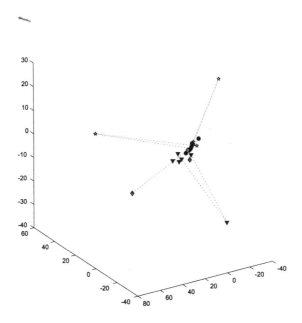

Figure 14.5. Multidimensional Scaling performed on the original (597×25) data set. (see color plates)

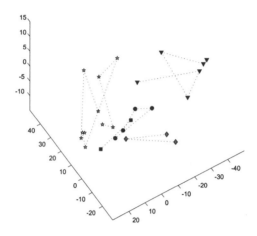

Figure 14.6. Multidimensional Scaling applied to the quaternary quantized selected data, using only the best 40 feature genes. (see color plates)

3. Confirmation of marker selection on a large scale

Although the use of supervised methods permits the identification of potential genes for clinical application, such candidate genes require validation by other methods prior to clinical implementation. Validation of newly identified target candidates requires analysis of hundreds of individual human tumors. Tissue-based molecular analysis is traditionally dependent on the serial examination of candidate gene expression at the DNA, mRNA or protein level by fluorescence *in situ* hybridization (FISH), chromagen-based *in situ* hybridization (CISH) or routine immunohistochemistry (IHC) performed on individual whole-block tissue sections. Although this approach is effective, the individual analysis of hundreds of specimens is pedestrian and highly labor-intensive. Moreover, inherent variability in experimental conditions from batch to batch of specimens analyzed is unavoidable and can have a deleterious effect on interpretation of results. An even more serious drawback is the fact that archival human tissue resources are finite in amount and a typical paraffin tissue block will be exhausted after only about a hundred tissue sections or so (frequently far fewer), thus severely restricting the number of studies that can be performed as new markers are identified through genomics strategies. This limitation of traditional tissue section-based analysis is particularly acute for collaborative cohort studies and cooperative clinical trials in which much effort has been expended in the meticulous collection of clinical data and survival statistics, and in the collective pooling of rare tumor samples for study. When the paraffin block tissue bank is exhausted, further correlative studies using newly-discovered markers cannot be performed and the rich clinical data base lies underutilized or fallow.

In recognition of the problems and limitations of traditional whole-block tissue section techniques, the tissue microarray (TMA) was developed, which allows as many as 1000 or more paraffin-embedded tissue cores from different specimens to be reassembled into a single ordinary paraffin block (the recipient or microarray block) for parallel analytical studies of molecular alterations at the DNA, RNA, or protein level (Wang *et al.*, 2002). TMA technology permits rapid assessment and validation of new molecular markers in a large number of tumors, with maximal conservation and utilization of irreplaceable human archival tissue. The major applications of TMA technology to date have been in basic and clinical translational research, particularly in the confirmation of tumor markers identified by gene expression profiling methods.

Additional methods exist for the validation of expression microarray data, including the real-time PCR-based assays alluded to above, focussed on measurement of mRNA rather than gene copy number. Such assays are especially useful for the validation of expression changes in genes for which appropriate antibodies do not yet exist.

4.   Conclusions

The practice of clinical genomics has patient care applications as the ultimate goal. Gene expression profiling studies have clearly demonstrated the utility of genomics in the areas of molecular classification, identification of novel subgroups (so-called "diseases within disease"), identification of new markers for diagnosis, and identification of novel targets for therapeutic intervention. Genomics data analyses have further demonstrated that an approach based on single gene analysis is not sufficiently robust for diagnostic applications, especially when multiple subtypes of diseases are involved. In contrast, combinations of genes provide much more useful information. Although gene expression profiling projects normally include the analysis of hundreds or thousands of genes in a single experiment, many of the genes thus analyzed have very little or no disease classification potential. Such genes should be excluded from diagnosis chips, which optimally will contain only a small subset of genes. Focused chips of this type are also more amenable to the strict quality controls and conditioning that are required for clinical applications. If the analysis is extended to the protein level via immuno-histochemistry, the number of genes to be tested should similarly not be too large if practical clinical utility is sought. In this paper, we have described a methodology for selecting a subset of genes that in combination have the greatest discriminating power for the four major subtypes of human gliomas. The 40 genes thus identified may potentially constitute a clinically useful diagnostic chip.

Acknowledgments

This work was partially supported by a grant from Texas Higher Education Coordinating Board.

References

Aldape K., Ginzinger D. G., Godfrey, T. E. (2002) Real-time quantitative polymerase chain reaction: a potential tool for genetic analysis in neuropathology. *Brain Pathology*, 12, 54-66.

Borg, I. and Groenen, P. (1997) *Modern multidimensional scaling: Theory and applications,* Springer, New York.

Caskey, L.S., Fuller, G.N., Bruner, J.M., Yung, W.K., Yung, W.K., Sawaya, R.E., Holland, E.C., and Zhang, W. (2000) Toward a molecular classification of the gliomas: histopathology, molecular genetics, and gene expression profiling. *Histology and Histopathology*, 15, 971-981.

Cairncross JG, Ueki K, Zlatescu MC, Lisle DK, Finkelstein DM, Hammond RR, Silver JS, Stark PC, Macdonald DR, Ino Y, Ramsay DA, Louis DN. (1998) Specific genetic predictors of chemotherapeutic response and survival in patients with anaplastic oligodendrogliomas. *Journal of the National Cancer Institute*, 90, 1473-1479.

Cover, T. M. and Thomas, J. A. (1991) *Elements of Information Theory,* John Wiley, New York.

Dudoit, S., Fridlyand, J., and T. P. Speed. (2000) *Comparison of Discrimination Methods for the Classification of Tumors Using Gene Expression Data.* Dept. of Statistics, University of California, Berkeley, Technical Report 576.

Fix, E. and Hodges, J. (1951) *Discriminatory analysis, nonparametric discrimination: consistency properties.* Technical Report, Randolph Field, Texas: USAF School of Aviation Medicine.

Fuller, C. E., Perry, A. (2002) Fluorescence in situ hybridization (FISH) in diagnostic and investigative neuropathology. *Brain Pathology*, 12, 67-86.

Fuller, G.N., Rhee, C.H., Hess, K.R., Caskey, L.S., Wang, R., Bruner, J.M., Yung, W.K., Zhang, W. (1999) Reactivation of insulin-like growth factor binding protein 2 expression in glioblastoma multiforme: a revelation by parallel gene expression profiling. *Cancer Research*, 59, 4228-4232.

Fuller, G.N., Hess, K.R., Rhee, C.H., Yung, W. K. A., Sawaya, R. A., Bruner, J. M., Zhang, W. (2002) Molecular classification of human diffuse gliomas by multidimensional scaling analysis of gene expression profiles parallels morphology-based classification, correlates with survival, and reveals clinically-relevant novel glioma subsets. *Brain Pathology*, 12, 108-116.

Gersho, A. and Gray, R. M. (1992) *Vector quantization and signal compression*. Kluwer, Boston.

Golub, T. R., Slonim, D. K., Tamayo, P., Huard, C., Gaasenbeek, M., Mesirov, J. P., Coller, H., Loh, M. L., Downing, J. R., Caligiuri, M. A., Bloomfield, C. D., Lander, E. S. (1999) Molecular classification of cancer: class discovery and class prediction by gene expression monitoring. *Science*, 285:531-537.

Kim, S., Dougherty, E. R., Barrera, J., Chen, Y., Bittner, M., and Trent, J. M. Strong feature sets from small samples. *Journal of Computational Biology*, in press.

Kleihues P. and Cavenee, W. K. (2000) *Pathology and Genetics of Tumours of the Nervous System*. IARC Press: Lyon.

Kohavi, R. and John, G. (1997) Wrapper for feature subset selection. *Artificial Intelligence*, 97, 273–324.

Mircean, C., Tabus, I., and Astola, J. (2002) Quantization and distance function selection for discrimination of tumors using gene expression data. *SPIE 2002, BiOS 2002 Symposium*, 19-25 January 2002, San Jose, CA.

Sallinen, S-L,, Sallinen, P.K., Haapasalo, H.K., Helin, H.J., Helen, P.T., Schraml, P., Kallioniemi, O-P., Kononen, J. (2000) Identification of differentially expressed genes in human gliomas by DNA microarray and tissue chip techniques. *Cancer Research*, 60, 6617-6622.

Sasaki, H., Zlatescu, M. C., Betensky, R. A., Johnk, L. B., Cutone, A. N., Cairncross, J. G., and Louis, D. N. (2002) Histopathological-Molecular Genetic Correlations in Referral Pathologist-Diagnosed Low-Grade Oligodendroglioma, *Journal of Neuropathology and Experimental Neurology*, 61, 58-63.

Shmulevich, I. and Zhang, W. Binary Analysis and Optimization-Based Normalization of Gene Expression Data. *Bioinformatics*, in press.

Smith, J.S., Perry, A., Borell, T.J., Lee, H.K., O'Fallon, J., Hosek, S.M., Kimmel, D., Yates, A., Burger, P.C., Scheithauer, B.W., Jenkins, R.B. (2000) Alterations of chromosome arms 1p and 19q as predictors of survival in oligodendrogliomas, astrocytomas, and mixed oligoastrocytomas. *Journal of Clinical Oncology*, 18, 636-645.

Stone, C. J. (1977) Consistent nonparametric regression (with discussions). *The Annals of Statistics* 5, 595-645.

Nigro, J.M., Takshashi, M.A., Ginzinger, D.G., Law, L., Passe, S., Jenkins, R.B., Aldape, K. (2001) Detection of 1p and 19q loss in oligodendroglioma by quantitative microsatellite analysis, a rea-time quantitative polymerase chain reaction assay. *American Journal of Pathology*, 158, 1253-1262.

Wang, H., Wang, H., Zhang, W., Fuller, G. N. (2002) Tissue microarrays: applications in neuropathology research, diagnosis, and education. *Brain Pathology*, 12, 95-107.

Zhang, W., Wang, H., Song S. W., Fuller, G. N. (2002) Insulin-like growth factor binding protein 2: gene expression microarrays and the hypothesis-generation paradigm. *Brain Pathology*, 12, 87-94.

Chapter 15

APPLICATION OF DNA MICROARRAY TECHNOLOGY TO CLINICAL BIOPSIES OF BREAST CANCER

Lajos Pusztai[1], W. Fraser Symmans[2], Thomas A. Buchholz[3], Jim Stec[4], Mark Ayers[4], Ed Clark[4], Funda Meric[5], David Stivers[6], and Kenneth Hess[6]

Departments of [1] Breast Medical Oncology, [2] Pathology, [3] Radiation Oncology, [5] Surgery, and [6] Biostatistics

The University of Texas M. D. Anderson Cancer Center, Houston, Texas, USA

[4] Millennium Predictive Medicine Inc., Cambridge, Massachusetts, USA

1. The promise of microarray technology in clinical medicine

DNA microarrays represent an important new tool to analyze cells and human tissues. The technology enables one to measure the expression of several thousands of mRNA species in a biological specimen. The gene density of currently used microarrays range from several hundred genes to over 30,000 human sequences. With the completion of the sequencing of the human genome, it has become apparent that the maximum number of human genes is around 30-40,000. Since the number of expressed genes in any particular human cell type is likely to be a lot less than that, it is technically possible to monitor almost the entire transcriptosome, the collection of all mRNAs present, in a biological sample. The ability to measure the expression of several thousands of genes simultaneously during various experimental conditions is a technological breakthrough that may revolutionize the study of cell biology and medicine. It may allow cell biologists to begin to reconstruct the complexity of biological systems by studying all the components simultaneously, at least at the level of mRNA. In medicine, this may be the first technique towards functional diagnostics, a technique that can obtain information on the functional status of a tissue and provide insight into the molecular activities that underlie distinct biological behavior of morphologically similar tumors. In this chapter, we outline potential applications of DNA microarrays for clinical research through examples from our own work and by others on human breast cancer.

2.   Breast cancer

Breast cancer is a major health problem in the developed world. The disease is often curable when discovered and treated at early stages, but once metastasis is detected in distant organs, cure is no longer attainable with current treatment modalities. Some micro-metastatic disease with metastatic deposits too small to be seen with current imaging techniques may sometimes be cured with systemic hormonal or chemotherapy (Loprinzi et al., 2001). Why some women with breast cancer develop metastatic disease whereas others do not, or why, some micrometastases can be cured with chemo- or hormonal therapy remains unknown. Furthermore, the clinical course of metastatic breast cancer is variable, some cancers progress slowly and respond to multiple lines of therapy, others show very aggressive behavior and resistance to all known treatments. Breast cancer is a tragic and fascinating example of the clinical heterogeneity that is observed with many other cancers. Seemingly similar tumors possess very different biological behavior.

It has long been suspected that the distinct biological behavior is due to molecular differences between histologically similar tumors. A large number of individual molecules have been evaluated as possible prognostic markers that could predict clinical outcome of breast cancer. For example, some data suggest that mutations in p53, over-expression of HER-2 or Epidermal Growth Factor receptors (EGFR) predict for a higher risk of recurrence after surgery and greater probability of death from breast cancer. However, none of these possess sufficient independent predictive value to have become clinically useful (Bast et al., 2001). Similarly, a variety of molecules have been evaluated as potential predictors of response to treatment. It has been proposed that the presence of bcl-2, MDR-1 or HER-2, among many others may confer resistance or sensitivity to certain chemotherapies (MacGrogan et al., 1996, Makris et al., 1997). Unfortunately, none of these have established clinical utility and the role of individual molecules as predictors of response to chemotherapy remains highly controversial. On the other hand, the presence of estrogen receptor (ER) and HER-2 are good, but far from perfect predictors of response to antiestrogen (i.e. tamoxifen) or anti-HER-2 (trastuzumab) therapy, respectively. A substantial proportion of cancers that are ER-positive will not respond to hormonal therapy. A full 60% of patients with HER-2-overexpressing cancer may not respond the trastuzumab treatment (Mass *et al.*, 2001). Clearly, refinements in our predictive tests are needed.

Despite the explosion in identifying new genes and understanding their function in cell growth, metastasis formation, angiogenesis and cell death, the clinical predictive and prognostic value of individual molecules remains limited. This may reflect our incomplete understanding of these processes. Consider, for example that only 4-6000 human genes have some rudimentary functional annotation, the remaining 20+ thousand is completely unknown. The limited utility of individual molecules to predict clinical outcome of cancer may also be due to the fact that these agents act in concert and form complex, perhaps redundant, interactive pathways. Therefore, individual components may only contribute minimal information on the activity of the entire system. DNA microarrays provide a much-needed technical tool to address both of these problems and has a potential to revolutionize clinical predictive and prognostic tests and the process of drug discovery.

3.   Gene finding in human tissues with microarrays

Transcriptional profiling is often used to attempt to identify novel genes that contribute to a disease process or biological function. For example, in the context of breast cancer, normal breast tissue could be compared with cancer to identify cancer-associated genes. These genes may serve as new targets for drug development. Cancers with an indolent course could be compared with rapidly progressive tumors to identify genes associated with poor prognosis. The technology could also be applied to learn about changes in the transcriptosome during exposure to chemotherapy. Chemotherapy-regulated genes likely represent signaling and effector pathways that participate in repair functions or programmed cell death. Identification of these genes could lead to development of new predictive tests or drugs that modulate response to therapy. However, the identification of such biologically important genes is complex enough in cell lines, but is even more challenging in human (or animal) tissues.

3.1   Impact of tissue and biological heterogeneity on microarray data interpretation

The identification of individual genes of interest from a large pool of transcripts is often critically dependent on differential expression between two specimens or groups of specimens. This raises several questions. What fold difference in expression values is sufficient to define

	Cancer Cells (%)		Lymphocytes (%)		Stromal Cells (%)	
	CBX	FNA	CBX	FNA	CBX	FNA
1	55	85	5	12	40	3
2	55	85	25	11	20	5
3	40	75	45	25	15	0
4	40	80	20	15	40	5
5	35	85	55	12	10	3
6	40	50	20	40	40	10
7	55	95	15	1	30	4

Table 15.1. Microscopic counts (%) of different cell types in matched pairs of core biopsy (CBX) and fine needle aspiration (FNA) samples from seven breast cancers used in a DNA microarray study.

a candidate gene for further research (Wittes *et al.*, 1999)? What frequency of differential expression should be considered relevant when multiple specimens are compared? What confounding factors can result in differential expression of a gene unrelated to the biological question under study? There are many factors that can contribute to differential expression of a gene between two tissue specimens obtained from different individuals, these include differences in age, race, sex and co-morbid illnesses (King *et al.*, 2001). Also, differences in tissue composition may vary from specimen to specimen even within the same organ. Cancer tissues are made up of a variety of cellular components including neoplastic cells, lymphocytes, stromal cells such as fibroblasts, adipocytes and blood vessels. Each tissue sample has slightly different proportions of these various cellular components. This tissue heterogeneity from sample to sample is illustrated in Table 15.1.

The table compares the cellular composition of fine needle aspiration (FNA) and core needle biopsies (CNB) of breast cancer obtained from the same patient from the same part of the tumor at the same time. The tumor cell content of an FNA biopsy may range from 50% to 95%. Furthermore, FNAs and CNBs have consistently different cellularity profiles, FNAs are enriched in cancer cells and depleted in stromal cells compared to CNB (Symmans *et al.*, 2001). Considering that microarray results can not identify the cellular source of a given mRNA, variation in cellular composition between specimens contributes an unavoidable noise to "differential expression" results. What additional difference age, race, concurrent medications or co-morbid illnesses contribute to gene expression profiles is poorly understood.

The numerous factors that can lead to differential expression of a gene between two tissue specimens, even in the absence of any experimental error, makes gene finding with microarrays a challenge. To illustrate this, a list of the top 10 differentially expressed genes between matching pairs of FNA and CNB from 8 different patients is presented in Figure 15.1. Genes overexpressed in a core biopsy compared to the matching FNA could be considered as molecules that characterize stromal cells adjacent to breast cancer and may hold biologically important information. However, as Figure 15.1 indicates rank order lists of differentially expressed genes tend to differ significantly among specimens obtained from different individuals. Looking at the 10 genes with highest CNB/FNA ratios for each case, the 8 lists together contain 59 distinct sequences. Only 9 genes appear on more than one list. Two genes are on three lists, and 7 genes are on four lists. When the list is expanded to top 50 the proportion of overlapping genes increases which indicate that the magnitude of differential expression is quite variable from sample to sample. The individual difference in these rank order lists is due to multiple factors. Experimental noise including error during hybridization or image acquisition contributes, but biological heterogeneity (age, race, medications, tissue sampling) is likely to play a major role. It is important to realize that the discrepancy seen in these lists do not necessarily reflect poor methodology or an unsuccessful experiment.

Picking individual genes of importance reliably may be more difficult than identifying a group of genes that contribute to a "molecular fingerprint". The assumption is that while the expression of individual genes may be prone to variations unrelated to the experimental question, looking at a larger number of genes across multiple specimens, a pattern may arise that may be more reliable than the individual components alone. One method to identify a "molecular fingerprint" is to define a set of genes that contribute most of the gene expression differences between two groups of samples. These genes could be defined by calculating two-sample t-scores for each gene. After choosing a significance level, a list of genes can be generated that is either positively or negatively associated with a group of samples. This approach yields a short list of genes often containing no more than a few dozen to few hundred sequences which are often extensively different in rank order from the individual lists presented in Figure 15.1. Genes that make up these "molecular fingerprints" are also reasonable candidates for further study to assess biological importance. However, many of the same uncertainties apply to individual components of the "molecular fingerprint" as described above.

In conclusion, microarray profiling is a useful method to identify individual disease- or biological process-associated genes in human tissues. However, due to the inherent biological complexity of tissues and sam-

Case 1	Case 2
collagen, type III, alpha 1	**DKFZP434B168 protein**
matrix metalloproteinase 2	AA757466
selenoprotein P	translation initiation factor 3, subunit 3
Duffy blood group	EST (Hs.30567)
immunoglobulin kappa variable 3D-15	EST (Hs.58351)
collagen, type III, alpha 1	EST (Hs.229675)
EGR1 binding protein 2	myosin, light polypeptide 4
EST (Hs.174404)	**ribosomal protein L34**
collagen, type III, alpha 1	collagen, type VI, alpha 1
immunoglobulin lambda joining 3	fibronectin 1

Case 3	Case 4
serum amyloid A4	**DKFZP434B168 protein**
apolipoprotein D	lysosomal-associated protein, transmembrane 4a
apolipoprotein D	**immunoglobulin lambda joining 3**
polymerase I and transcript release factor	**S-adenosylhomocysteine hydrolase**
collagen, type III, alpha 1	**osteoblast specific factor 2**
AI820704::R54983::R55428::AI732283	EST (Hs.93102)
putative lymphocyte G0/G1 switch gene	**collagen, type III, alpha 1**
lactotransferrin	osteoblast specific factor 2
adipocyte fatty acid binding protein 4	small inducible cytokine subfamily B, member 10
immunoglobulin heavy constant gamma 3	**collagen, type III, alpha 1**

Case 5	Case 6
adipocyte fatty acid binding protein 4	**secreted phosphoprotein 1 (osteopontin)**
EST (Hs.108211)	EST (Hs.153958, Hs.699)
putative lymphocyte G0/G1 switch gene	**S-adenosylhomocysteine hydrolase**
EST (Hs.182171)	EST (Hs.275245, Hs.252723)
hypothetical protein FLJ10718	ribonuclease, RNase A family, 1 (pancreatic)
EST (Hs.127286)	MHC, class II, DQ alpha 1
EST (Hs.121824)	cytochrome c oxidase subunit Vic
cDNA FLJ22425 fis, clone HRC08686	zinc finger protein 313
AI335863	hypothetical protein DKFZp761H221
EST (Hs.189299)	EST (Hs.130251)

Case 7	Case 8
RAN binding protein 9	decidual protein induced by progesterone
ATP citrate lyase	**ribosomal protein L34**
osteoblast specific factor 2	**adipocyte fatty acid binding protein 4**
secreted phosphoprotein 1 (osteopontin)	serum amyloid A4
EST (Hs.27769)	serum amyloid A1
osteoblast specific factor 2	AHNAK nucleoprotein (desmoyokin)
hydroxyacyl-Coenzyme A dehydrogenase	fatty acid binding protein 5
cDNA FLJ13598 fis, clone PLACE1009921	aquaporin 1
small breast epithelial mucin	MHC, class II, DR beta 1
insulin-like growth factor binding protein 7	H19, imprinted maternally expressed

Figure 15.1. The top 10 differentially expressed genes in 8 individual pairs of core needle (CNB) and fine needle (FNA) biopsies. The biopsies were profiled with a 25,000 gene cDNA array and results compared in pairs for 8 individuals. Ratios of gene expression were calculated for each sequence by dividing the normalized expression value observed in a CNB by that of a matching FNA. When the same gene appears twice on a list, it indicates that the sequence was represented more than once on the array and both duplicate spots showed high level of differential expression. The results demonstrate that most of the genes overrepresented in CNB relative to FNA are sequences that are commonly expressed in stromal cells including fibroblasts or fat cells. This is consistent with the cellular differences between the tissues obtained by these two distinct sampling methods. The individual lists differ greatly from one another containing very few overlapping genes. Genes that appear in more than one list are bolded.

pling error, even if the technology itself were noise-free, differentially expressed genes can not be expected to be confirmed by other methods a 100% of the time. Single genes almost never separate two *a priori* defined closely related groups of biological specimens perfectly, unless the gene itself (such as ER status) defines the groups. Most microarray experiments report positive confirmation rates for putative differentially expressed genes ranging from 60-80% when tested on a larger set of specimens and using a different method (Roche *et al.*, 2001; Li *et al.*, 2001). What constitutes acceptable confirmation when a cancer-associated gene is tested in many cancerous and normal tissues also remains unclear. For example, a microarray experiment identifies gene X as overexpressed in cancer compared to normal tissues in 6 out of 10 cases. A confirmatory experiment finds that 3 of 10 cancers indeed overexpress this molecule compared to an arbitrary cutoff based on average expression in normal tissue using a different method (immunohistochemistry or RT-PCR). Does this result confirm or refute the initial observation? It is important to remember that the sensitivity of cDNA microarrays to detect relative difference in mRNA expression may be quite different compared to RT-PCR. It is certainly is quite different from immunohistochemistry that detects proteins in a semi quantitative manner. Because of the biological heterogeneity that contributes to expression heterogeneity of any single gene, the functional relevance of differentially expressed genes to the biological question under study also remains inherently uncertain. In the above example, a confirmation rate of 30% (3/10) may equally represent a fundamentally important gene for a third of breast cancers (like for example HER-2) or may simply be a biological variation without pathobiological importance. Differentially expressed genes identified by transcriptional profiling may be best regarded as the starting points for further experiments.

4.   Molecular classification of breast cancer based on mRNA expression profiles

Global mRNA expression profiles provide a potentially useful tool to classify cancer into biologically and clinically relevant groups. It is expected that distinct "molecular fingerprints" will be associated with clinically important characteristics such as metastatic ability and response to therapy (Khan *et al.*, 2001, Su *et al.*, 2001). When biological specimens are compared to one another to determine similarity based on transcriptional profiles, two very commonly used mathematical techniques are hierarchical clustering and multidimensional scaling (Eisen *et al.*, 1998). The results are displayed as branching trees where adjacent

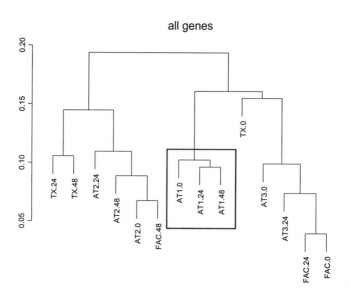

Figure 15.2. Hierarchical clustering of serial core biopsies obtain from 5 different individuals. Four individuals had 3 biopsies (TX, FAC, AT1, AT2) and one had 2 biopsies only (AT3). Unsupervised clustering was performed using all genes included in the experiment. Each hybridization was performed in duplicate and the average of the two experiments was used for each gene for further analysis. Using this unfiltered set of genes ($n = 25,000$), the relationship between the specimens is obscured by noise. The 3 individual biopsies from the same person only cluster together in one case, AT1.

branches represent more closely related specimens, or as points in two- or three-dimensional space. However, the results of these analyses are quite sensitive to the genes that are fed into the clustering program. This is both a potential source of error and an opportunity for clinical application.

4.1   Impact of selection of genes on clustering outcomes

Cluster tree diagrams and multidimensional scaling figures are visually appealing and intuitively easy to interpret. However, results of clustering are sensitive to the selection of genes that are used for the clustering algorithm. Genes expressed at low levels, close to the detection limit of the technology, contribute substantial noise. On the other hand, genes that are uniformly expressed across all specimens, this often include the majority of expressed genes, do not contribute to the separation of molecular profiles. To determine what the threshold is for reliable detection of genes with low expression, or what constitutes uniform expression across all specimens is not without challenges. Thresholds for gene expression intensity are often linked to the mean or median expression of all genes detected on a membrane array or restricted to percentiles. Differential expression among specimens is often defined by a coefficient of variation (CV). Figures 15.2 and 15.3 illustrate how the cluster trees change depending on the gene filtration criteria in five cases of serial biopsies. Figure 15.2 is the result of hierarchical clustering using 25,000 human sequences without any filtering of genes. Figure 15.3 shows results generated from the same data set after gene filtering and using only genes ($n = 20,849$) with absolute normalized expression values > 0.5 and low variability in replicate experiment. With more stringent selection of genes, the background noise is removed and it becomes apparent that serial biopsies from the same patient tend to cluster together despite concurrent treatment with chemotherapy. Differences between individual cases appear to be larger than differences within cases during serial sampling (Buchholz *et al.*, 2002). The price of this "cleaning up" exercise is that the process could also remove some genes of potential interest.

Depending on the biological question asked, different sets of genes are appropriate for cluster analysis after filtering for "noise". In a set of breast cancer needle biopsies we asked a question if global gene expression patterns correlate with known clinicopathologic parameters such as tumor size, nodal status, nuclear grade or hormone receptor status (Pusztai *et al.*, 2001). To determine this, multidimensional scaling was performed with all genes whose expression was above the reliable detection limit ($n = 13,924$) and clinical parameters were overlaid on the results. This sort of analysis can reveal the major organizing force of the transcriptosome and its relationship to known clinical parameters. Figure 15.4 illustrates that the cases appear to separate, although incompletely, along estrogen receptor status of the tumors.

However, there is more information hidden in the transcriptosome. Multiple "molecular fingerprints" are embedded in the global profile that could be used to refine molecular classification schemes based on transcriptional profiles. As described in the gene finding section, specimens can be grouped based on a clinical characteristic (i.e. ER positive versus ER negative cancers) and genes can be defined that differentiate the two groups. Using only the differentially expressed genes, a more precise separation of the groups is expected. This is illustrated in Figure 15.5. Using approximately 100 genes that contribute most of the difference to the distinct gene expression profiles of ER negative compared to ER pos-

genes above 0.5 with low replicate variability

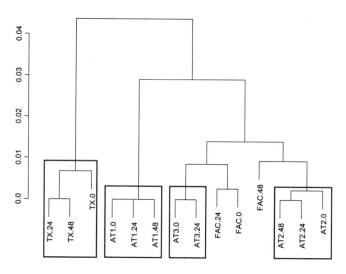

Figure 15.3. Hierarchical clustering of serial core biopsies obtain from 5 different individuals. For this figure, the same primary data set is used as presented in Figure 15.2. Unsupervised clustering was performed after discarding genes with normalized expression values <0.5 and genes that showed high level of variability in replicate hybridization experiments. Using this filtered set of genes ($n = 20,849$), the relationship between the specimens becomes clearer. Individual biopsies from 4 of the 5 cases cluster together in terminal branches of the diagram.

itive breast cancers, a perfect separation of specimens can be achieved. However, it has to be kept in mind that the list of genes contributing to the fingerprint was generated from the same cases that were subjected to the multidimensional scaling, often referred to as discovery or training set, therefore a good separation is expected. The real discriminating power of such a "fingerprint of ER status" can only be determined by testing the same sets of genes prospectively on independent specimens. Figure 15.6 illustrates that when 9 new cases were added and the analysis repeated, all new cases clustered correctly with the ER negative or ER positive groups corresponding to their clinical ER status. Distinct transcriptional profiles associated with hormone receptor positive compared to hormone receptor negative breast cancer has also been reported by others (Gruvberger *et al.*, 2001). Similar strategies have been successfully applied to distinguish hereditary breast cancer from sporadic tumors and to distinguish normal breast epithelium from cancer based

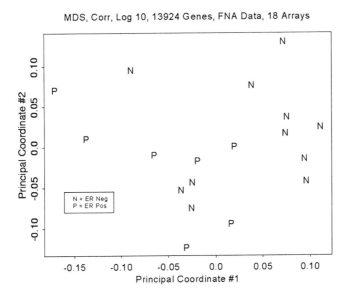

Figure 15.4. Multidimensional scaling of transcriptional profiles of fine needle biopsies of 18 breast cancers. Estrogen receptor negative cases are denoted with N and hormone receptor positive cases with P.

on expression profiles (Hedenfalk *et al.*, 2001, Perou *et al.*, 1999). Even a novel classification scheme of breast cancer to luminal and basal type has been proposed based on gene expression patterns (Perou *et al.*, 2000, Lonning *et al.*, 2001).

The most exciting promise of this molecular classification of tumors is that it might predict clinical outcomes that can not currently be predicted with existing routine pathologic techniques. Whether the paradigm of analysis outlined for ER status will hold up to predict other clinical characteristics remains to be tested. Clearly, ER status has a profound effect on the transcriptosome of breast cancer and unsupervised clustering without extensive gene selection already separates tumors into two groups that roughly correspond to ER status. Sets of genes that determine resistance to certain type of chemotherapy or predispose to metastatic ability may be more limited and may not influence the transcriptosome to the same degree as ER status does. In this case, separation between groups with predicted good versus bad response may not be so robust as seen with the "fingerprint of ER".

However, the precision of a prognostic or predictive test need not be close to 100% in order to be clinically useful. In fact, what may be considered clinically useful predictive power will depend on clinical circumstances (Hayes *et al.*, 1998). For example, combined modality adjuvant systemic therapy for breast cancer may increase long term survival and cure by as much as 30-50% (Loprinzi *et al.*, 2001). The same treatment, however, fails to cure patients with macroscopic metastatic disease. Therefore, women at the time of diagnosis after completion of their surgery face a gamble of life and death. They may already be cured from cancer by surgery alone, or they may harbor micrometastatic disease. If they harbor micrometastasis, some of these may be eradicated by 4-6 months of adjuvant chemotherapy and/or further hormonal treatment. On the other hand, if they miss this opportunity to maximize the chance of cure and the cancer recurs, cure is no longer attainable. It has been reported that in this situation a substantial number of women in the USA are willing to undergo adjuvant chemotherapy to gain 1-2% absolute improvement in survival (Ravdin *et al.*, 1998). Indeed, current recommendations of adjuvant therapy by the National Cancer Institute

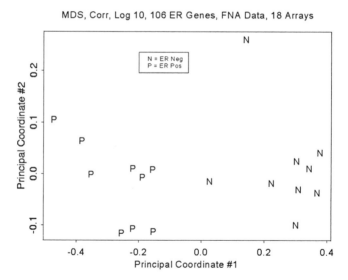

Figure 15.5. Multidimensional scaling of transcriptional profiles of fine needle biopsies of 18 breast cancers. Estrogen receptor negative cases are denoted with N and hormone receptor positive cases with P. The same set of primary data is used as in Figure 15.4, but multidimensional scaling was performed by using only 106 of the most differentially expressed genes between ER negative and ER positive cancers. As expected, a clear separation of cases is seen.

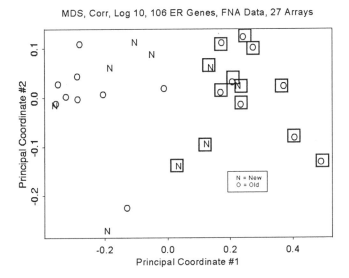

Figure 15.6. Multidimensional scaling of transcriptional profiles of fine needle biopsies of 27 breast cancers. The same 106 genes that were identified in the first 18 cases (discovery set) as the molecular fingerprint of ER status are tested on 9 new cases. The cases included in Figures 15.4 and 15.5 are denoted with O (old) and the 9 new cases for validation are identified by N. All ER positive cases are marked with a square.

and the National Comprehensive Cancer Network call for considering adjuvant chemotherapy for most breast cancers over 1 cm of size even though the long term risk of relapse from small lymph node negative tumors may not be greater than 20% (http://consensus.nih.gov, Carlson *et al.*, 2000). In this clinical situation, a prognostic test that aims to separate women who are cured by surgery alone from those who will relapse will need to show a very high degree of predictive precision. A potentially life saving treatment may not be denied to a patient mistakenly classified into the low risk group. A recent study reported an 83% correct prediction of relapse based on transcriptional profiles (van t Veer *et al.*, 2001). Whether this is sufficiently high to be accepted for clinical use or if further improvements can be made remains to be seen. In contrast, a test that aims to optimize the use of standard adjuvant chemotherapy may not need to be close to perfect in order to be clinically useful and acceptable. There are a number of chemotherapy regimens that are considered acceptable adjuvant treatment for breast cancer, these include

6 courses of 5-FU, adriamycin, cyclophosphamide (FAC/CAF) or cyclophosphamide, methotrexate, 5FU (CMF) or 4 courses of adriamycin cyclophosphamide (AC) alone or followed by 4 additional treatments with paclitaxel (AC/T) (Carlson *et al.*, 2000). When these regimens are used preoperatively, some patients (13-25%) undergo complete eradication of their tumor in the breast by the time they complete chemotherapy. Other tumors show minimal or no response to cytotoxic treatment indicating substantial individual differences in sensitivity to any of these regimens. The role of paclitaxel in adjuvant chemotherapy of breast cancer is particularly intriguing. Two large randomized studies have failed to show improvement in overall survival after adding 4 courses of paclitaxel to 4 courses of AC (http://consensus.nih.gov). Unplanned subgroup analysis, however suggests that a minority of women may benefit from this more prolonged treatment. Considering the gamble of life and death that women and their physicians face when deciding systemic adjuvant therapy for breast cancer and the low threshold for accepting a potentially life saving treatment, a predictive test that could predict for example a 50% chance of benefiting from addition of paclitaxel may be acceptable to many.

5. Pathway finding by expression analysis of human tissues

The third, and most exciting promise of DNA microarray technology, in addition to identifying novel genes and molecular classification schemes, is to provide a mechanistic insight into the complexity of cellular processes. This is the area where the least amount of progress has been made so far. A commonly employed experimental strategy is to expose cells to a certain stimulus and obtain transcriptional profiles at several time points during the experiment (Iyer V.R., 1999). The expectation is that the global transcriptional response will reveal a grand picture of molecular pathways as they interact and respond to a perturbation. To discover previously unknown molecular pathways or even extend and add putative new molecules to known pathways by analyzing serial gene expression changes is difficult. Transcriptional profiling produces descriptive results. Causal relationship and the direction of causality between any two or more genes whose expression changes concordantly is impossible to determine from profiling experiments alone. In other words, which gene caused what, and whether it caused it directly or indirectly through several other detected or undetected genes is difficult, if not impossible, to decipher from expression analysis alone. Nevertheless, identifying genes whose expression is coordinated can provide a working hypothesis for further experiments. However, a single time series experiment will not establish a rule for coordinated gene expression change, whereas multiple repeat experiments may. Interestingly, independent repeat experiments often do not reproduce the same global expression changes in terms of individual genes perturbed and magnitude of change observed. This is sometimes interpreted as a problem with the technology. However, repeat experiments can never possibly produce the exact same result. This is particularly true when the results contain great details as a "several-thousand-gene" expression profile does. An overarching general common response should nevertheless be discernible and should identify a consistent pattern of change that reflects "the biological response pathways". To see such a pattern is a huge computational and bioinformatics challenge. Consider that a 2- to 3-fold change in any gene may be functionally important but unrelated to the experimental question. It could be caused by subtle differences in cell culture conditions. The same change may also be functionally unimportant caused by random fluctuation in the expression level of that particular gene. How tightly the expression of individual genes is regulated and what functional thresholds exist for levels of gene expression is unknown. Of course, the same 2- to 3-fold change could be very relevant for the experiment. For a single molecule these possibilities can easily be gauged by doing a large number of multiple repeat experiments. However this problem is not easily solved when several thousand genes are monitored simultaneously. With increasing understanding of the function of more and more genes as well as the interactions among them, pathway finding should become easier.

6.   Summary

This paper has not addressed the numerous technical pitfalls of microarray technology. Any conclusion made from an experiment is only as reliable as the results on which it is based. With the widespread use of DNA arrays numerous technical problems have also become more public. Designing the arrays is a science on its own, verification of sequences placed on an array and optimization of hybridization conditions are important. Outcome is also critically dependent on good quality of RNA from the tissue. Acquisition of images and methods to deal with spill over signal ("bleeding") when using radiolabeled probes must be optimized. The statistical analysis of results needs expert input from biostatisticians dedicated and interested in the field. Microarray experiments are relatively complex to carry out and analyze, nevertheless, the technology is mature enough to produce valuable results. Among the many applications and promises of the technique, the most likely to yield practical results first are the attempts to classify cancer based on its expression profile. Several reports demonstrated successful diagnostic/prognostic application of DNA microarrays to various cancers (Khan *et al.*, 1998; Anbazhagan *et al.*, 1999; Alizadeh *et al.*, 2000; Takahashi *et al.*, 2001; van t Veer *et al.*, 2001). It is important to realize that these clustering functions can be carried out with select groups of genes without understanding any of the functions of the molecules involved in the classification. Microarray results are also yielding a large number of promising novel targets for drug development. Numerous inevitable false leads hinder this gene finding process but true gems are likely to be found. The field where microarrays have not yet delivered the expected answers, is its most exciting future potential. The potential to gain insight into the vast complexity of biological systems. This is a major bioinformatics challenge that is compounded by our rudimentary understanding of the function of most human genes. However, major advances are likely to happen in both areas within the coming decade and this, combined with DNA microarray technology, may indeed result in revolutionary new insights in biology – a goal that is well worth working for.

References

Alizadeh A.A., eisen, M.B., Davis, R.E., *et al.*, (2000) Distinct types of diffuse large B-cell lymphoma identified by gene expression profiling. Nature, 403, 503-511.

Anbazhugan, R., Tihan, T., Bornman D.M., *et al.* (1999) Classification of small cell lung cancer and pulmonary carcinoid by gene expression profiles. Cancer Research, 59, 5119-5122.

Bast, Jr., R. C., Ravdin, P., Hayes, D. F., Bates, S., Fritsche, Jr., H., Jessup, J. M., Kemeny, N., Locker, G. Y., Mennel, R. G., Somerfield, M. R. (2001) 2000 Update of recommendations for the use of tumor markers in breast and colorectal cancer: Clinical practice guidelines of the American Society of Clinical Oncology. Journal of Clinical Oncology, 19(6), 1865-1878.

Buchholz, T. A., Stivers, D., Stec, J., Ayers, M., Brown, J. L., Pusztai, L.. (2002) Global gene expression changes during neoadjuvant chemotherapy of human breast cancer. American Radium Society 84th Annual Meeting, Las Croabas, Puerto Rico.

Carlson, R.W., Anderson, B.O., Bensinger, W., et al (2000) NCCN Practice guidelines for breast cancer. Oncology 14 (11A) 33-49.

Eisen, M. B., Spellman, P. T., Brown, P. O., Botstein, D. (1998) Cluster analysis and display of genome-wide expression patterns. Proc Natl Acad Sci, USA, 95, 14863-14868.

Gruvberger, S., Ringner, M., Chen, Y., Panavally, S., Saal, L. H., Borg, A., Ferno, M., Peterson, C., Meltzer, S. (2001) Estrogen receptor status in breast cancer is associated with remarkably distinct gene expression patterns. Cancer Research, 61, 5979-5984.

Hayes, D. F., Trock, B., Harris, A. L. (1998) Assessing the clinical impact of prognostic factors: When is "statistically significant" clinically useful? Breast Cancer Research and Treatment, 52, 305-319.

Hedenfalk, I., Duggan, D., Chen, Y., Radmacher, M., Binnter, M., Simon, R., Meltzer, P., Gusterson, B., Esteller, M., Kallioniemi, O. P., Wilfind, B., Borg, A., Trent, J. (2001) Gene-expression profiles in hereditary breast cancer. N Engl J Med, 344(8), 539-548.

Iyer, V.R., Eisen, M.B., Ross, D.T., et al (1999) The transcriptional program in response of human fibroblasts to serum. Science, 283, 83-87.

Kahn, J., Simon, R., Bittner, M., et al. (1998) Gene expression profiling of alveolar rhabdomyosarcoma with cDNA microarrays. Cancer Research 58, 5009-5013.

Khan, J., Wei, J. S., Ringner, M., Saal, L. H., Ladanyi, M., Westerman, F., Berthold, F., Schwab, M., Antonescu, C. R., Peterson, C., Meltzer, P. S. (2001) Classification and diagnostic prediction of cancers using gene expression profiling and artificial neural networks. Nature Medicine, 7(6), 673-660.

King, H. C., Sinha, A. A. (2001) Gene expression profile analysis by DNA microarrays. JAMA, 286(18): 2280-2288.

Li, Z., Tsimelzon, A, Immaneni, A., et al., (2001) Genetic alterations in the progression of non-invasive to invasive breast cancer. Breast Cancer Research and Treatment, 69(3), 285 (Abst 23).

Lonning, P. E., Sorlie, T., Perou, C. M., Brown, P. O., Botstein, D., Borresen-Dale, A-L. (2001) Microarrays in primary breast cancer – lessons from chemotherapy studies. Endocrine-Related Cancer, 8, 259-263.

Loprinzi, C. L., Thome, S. D. (2001) Understanding the utility of adjuvant systemic therapy for primary breast cancer. Journal of Clinical Oncology, 19(4), 972-979.

MacGrogan, G., Mauriac, L., Durand, M., Bonichon, F., Trojani, M., de Mascarel, I., Coindre, J. M. (1996) Primary chemotherapy in breast invasive carcinoma: predictive value of the immunohistochemical detection of hormonal receptors, p53, c-erB-2, MiB1, pS2 and GSTπ, British Journal of Cancer. 74(9),1458-65.

Makris, A., Powles, T. J., Dowsett, M., Osborne, C. K., Trott, P. A., Fernando, I. N., Ashley, S. E., Ormerod, J.C., Titley, R.K., Gregory, R. K., Allred, D. C. (1997) Prediction of response to neoadjuvant chemoendrocrine therapy in primary breast carcinomas. Clinical Cancer Research, 3, 593-600.

Mass, R., Press, M., Anderson, S., Slamon, D. (2001) Improved survival benefits from Herceptin (trastuzumab) and chemotherapy in patients selected by fluorescence in situ hybridization. Breast Cancer Research and Treatment, 69(3), 213.

Perou, C. M., Jeffrey, S. S., Rijn, M., Rees, C. A., Eisen, M. B., Ross, D. T., Pergamenschikov, A., Williams, C. F., Zhu, S. X., Lee, J. C. F., Lashkari, D., Shalon, D., Brown, P. O., Botstein, D. (1999) Distinctive gene expression patterns in human mammary epithelial cells and breast cancers. Proc Natl Acad Sci, UA, 96, 9212-9217.

Perou, C. M., Serlie, T., Eisen, M. B., Rijn, M., Jeffrey, S. S., Rees, C. A., Pollack, J. R., Ross, D. T., Johnsen, H., Akslen, L. A., Fluge, O., Pergamenschikov, A., Williams, C., Zhus, S. X., Lenning, P. E., Berresen-Dale, A. L., Brown, P. O., Botstein, D. (2000) Molecular portraits of human breast tumours. Nature, 406, 747-752.

Pusztai, L., Symmans, W. F., Ayers, M., Stec, J., Clark, E., Hess, K., Sneige, N., Esteva, F. J., Banu, A., Rosales, M. M., Hortobagyi, G. N. (2001) Correlation between gene expression profiles obtained from single passage fine needle aspirations (FNA) of breast cancer and clinicopathologic parameters. Clinical Cancer Research, 7, 3692s (Abst 196).

Ravdin, P. M., Siminoff, I. A., Harvey, J. A., (1998) Survey of breast cancer patients concerning their knowledge and expectations of adjuvant therapy. Journal of Clinical Oncology, 16(2), 515-521.

Roche, P.C., Mertens, M.L., Couch F.J., et al (2001) Utilization of gene expression profiling to identify markers of disease recurrence in node-negative breast cancer. Breast Cancer Research and Treatment, 69(3), 285 (Abst 17).

Su, A. I., Welsh, J. B., Sapinoso, L. M., Kern, S. G., Dimitrov, P., Lapp, H., Schultz, P. G., Powell, S. M., Moskaluk, C. A., Frierson, Jr., H. F., Hampton, G. M. (2001) Molecular classification of human carcinomas by use of gene expression signatures. Cancer Research, 61, 7388-7393.

Symmans, W. F., Pusztai, L., Ayers, M., Clark, E., Stec, J., Buchholz, T. A., Sneige, N., Hortobagyi, G. N. (2001) RNA yield from needle biopsies for cDNA microarray analysis of breast cancer prior to neoadjuvant chemotherapy. Breast Cancer Research and Treatment, 69(3), 285 (Abst. 518).

Takahashi, M., Rhodes, D.R., Furge, K.A., Kanayama, H-O., et al., (2001) Gene expression profiling of clear cell renal cell carcinoma:

gene identification and prognostic classification. Proc. Natl. Acad. Sci USA., 98, 9754-9759.

van t Veer, L. J., van de Vijver, M. J., Dai, H., He, Y. D., Hart, A. A. M., Mao, M., Roberts, C., Bernards, R., Linsley, P. S., Friend, S. H. (2001) Expression profiling predicts poor outcome of disease in young breast cancer patients. European Journal of Cancer, 37, Suppl. 6, 271 (Abst. 1000).

Wittes, J., Friedman, H.P. (1999) Searching for evidence of altered gene expression: A comment on statistical analysis of microarray data. Journal of National Cancer Institute, 91, 400-401.

Chapter 16

ALTERNATIVE SPLICING: GENETIC COMPLEXITY IN CANCER

Sonya W. Song[1], Gilbert J. Cote[2], Chunlei Wu[1], and Wei Zhang[1]

[1] *Cancer Genomics Laboratory, Department of Pathology*
[2] *Department of Endocrine Neoplasia & Hormonal Disorders Research*
The University of Texas M. D. Anderson Cancer Center, Houston, Texas, USA

1. Introduction

It was found late in the 1970s that a number of DNA segments called exons, interrupted by segments called introns, come to constitute the final gene transcription product, messenger RNA (mRNA), through a post-transcriptional process called RNA splicing (Berget *et al.*, 1977; Chow *et al.*, 1977; Gilbert 1978). Only a few years after the initial discovery of exons and RNA splicing, it was found that there is an additional level of gene regulation, alternative RNA splicing. While an intriguing finding at the time, there was really no way to completely predict the impact that this finding would have on our understanding of gene expression. Early researchers deemed alternative splicing a specialized phenomenon that likely only affected the regulation of a minority of genes. However, as the genome initiative pressed forward, the number of genes found to undergo some type of alternative splicing gradually increased, and the expressed sequence tags (EST) database grew, so did the realization that a significant number of genes do indeed utilize some form of alternative RNA processing to generate genetic diversity. With the complete sequencing of the genome, it became apparent that much of human diversity actually stems from alternative RNA splicing. In fact, it is currently estimated that as many as 60% of all genes are subject to alternative RNA splicing thereby allowing the generation of hundreds of thousands of different mRNAs from the 30-40 thousand predicted genes and thus adding another layer of complexity to the human genome. Since alternative splicing appears to be associated with important biological and pathological processes, understanding how any given cell ultimately forms its final mRNA complements from a well-defined pool of RNA precursors provide insight into the genesis of disease and strategies for treatment. This chapter will focus on how alternative splicing adds further complexity to the interpretation of gene structures and their protein products. We also describe the biological importance of alternative splicing to the development of normal and cancer cells.

2.   Complexities of Gene Finding and Proteomics from Alternative Splicing

2.1   Computational tools for gene finding and alternative splicing analysis

Completion of the human genome sequencing project has heightened the urgent need for computational tools that systematically explore the genetic information embedded in the raw genomic sequences. Genome annotation was thus born to identify genes and their structures on the complete genomes. The computational methods developed for gene finding in eukaryotes are commonly used for predicting protein-coding regions in genomic DNA. These computational programs generally use three major types of genetic information, with algorithms applied to combine the information into a coherent gene prediction.

The three types of genetic information used for predicting gene structures include signals, content measures, and similarity measures (Stormo, 2000). *Signals* in DNA sequences include donor and acceptor sites that are used to identify potential splice junctions; start and stop codons that assist in categorizing exons into four classes (single exon genes, initial exons, terminal exons, and internal exons) (Guigo *et al.*, 1992); promoter-related sites, such as TATA boxes, transcription factor binding sites, and CpG islands (Uberbacher *et al.*, 1996); and polyadenylation signals to define the 3' end of the gene. *Content measures* are applied to distinguish the coding region from the noncoding region since these two regions have statistically different properties (Staden and MacLachlan, 1982; Fickett, 1982; Fickett and Tung, 1992; Brunak, *et al.*, 1991; Uberbacher and Mural, 1991). *Similarity measures* are also frequently used in gene recognition. The spliced alignment algorithm is thus designed to compare the similarity of human DNA sequences to already known gene sequences from other species and/or to cDNA, EST, or mRNA sequences published in the GeneBank database (Mironov *et al.*, 1998; Gelfand *et al.*, 1996; Mott, 1997).

A variety of computational programs for gene annotation have been developed to combine the information described above. However, because of the complexity of gene organization in eukaryotes and the com-

Programs (functions)	URL	References
GRAIL (g.f.[1])	http://compbio.ornl.gov/Grail-1.3/	Uberbacher and Mural et al., 1996
GeneID (g.f.[1])	mailto:geneid@darwin.bu.edu http://igs-server.cnrs-mrs.fr/igs/banbury/geneid-email.html	Parra et al., 2000
BLOCK (g.f. [1]/ g.f[2])	http://blocks.fhcrc.org/	Henikoff et al., 2000
HMMGene (g.f.[1])	http://www.cbs.dtu.dk/services/HMMgene/	Krogh, 1997
Fgenes (g.f.[1])	http://genomic.sanger.ac.uk/gf/gf.shtml	Solovyev et al.,1995
TRF (r.[3])	http://c3.biomath.mssm.edu/trf.test.html	Benson et al., 1999
CoreInspector (p.r.[4])	http://www.gsf.de/biodv/	Frech et al., 1997
MCPromoter (p.r.[4])	http://promoter.informatik.uni-erlangen.de/	Ohler et al., 2001
GeneWise (p.s.[5], g.f.[2])	http://www.sanger.ac.uk/Software/Wise2/	Birney et al., 2000
Genie (g.f.[1], p.r.[4])	http://www.fruitfly.org/seq_tools/genie.html	Reese et al., 2000
MAGPIE (g.f.[1],p.r.[4], a.[6],r.[3], g.f.[2])	http://genomes.rockefeller.edu/magpie/	Gaasterland et al., 1996

Table 16.1. Selected annotation programs and their internet websites. ([1]gene finding; [2]gene function; [3]repeat; [4] protein recognition; [5] protein similarity; [6]EST/cDNA alignment)

binatorial possibilities for exon assembly, the accuracy with which these computational tools can predict genes is relatively low. Specifically the correlation between predicted and actual genes is around 70%, with just 40% to 50% of exons predicted correctly (Burset and Guigo, 1996). The Genome Annotation Assessment Project (GASP) was purposely organized to evaluate the quality of current computational gene annotation tools when applied to the *Drosophila melanogaster* genome sequences (Reese *et al.*, 2000). Eleven annotation programs that are most commonly used were selected for the project (Table 16.1).

The GASP work demonstrated that the most of the gene finders correctly identified over 95% of coding nucleotides and more than 40% of the genes with correct intron/exon structures. It was also found that the homology-based annotation tools recognized and associated functions with almost half of the genes, whereas the promoter predictors generated high false-positive findings. However, one must still keep in mind that the human genome is far more complex than that of the *Drosophila* with significant differences in the gene architecture.

The review of these computational tools also revealed that none of the programs directly addressed the issue of alternative splicing in gene annotation. This is a concern because as Reese *et al.* (2000) found in their work, the unintentional inclusion of alternative spliced transcripts from a single gene in a standard data set confounded the scoring process. Even when Gelfand *et al.* (1996) and Mironov *et al.* (1998) used the spliced sequence alignment algorithms for gene prediction, they did not incorporate alternative splicing in their gene prediction.

However, to be most reliable in predicting complex exon assemblies, the computational tools must take into consideration of the mechanisms and sequence signals required for splicing. Although adding the concept of alternative splicing to the gene prediction process seemingly increases the complexity of gene finding, it would actually eventually improve the accuracy of our prediction of true gene products since it would rule out many false possibilities of exon assembly. Of course, there is a long way to go before we can incorporate alternative splicing into our prediction program, since we still have a very limited understanding of the mechanism of alternative splicing.

Some effort has been focused on trying to accurately estimate the prevalence of alternative processing and to define any unique characteristics of these genes. Earlier guesses placed the frequency of alternative RNA splicing at ~5% (Sharp, 1994). However, this figure has been revised with the growth of DNA and EST databases which has allowed more recent studies to be performed using data sampling. For example, in a study by Hanke (1999), a set of 475 human proteins was aligned against the six-frame translations of ~1.3 million ESTs using TBLASTN. These researchers found alternative ESTs containing common coding regions in 162 (34%) of these proteins. In a parallel study by Mironov *et al.* (1999), a computational program called Procrustes-EST was developed with an algorithm that could inventory exon-intron boundaries of known genes with the TIGR Human Gene Index. This sampling method yielded a similar alternative splicing frequency of ~35%. Others (Brett *et al.*, 2000; Modrek *et al.*, 2001) also found that at least 30% to 40% of human genes may be subject to alternative splicing. With the completion of the sequencing of the human genome, this number has continued to grow. In a recent survey, participants estimated that as much as 60% of all human genes produce alternative mRNAs and thought the number was likely to be higher. More importantly, nearly 75% of alternatively spliced genes derived from chromosomes 19 and 22 have altered protein-coding region, rather than noncoding 5' and 3' UTRs (International Human Genome Sequencing Consortium, 2001).

Database	URL	References
ASDB	http://devnull.lbl.gov:8888/ alt/index.html	Gelfand et al. 1999
Alternative Exons	http://cgsigma.cshl.org/ new_alt_exon_db2 http://elmo.ims.u-tokyo.ac.jp/altspl/	Stamm et al. 2000
ASMAMDB	http://166.111.30.65/ ASMAMDB.html	Ji et al. 2001
HASDB	http://www.bioinformatics.ucla.edu/ ~splice/HASDB/	Modrek, 2001
ISIS	http://isis.bit.uq.edu.au	Croft et al. 2000
EST-confirmed splice sites	http://industry.ebi.ac.uk/ ~thanaraj/splice.html	Thanaraj, 1999

Table 16.2. Alternative Splicing - Related Databases.

A few groups have begun the daunting task of documenting alternative spliced genes (Table 16.2), although a comprehensive public database is still a long ways away. The first was the Alternative Splicing Database (ASDB) initiated by Gelfand *et al.* (1999). It allows searches by either protein or DNA sequences. Stamm *et al.* (2000) compiled a comprehensive database of alternative exons. These sites have information from multiple species and permit searches by a variety of parameters, including predicting enhancer/silencer elements in known sequences. The most recent databases include the AsMamDB which is a growing compilation of mammalian genes that undergo alternative splicing (Ji *et al.*, 2001), and the Human Alternative Splicing Database (HASDNB) (Modrek *et al.*, 2001). The former contains a rather small set of genes, but these have been grouped by their cytobands and expression patterns in an effort to help look for regulatory mechanisms. The latter database has extended these efforts and contains a much larger dataset. The ISIS database developed by Croft *et al.* (2000) contains the information of over 170,000 intron sequences and shows the high frequency of alternative splicing in human genes. A dataset of EST-confirmed splice sites constructed by Thanaraj in 1999 classifies splice sites as donors, acceptors, genuine, and alternative sites. The information provided by these databases is of considerable value to the computational gene finding, in that it will eventually improve the accuracy of gene prediction.

2.2   Complexity of functional genomics - protein diversity from alternative splicing

The recognition that alternative RNA splicing occurs frequently and serves as a key mechanism for several cell-regulatory events has highlighted the importance of this process. In fact, it is becoming increasingly clear that alternative splicing enhances diversification of the proteomic world and creates more challenges for functional genomics.

As noted above, prior to the sequencing of the human genome, about 35% to 40% of human genes were believed to produce alternatively splicing mRNAs (Mironov *et al.*, 1999). Now this number is thought to have been significantly underestimated (Black, 2001; Graveley, 2001) because the ESTs used for the analysis were taken from a limited number of tissues or developmental states, thus only covering a limited portion of transcripts generated from a single gene. In addition, most of the ESTs used represented only the 5' or 3' end and thus ignored alternative splicing that occurred in the coding region of mRNA transcripts. This underestimation has been borne out by recent studies showing that as much as 60% of genes may produce alternative RNA isoforms (International Human Genome Sequencing Consortium, 2001). This, too, may not be an accurate representation, however, as existing ESTs expressed from only two chromosomes were sampled.

Perhaps of greater significance is the shear number of alternative mRNAs that may be produced from the genome. On average, a human gene can encode two to four transcript variants (Galas, 2001). In some extreme instances, a single gene can even generate hundreds to thousands of mRNA isoforms, such as the neurexin gene which encodes over 1000 different neurexin mRNAs, and the Down syndrome cell adhesion molecule (Dscam) which generates over 38,000 different transcripts (Graveley, 2001). Thus, alternative splicing could yield hundreds of thousands of proteins, far more than the 30,000 estimated human genes (Claverie, 2001). This large number of proteins means that the proteome is considerably more complex than the genome, an observation that is more consonant with human complexity than that of the complexity of the human genes themselves.

2.3   Microarray technology applied to the study of alternative splicing and confirmation of exon prediction

Most microarray chips, either cDNA arrays or oligoarrays, are generally designed to detect gene expression without differentiating between the expression of different mRNA splicing isoforms because these chips cannot detect the differential expression patterns of isoforms present in a cell. These expression microarrays thus do not provide insight into an important molecular profile generated by alternative splicing variants.

However, the microarray can be tailored to evaluate some alternative spliced transcripts under certain circumstances. Technically, this is rather straightforward. For example, if a gene has two splicing variants, one of full length and one missing an exon, two probes, one representing a common exon sequence and the other missing exon sequence, can be deposited on the microarray. To determine the relative levels of the two variants in two cell populations, the relative hybridization intensities of the two spots can be compared. However, differentiating among different variants will be a daunting task if multiple variants with different exon skipping are present in cells. In this situation, the averaging effect of the mixture will be a hindrance to distinguishing individual forms.

The microarray, however, can also be useful in the determination of exons. One example is the "exon" and "tiling" arrays devised by Shoemaker *et al.* (2001) that were used to validate and refine computational gene predictions and define full-length transcripts. The exon arrays are composed of 50 to 60 base-long oligonucleotide probes derived from predicted exons. When these probes are hybridized with fluorescently labeled cDNAs from particular tissues, authentic exons can be identified, gene boundaries defined, and different exon assemblies aligned. "Tiling" arrays which achieve a higher-resolution view of the gene structure were designed by overlapping oligonucleotides to cover an entire genomic sequence of interest. Such an array can reveal exons not identified by algorithms and thereby refine the gene structures shown by the "exon" arrays, thus providing information about the alternative splicing of the interested genes.

3. Alternative Splicing – A Common Regulatory Mechanism of Gene Expression in Higher Eukaryotes

3.1 Discovery of alternative splicing and its regulatory patterns in gene expression

The concept of alternative splicing was born when Gilbert (1978) first proposed the question "why genes in pieces." Following that, two human genes, those encoding immunoglobulin (Choi *et al.*, 1980) and calcitonin/calcitonin gene-related peptide (Amara *et al.*, 1982), were identified to undergo alternative RNA splicing. Over the past 20 years, an increasing number of genes have been found to produce alternative splicing mRNA isoforms, with most of their variants identified by accident. However, the greatest recognition of the importance of alternative splicing came when the human genome sequencing project and its initial sequence analysis were completed as described previously. It is now known that alternative splicing events can be constitutive or regulated in response to developmental, physiological, or tissue-specific changes. These features make alternative splicing a powerful and versatile mechanism for strictly regulating gene expression, as well as for modulating the functions of their gene products.

One of the best-studied examples of the resultant impact of alternative RNA splicing is sex determination in *Drosophila*. Sex determination begins with a chromosome counting in the diploid fly embryo (female, XX; or male, XY). The key factors that implement sex determination once the decision is made are proteins that regulate alternative RNA splicing. For example, Sex-lethal (Sxl) and doublesex (dsx), two key players in sex identity, regulate a male- or female-specific RNA splicing of specific genes respectively, which leads to the creation a male or female fly. For a detailed description of the mechanisms involved in the development of sex identity in *Drosophila*, please review the papers of Lopez (1998) and Lou *et al.*(2001).

While it is clear that some type of alternative RNA splicing affects every cell type in the body, the central nervous system seems to stand apart in this regard. There are examples of tissue-specific alternative splicing, in which highly differentiated neuron cells produce thousands of alternative spliced mRNAs that are translated into diversified protein counterparts. These protein isoforms vary in their structures and functions, which leads to the fine-tuning of their physiological roles in learning and memory, neuron cell recognition, neurotransmission, ion channel function, and neurotransmitter receptor specificity. Grabowski *et al.* (1998 and 2001) have provided a comprehensive discussion of the alternative splicing events in the nervous system.

3.2   Different types of alternative RNA processing and their functional consequences

Genetic and biochemical approaches have identified that even simple alternative splicing events occur as a result of a complex interplay of *cis*-acting regulatory elements located in pre-mRNA sequences and *trans*-acting factors expressed in the cell. First, there exists a series of additional elements that modulate the selection of 5' and 3' splice sites, as well as polyadenylation sequences. These elements are functionally classified as either enhancers or silencers, although neither class has a specific consensus sequence associated with it. Lou and Gagel (2001) summarized these elements and some of their characteristics. Trans-acting factors involved in the regulation of alternative splicing are also typically classified into two groups based on the presence or absence of the C-terminal arginine-serine-rich domain. The SR group of proteins all have a highly charged C-terminal domain that is thought to be involved in protein/protein interactions to either recruit additional proteins to the RNA or to bridge gaps that exist between proteins bound at different RNA elements (splice sites or regulatory sequences). The non-SR proteins perform a variety of functions, from the specific binding of the RNA regulatory sequence to the enzymatic modification of regulatory proteins (Smith and Valcarcel, 2000; Wagner and Garcia-Blanco, 2001; Wollerton *et al.*, 2001; Kramer, 1996; Neubauer *et al.*, 1998; Hastings and Krainer, 2001). We are truly just beginning to uncover the hundreds of proteins involved in both constitutive and alternative RNA splicing.

No matter what the elements or factors are, the specific types of alternative processing events can be categorized into a relatively few types of alternative RNA splicing (Lou and Gagel, 2001): (1) the inclusion or exclusion of one or more exons; (2) 5'- or 3'-terminal exon changes, resulting in the alternative usage of promoters or alternative polyadenylation; (3) intron removal or retention; and (4) the alternative usage of splice sites. However, the true complexity of RNA processing is appreciated when it is recognized that many transcripts of multiple exons are involved and that sometimes more than one type of alternative RNA processing occurs in a single pre-mRNA.

As a result, the different types of alternative splicing have effects on the gene products. They can turn on or off gene expression by excluding or including stop codons (Bopp *et al.*, 1991), or they can change the structures or functions of proteins by inserting or deleting some amino acids (Vogan *et al.*, 1996; Schwarzbauer, 1991). They can even just alter the stability of the RNA by changing noncoding regions (Sureau *et al.*, 2001). In some cases, the functional changes in the proteins are directly correlated with human diseases (Epstein *et al.*, 1994; Bruening *et al.*, 1992; Goedert *et al.*, 2000; Le Corre *et al.*, 2000; Nagai *et al.*, 1998; Sandbrink *et al.*, 1993; Dredge *et al.*, 2001; Mayer *et al.*, 2000; Penninger *et al.*, 2001) and are frequently associated with human cancers (Bala and Peltomaki, 2001).

4.   Alternative Splicing in Cancer and its Clinical Values in Cancer Treatment

4.1   Dysregulation of alternative splicing in cancer

The genes encoding calcitonin and fibronectin were the first genes identified to have aberrant RNA splicing associated with tumorigenesis (Steenbergh *et al.*, 1984; Castellani *et al.*, 1986). Since then, more than forty genes have been identified that generate tumor-specific splicing iso-forms, allowing certain altered splicing products to be correlated with cancer development. The role that many of these aberrant products play in the induction or maintenance of a given tumor type is debatable; however, many of these proteins have defined roles in cell growth. The genes are functionally variable and include those that encode growth factors, receptors and kinases (e.g. fibroblast growth factor and its receptor (FGF/FGFR), vascular endothelial growth factor (VEGF), epidermal growth factor receptor (EGFR), estrogen receptor (ER), platelet-derived growth factor (PDGF), HER2) (Valve *et al.*, 2001; Neufeld *et al.*, 1999; Aigner *et al.*, 2001; Eley *et al.*, 1998; Pfeffer *et al.*, 1996); transcription factors (e.g. WT1) (Lee and Haber, 2001); membrane proteins (e.g. CD44) (Cooper, 1995; Zhao *et al.*, 1998); cell cycle genes (e.g., cyclin D1) (Bala and Peltomaki, 2001), and genes encoding proteins involved in apoptosis (e.g., Bcl-2) (Cote, *et al.*, 2001; Almog *et al.*, 2000).

Most impressive has been the study of the alternative splicing of apoptosis genes in cancer, from genes encoding death receptors (e.g. Fas, TRICK2), to genes encoding caspases (executive killers), to antiapoptotic genes (e.g. Bcl-2, Bax) (Jiang and Wu, 1999). The altered splicing in tumor cells can turn gene products from membrane-bound to soluble, from active to inactive/less active, or from preventing to promoting proteins. Examples are Fas and FasTMDel (Ruberti *et al.*, 1996), ICEα and ICEδ (Alnemri et al, 1995); and Bcl-x and Bcl-xS (Xerri *et al.*, 1998).

While changes in RNA splicing are clearly associated with many types of cancers, the specific underlying mechanisms responsible for these changes are less clear. In most cases investigators have merely reported the correlation of specific splicing variants with particular cancers. Point mutations of donor/acceptor splicing sites or enhancer/silencer elements of genes cause aberrant splicing in some cases (Lee and Haber, 2001; Scheffer *et al.*, 2000; Hartikainen *et al.*, 2000; Bromidge *et al.*, 2000; Cohen *et al.*, 1989). In fact, Krawczak *et al.* (1992) estimated that 15% of point mutations that result in human genetic disease create an RNA splicing defect. A specific example of point mutations that alter RNA splicing and lead to cancer formation are those involving the *p53* gene

(Varley, 1997; Frebourg, 1995; Jolly *et al.*, 1994; Avigad *et al.*, 1997). It is now clear that altered splicing leads to a loss of *p53* gene function.

How RNA splicing is altered in the absence of discernible mutations is less clear. It is thought that alterations in trans-acting factor composition are responsible, though few studies investigating this have been done. In a well- characterized mouse model of mammary tumorigenesis, it was shown that the ratio and levels of specific SR proteins are altered (Stickeler *et al.*, 1999). The spectrum of specific splicing events affected, however, was not examined. In tumor cell culture models, modulation of splicing factor expression has also been shown to alter cell growth. For example overexpression of the SR protein SC35 in cancer decreased the ratio of caspase-2S to caspase-2L transcripts, resulting in increased apoptosis. However, the overexpression of heterogeneous nuclear ribonucleoprotein (hnRNP) A1 generated opposite outcomes (Jiang *et al.*, 1998). Finally decreased expression of the polypyrimidine tract-binding (PTB) protein has been found in malignant glioblastoma multiforme and correlated with changes in the splicing of the receptor FGFR-1 (Jin *et al.*, 2000).

Even though we are only beginning to comprehend the mechanisms involved in the regulation of alternative RNA splicing in cancer, the strong correlation of altered splicing with cancer occurrence and/or progression has shown the potential value of alternative splicing in cancer diagnosis and treatment.

4.2 Clinical application of alternative splicing in cancer

The detection of tumor-specific splicing variants has been proposed as a means of cancer diagnosis or prognosis because of their tight correlation. Although extensive studies have not been conducted, the potential clinical values of some genes have been reported. For example, *KLK8* (neuropsin/ovasin-serine protease) type 3 and 4 variants have been suggested as an independent and favorable marker for ovarian cancer because *KLK8* types 3 and 4 variants encode secreted proteins that are easily detected in patients' serum (Magklara *et al.*, 2001). In addition, Cdc25B mRNA levels were significantly correlated with survival in colorectal carcinoma patients, implying the involvement of cdc25B in the progression of disease (Hernandez *et al.*, 2001). The absence of the CD44v6 variant in non-small cell lung cancer was also associated with significantly longer survival in patients, suggesting its prognostic value in patients with this cancer (Nguyen *et al.*, 2000). The presence of MDM2 short forms was found to be indicative of high-grade malignancy in patients with breast cancer (Hori et al., 2000). The challenge now is to develop accurate methods to monitor these differences, although newer methods of RNA amplification, clinical grade assays, and real-time quantitative polymerization chain reaction have already overcome a primary obstacle of obtaining sufficient tissue for RNA isolation and analysis.

Besides the use of alternative splicing in diagnosis there is the very real possibility of targeting therapies to specific alternative splicing events. Two general approaches have been discussed to date. Since these methods target tumor-specific splicing isoforms, they would selectively kill only cancer cells and not affect healthy cells. One experimental approach has been the use of antisense oligonucleotides to target specific splicing signals that are utilized to produce splicing variants in cancer (Mercatante and Kole, 2000; Nagel et al., 1993; Karras et al., 2000; Sierakowska et al., 2000). The Kole laboratory has now extended this approach to target the alternative 5' splice sites of Bcl-x in order to shift its splicing pattern from the antiapoptotic Bcl-xL isoform to the proapoptotic Bcl-xS isoform, then eventually increasing the apoptosis of tumor cells (Mercatante et al., 2001). This group has also applied the same strategy to block the aberrant splice sites of beta-globin genes and to restore the use of correct splice sites. An antisense oligonucleotide was also designed to selectively inhibit the tumor-specific expression of the human interleukin-5 receptor-alpha chain mRNA isoform and ultimately harm cancer cells (Karras et al., 2000).

The other broad therapeutic approach is to target the final aberrant protein products rather than the RNAs. For example, the high-affinity monoclonal antibody specific for CD44v6 was generated by Heider and his colleagues (1996) and found to be fast and selectively taken up by tumor cells, thus serving as a promising target for the immunotherapy of squamous cell carcinomas.

5.   Summary

Perhaps the greatest surprise that emerged from the sequencing of the human genome was just how few genes were found. Instead it appears that genetic and ultimately protein diversity is achieved through alternative RNA processing. Several examples of changes in the splicing of a single gene transcript that greatly affect cellular function have now been recognized. However, it is also clear that the regulated alternative splicing of gene families is likely to play a key role in cellular differentiation pathways, tissue development, and perhaps neoplastic transformation. Unfortunately, current oligo and cDNA arrays may be missing some of the important changes that occur in mRNA isoform levels. The challenge then is to develop better arrays that detect mRNA isoforms derived through alternative splicing. Therefore, it is now important to better understand the mechanisms involved in the regulation of splicing so that perhaps we can better predict when alternative splicing may occur. In the meantime, array analysis does provide an important tool for identifying those genes subject to alternative processing in order to begin compiling databases of mRNA isoforms.

References

Aigner, A. , Juhl, H. , Malerczyk, C. , Tkybusch, A. , Benz, C. C. , Czubayko, F. (2001) Expression of a truncated 100 kDa HER2 splice variant acts as an endogenous inhibitor of tumour cell proliferation. Oncogene, 20, 2101-11.

Almog, N. , Goldfinger, N. , Rotter, V. (2000) p53-dependent apoptosis is regulated by a C-terminal alternatively spliced form of murine p53. Oncogene, 19, 3395-403.

Alnemri, E. S. , Fernandes-Alnemri, T. , Litwack, G. (1995) Cloning and expression of four novel isoforms of human interleukin-1 beta converting enzyme with different apoptotic activities. J Biol Chem, 270, 4312-7.

Amara, S. G. , Jonas, V. , Rosenfeld, M. G. , Ong, E. S. , Evans, R. M. (1982) Alternative RNA processing in calcitonin gene expression generates mRNAs encoding different polypeptide products. Nature, 298, 240-4.

Amara, S. G. , Jonas, V. , Rosenfeld, M. G. , Ong, E. S. , Evans, R. M. (1982) Alternative RNA processing in calcitonin gene expression generates mRNAs encoding different polypeptide products. Nature, 298, 240-4.

Avigad, S. , Barel, D. , Blau, O. , Malka, A. , Zoldan, M. , Mor, C. , Fogel, M. , Cohen, I. J. , Stark, B. , Goshen, Y. , Stein, J. , Zaizov, R. (1997) A novel germ line p53 mutation in intron 6 in diverse childhood malignancies. Oncogene, 14, 1541-5.

Bala, S. , Peltomaki, P. (2001) CYCLIN D1 as a genetic modifier in hereditary nonpolyposis colorectal cancer. Cancer Res, 61, 6042-5.

Benson, G. (1999) Tandem repeats finder: a program to analyze DNA sequences. Nucleic Acids Res, 27, 573-80.

Berget, S.M. , Moore C. , Sharp P.A. (1977) Spliced segments at the 5' terminus of adenovirus 2 late mRNA. Proc Natl Acad Sci USA, 74, 3171-5.

Birney, E. , Durbin, R. (2000) Using GeneWise in the Drosophila annotation experiment. Genome Res, 10, 547-8.

Black, D. L. (2000) Protein diversity from alternative splicing: a challenge for bioinformatics and post-genome biology. Cell, 103, 367-70.

Bopp, D. , Bell, L. R. , Cline, T. W. , Schedl, P. (1991) Developmental distribution of female-specific Sex-lethal proteins in Drosophila melanogaster. Genes Dev, 5, 403-15.

Brett, D. , Hanke, J. , Lehmann, G. , Haase, S. , Delbruck, S. , Krueger, S. , Reich, J. , Bork, P. (2000) EST comparison indicates 38% of human mRNAs contain possible alternative splice forms. FEBS Lett, 474, 83-6.

Bromidge, T. , Lowe, C. , Prentice, A. , Johnson, S. (2000) p53 intronic point mutation, aberrant splicing and telomeric associations in a case of B-chronic lymphocytic leukaemia. Br J Haematol, 111, 223-9.

Bruening, W. , Bardeesy, N. , Silverman, B. L. , Cohn, R. A. , Machin, G. A. , Aronson, A. J. , Housman, D. , Pelletier, J. (1992) Germline intronic and exonic mutations in the Wilms' tumour gene (WT1) affecting urogenital development. Nat Genet, 1, 144-8.

Brunak, S. , Engelbrecht, J. , Knudsen, S. (1991) Prediction of human mRNA donor and acceptor sites from the DNA sequence. J Mol Biol, 220, 49-65.

Burset, M. , Guigo, R. (1996) Evaluation of gene structure prediction programs. Genomics, 34, 353-67.

Castellani, P. , Siri, A. , Rosellini, C. , Infusini, E. , Borsi, L. , Zardi, L. (1986) Transformed human cells release different fibronectin variants than do normal cells. J Cell Biol, 103, 1671-7.

Choi, E. , Kuehl, M. , Wall, R. (1980) RNA splicing generates a variant light chain from an aberrantly rearranged kappa gene. Nature, 286, 776-9.

Chow L.T. , Gelinas R.E. , Broker T.R. , Roberts R.J. (1977) An amazing sequence arrangement at the 5' ends of adenovirus 2 messenger RNA. Cell 12,1-8.

Claverie, J. M. (1997) Computational methods for the identification of genes in vertebrate genomic sequences. Hum Mol Genet, 6, 1735-44.

Claverie, J. M. (2001) Gene number. What if there are only 30,000 human genes? Science, 291, 1255-7.

Cohen, J. B. , Broz, S. D. , Levinson, A. D. (1989) Expression of the H-ras proto-oncogene is controlled by alternative splicing. Cell, 58, 461-72.

Cooper, D. L. (1995) Retention of CD44 introns in bladder cancer: understanding the alternative splicing of pre-mRNA opens new insights into the pathogenesis of human cancers. J Pathol, 177, 1-3.

Cooper, T. A. , Mattox, W. (1997) The regulation of splice-site selection, and its role in human disease. Am J Hum Genet, 61, 259-66.

Cote, J. , Dupuis, S. , Jiang, Z. , Wu, J. Y. (2001) Caspase-2 pre-mRNA alternative splicing: identification of an intronic element containing a decoy 3' acceptor site. Proc Natl Acad Sci U S A, 98, 938-43.

Croft, L. , Schandorff, S. , Clark, F. , Burrage, K. , Arctander, P. , Mattick, J. S. (2000) ISIS, the intron information system, reveals the high frequency of alternative splicing in the human genome. Nat Genet, 24, 340-1.

Dredge, B. K. , Polydorides, A. D. , Darnell, R. B. (2001) The splice of life: alternative splicing and neurological disease. Nat Rev Neurosci, 2, 43-50.

Eley, G. , Frederick, L. , Wang, X. Y. , Smith, D. I. , James, C. D. (1998) 3' end structure and rearrangements of EGFR in glioblastomas. Genes Chromosomes Cancer, 23, 248-54.

Epstein, J. A. , Glaser, T. , Cai, J. , Jepeal, L. , Walton, D. S. , Maas, R. L. (1994) Two independent and interactive DNA-binding subdomains of the Pax6 paired domain are regulated by alternative splicing. Genes Dev, 8, 2022-34.

Fickett, J. W. (1982) Recognition of protein coding regions in DNA sequences. Nucleic Acids Res, 10, 5303-18.

Fickett, J. W. , Tung, C. S. (1992) Assessment of protein coding measures. Nucleic Acids Res, 20, 6441-50.

Fire, A. , Xu, S. , Montgomery, M. K. , Kostas, S. A. , Driver, S. E. , Mello, C. C. (1998) Potent and specific genetic interference by double-stranded RNA in Caenorhabditis elegans. Nature, 391, 806-11.

Fraser, A. G. , Kamath, R. S. , Zipperlen, P. , Martinez-Campos, M. , Sohrmann, M. , Ahringer, J. (2000) Functional genomic analysis of C. elegans chromosome I by systematic RNA interference. Nature, 408, 325-30.

Frebourg, T. , Barbier, N. , Yan, Y. X. , Garber, J. E. , Dreyfus, M. , Fraumeni J Jr, Li, F. P. , Friend, S. H. (1995) Germ-line p53 mutations in 15 families with Li-Fraumeni syndrome. Am J Hum Genet, 56, 608-15.

Frech, K. , Dietze, P. , Werner, T. (1997) ConsInspector 3.0: new library and enhanced functionality. Comput Appl Biosci, 13, 109-10.

Gaasterland, T. , Sensen, C. W. (1996) MAGPIE: automated genome interpretation. Trends Genet, 12, 76-8.

Galas, D. J. (2001) Sequence interpretation. Making sense of the sequence. Science, 291, 1257-60.

Gelfand, M. S. , Dubchak, I. , Dralyuk, I. , Zorn, M. (1999) ASDB: database of alternatively spliced genes. Nucleic Acids Res, 27, 301-2.

Gelfand, M. S. , Mironov, A. A. , Pevzner, P. A. (1996) Gene recognition via spliced sequence alignment. Proc Natl Acad Sci U S A, 93, 9061-6.

Gilbert, W. (1978) Why genes in pieces? Nature, 271, 501.

Goedert, M. , Ghetti, B. , Spillantini, M. G. (2000) Tau gene mutations in frontotemporal dementia and parkinsonism linked to chromosome 17 (FTDP-17). Their relevance for understanding the neurogenerative process. Ann N Y Acad Sci, 920, 74-83.

Gonczy, P. , Echeverri, G. , Oegema, K. , Coulson, A. , Jones, S. J. , Copley, R. R. , Duperon, J. , Oegema, J. , Brehm, M. , Cassin, E. , Hannak, E. , Kirkham, M. , Pichler, S. , Flohrs, K. , Goessen, A. , Leidel, S. , Alleaume, A. M. , Martin, C. , Ozlu, N. , Bork, P. , Hyman, A. A. (2000) Functional genomic analysis of cell division in C. elegans using RNAi of genes on chromosome III. Nature, 408, 331-6.

Grabowski, P. J. (1998) Splicing regulation in neurons: tinkering with cell-specific control. Cell, 92, 709-12.

Grabowski, P. J. , Black, D. L. (2001) Alternative RNA splicing in the nervous system. Prog Neurobiol, 65, 289-308.

Graveley, B. R. (2001) Alternative splicing: increasing diversity in the proteomic world. Trends Genet, 17, 100-7.

Guigo, R. , Knudsen, S. , Drake, N. , Smith, T. (1992) Prediction of gene structure. J Mol Biol, 226, 141-57.

Hanke, J. , Brett, D. , Zastrow, I. , Aydin, A. , Delbruck, S. , Lehmann, G. , Luft, F. , Reich, J. , Bork, P. (1999) Alternative splicing of human genes: more the rule than the exception? Trends Genet, 15, 389-90.

Hartikainen, J. M. , Pirskanen, M. M. , Arffman, A. H. , Ristonmaa, U. K. , Mannermaa, A. J. (2000) A Finnish BRCA1 exon 12 4216-2nt A to G splice acceptor site mutation causes aberrant splicing and frameshift, leading to protein truncation. Hum Mutat, 15, 120.

Hastings, M. L. , Krainer, A. R. (2001) Pre-mRNA splicing in the new millennium. Curr Opin Cell Biol, 13, 302-9.

Heider, K. H. , Sproll, M. , Susani, S. , Patzelt, E. , Beaumier, P. , Ostermann, E. , Ahorn, H. , Adolf, G. R. (1996) Characterization of a high-affinity monoclonal antibody specific for CD44v6 as candidate for immunotherapy of squamous cell carcinomas. Cancer Immunol Immunother, 43, 245-53.

Henikoff, J. G. , Henikoff, S. (2000) Drosophila genomic sequence annotation using the BLOCKS+ database. Genome Res, 10, 543-6.

Hernandez, S. , Bessa, X. , Bea, S. , Hernandez, L. , Nadal, A. , Mallofre, C. , Muntane, J. , Castells, A. , Fernandez, P. L. , Cardesa, A. , Campo, E. (2001) Differential expression of cdc25 cell-cycle-activating phosphatases in human colorectal carcinoma. Lab Invest, 81, 465-73.

Hori, M. , Shimazaki, J. , Inagawa, S. , Itabashi, M. , Hori, M. (2000) Alternatively spliced MDM2 transcripts in human breast cancer in relation to tumor necrosis and lymph node involvement. Pathol Int, 50, 786-92.

International Human Genome Sequencing Consortium. (2001) Initial sequencing and analysis of the human genome. Nature, 409, 860-921.

Ji, H. , Zhou, Q. , Wen, F. , Xia, H. , Lu, X. , Li, Y. (2001) AsMamDB: an alternative splice database of mammals. Nucleic Acids Res, 29, 260-3.

Jiang, Z. H. , Wu, J. Y. (1999) Alternative splicing and programmed cell death. Proc Soc Exp Biol Med, 220, 64-72.

Jiang, Z. H. , Zhang, W. J. , Rao, Y. , Wu, J. Y. (1998) Regulation of Ich-1 pre-mRNA alternative splicing and apoptosis by mammalian splicing factors. Proc Natl Acad Sci U S A, 95, 9155-60.

Jin, W. , McCutcheon, I. E. , Fuller, G. N. , Huang, E. S. , Cote, G. J. (2000) Fibroblast growth factor receptor-1 alpha-exon exclusion and polypyrimidine tract-binding protein in glioblastoma multiforme tumors. Cancer Res, 60, 1221-4.

Jolly, K. W. , Malkin, D. , Douglass, E. C. , Brown, T. F. , Sinclair, A. E. , Look, A. T. (1994) Splice-site mutation of the p53 gene in a family with hereditary breast-ovarian cancer. Oncogene, 9, 97-102.

Karras, J. G. , McKay, R. A. , Lu, T. , Dean, N. M. , Monia, B. P. (2000) Antisense inhibition of membrane-bound human interleukin-5 receptor-alpha chain does not affect soluble receptor expression and induces apoptosis in TF-1 cells. Antisense Nucleic Acid Drug Dev, 10, 347-57.

Kramer, A. (1996) The structure and function of proteins involved in mammalian pre-mRNA splicing. Annu Rev Biochem, 65, 367-409.

Krawczak, M. , Reiss, J. , Cooper, D. N. (1992) The mutational spectrum of single base-pair substitutions in mRNA splice junctions of human genes: causes and consequences. Hum Genet, 90, 41-54.

Krogh, A. (1997) Two methods for improving performance of an HMM and their application for gene finding. Proc Int Conf Intell Syst Mol Biol, 5, 179-86.

Le Corre, S. , Harper, C. G. , Lopez, P. , Ward, P. , Catts, S. (2000) Increased levels of expression of an NMDARI splice variant in the superior temporal gyrus in schizophrenia. Neuroreport, 11, 983-6.

Lee, S. B. , Haber, D. A. (2001) Wilms tumor and the WT1 gene. Exp Cell Res, 264, 74-99.

Lopez, A. J. (1998) Alternative splicing of pre-mRNA: developmental consequences and mechanisms of regulation. Annu Rev Genet, 32, 279-305.

Lou, H. , Gagel, R. F. (2001) Alternative ribonucleic acid processing in endocrine systems. Endocr Rev, 22, 205-25.

Magklara, A. , Scorilas, A. , Katsaros, D. , Massobrio, M. , Yousef, G. M. , Fracchioli, S. , Danese, S. , Diamandis, E. P. (2001) The human KLK8 (neuropsin/ovasin) gene: identification of two novel splice variants and its prognostic value in ovarian cancer. Clin Cancer Res, 7, 806-11.

Mayer, K. , Ballhausen, W. , Leistner, W. , Rott, H. (2000) Three novel types of splicing aberrations in the tuberous sclerosis TSC2 gene caused by mutations apart from splice consensus sequences. Biochim Biophys Acta, 1502, 495-507.

Mercatante, D., Kole, R. (2000) Modification of alternative splicing pathways as a potential approach to chemotherapy. Pharmacol Ther, 85, 237-43.

Mercatante, D. R. , Bortner, C. D. , Cidlowski, J. A. , Kole, R. (2001) Modification of alternative splicing of Bcl-x pre-mRNA in prostate and breast cancer cells. Analysis of apoptosis and cell death. J Biol Chem, 276, 16411-7.

Mironov, A. A. , Fickett, J. W. , Gelfand, M. S. (1999) Frequent alternative splicing of human genes. Genome Res, 9, 1288-93.

Mironov, A. A. , Roytberg, M. A. , Pevzner, P. A. , Gelfand, M. S. (1998) Performance-guarantee gene predictions via spliced alignment. Genomics, 51, 332-9.

Modrek, B. , Resch, A. , Grasso, C. , Lee, C. (2001) Genome-wide detection of alternative splicing in expressed sequences of human genes. Nucleic Acids Res, 29, 2850-9.

Mott, R. (1997) EST_GENOME: a program to align spliced DNA sequences to unspliced genomic DNA. Comput Appl Biosci, 13, 477-8.

Nagai, M. , Abe, K. , Okamoto, K. , Itoyama, Y. (1998) Identification of alternative splicing forms of GLT-1 mRNA in the spinal cord of amyotrophic lateral sclerosis patients. Neurosci Lett, 244, 165-8.

Nagel, K. M. , Holstad, S. G. , Isenberg, K. E. (1993) Oligonucleotide pharmacotherapy: an antigene strategy. Pharmacotherapy, 13, 177-88.

Neubauer, G. , King, A. , Rappsilber, J. , Calvio, C. , Watson, M. , Ajuh, P. , Sleeman, J. , Lamond, A. , Mann, M. (1998) Mass spectrometry and EST-database searching allows characterization of the multi-protein spliceosome complex. Nat Genet, 20, 46-50.

Neufeld, G. , Cohen, T. , Gengrinovitch, S. , Poltorak, Z. (1999) Vascular endothelial growth factor (VEGF) and its receptors. FASEB J, 13, 9-22.

Nguyen, V. N. , Mirejovsky, T. , Melinova, L. , Mandys, V. (2000) CD44 and its v6 spliced variant in lung carcinomas: relation to NCAM, CEA, EMA and UP1 and prognostic significance. Neoplasma, 47, 400-8.

Ohler, U. , Niemann, H. , Liao Gc, Rubin, G. M. (2001) Joint modeling of DNA sequence and physical properties to improve eukaryotic promoter recognition. Bioinformatics, 17 Suppl 1, S199-206.

Parra, G. , Blanco, E. , Guigo, R. (2000) GeneID in Drosophila. Genome Res, 10, 511-5.

Penninger, J. M. , Irie-Sasaki, J. , Sasaki, T. , Oliveira-dos-Santos, A. J. (2001) CD45: new jobs for an old acquaintance. Nat Immunol, 2, 389-96.

Pfeffer, U. , Fecarotta, E. , Arena, G. , Forlani, A. , Vidali, G. (1996) Alternative splicing of the estrogen receptor primary transcript normally occurs in estrogen receptor positive tissues and cell lines. J Steroid Biochem Mol Biol, 56, 99-105.

Reese, M. G. , Hartzell, G. , Harris, N. L. , Ohler, U. , Abril, J. F. , Lewis, S. E. (2000) Genome annotation assessment in Drosophila melanogaster. Genome Res, 10, 483-501.

Reese, M. G. , Kulp, D. , Tammana, H. , Haussler, D. (2000) Genie–gene finding in Drosophila melanogaster. Genome Res, 10, 529-38.

Ruberti, G. , Cascino, I. , Papoff, G. , Eramo, A. (1996) Fas splicing variants and their effect on apoptosis. Adv Exp Med Biol, 406, 125-34.

Sandbrink, R. , Banati, R. , Masters, C. L. , Beyreuther, K. , Konig, G. (1993) Expression of L-APP mRNA in brain cells. Ann N Y Acad Sci, 695, 183-9.

Scheffer, H. , Van Der Vlies, P. , Burton, M. , Verlind, E. , Moll, A. C. , Imhof, S. M. , Buys, C. H. (2000) Two novel germline mutations of the retinoblastoma gene (RB1) that show incomplete penetrance, one splice site and one missense. J Med Genet, 37, E6.

Schwarzbauer, J. E. (1991) Alternative splicing of fibronectin: three variants, three functions. Bioessays, 13, 527-33.

Sharp, P. A. (1994) Split genes and RNA splicing. Cell, 77, 805-15.

Shoemaker, D. D. , Schadt, E. E. , Armour, C. D. , He, Y. D. , Garrett-Engele, P. , McDonagh, P. D. , Loerch, P. M. , Leonardson, A. , Lum, P. Y. , Cavet, G. , Wu, L. F. , Altschuler, S. J. , Edwards, S. , King, J. , Tsang, J. S. , Schimmack, G. , Schelter, J. M. , Koch, J. , Ziman, M. , Marton, M. J. , Li, B. , Cundiff, P. , Ward, T. , Castle, J. , Krolewski, M. , Meyer, M. R. , Mao, M. , Burchard, J. , Kidd, M. J. , Dai, H. , Phillips, J. W. , Linsley, P. S. , Stoughton, R. , Scherer, S. , Boguski, M. S. (2001) Experimental annotation of the human genome using microarray technology. Nature, 409, 922-7.

Sierakowska, H. , Agrawal, S. , Kole, R. (2000) Antisense oligonucleotides as modulators of pre-mRNA splicing. Methods Mol Biol, 133, 223-33.

Smith, C. W. , Valcarcel, J. (2000) Alternative pre-mRNA splicing: the logic of combinatorial control. Trends Biochem Sci, 25, 381-8.

Solovyev, V. V. , Salamov, A. A. , Lawrence, C. B. (1995) Identification of human gene structure using linear discriminant functions and dynamic programming. Proc Int Conf Intell Syst Mol Biol, 3, 367-75.

Staden, R. , McLachlan, A. D. (1982) Codon preference and its use in identifying protein coding regions in long DNA sequences. Nucleic Acids Res, 10, 141-56.

Stamm, S. , Zhu, J. , Nakai, K. , Stoilov, P. , Stoss, O. , Zhang, M. Q. (2000) An alternative-exon database and its statistical analysis. DNA Cell Biol, 19, 739-56.

Steenbergh, P. H. , Hoppener, J. W. , Zandberg, J. , Van de Ven, W. J. , Jansz, H. S. , Lips, C. J. (1984) Calcitonin gene related peptide coding sequence is conserved in the human genome and is expressed in medullary thyroid carcinoma. J Clin Endocrinol Metab, 59, 358-60.

Stickeler, E. , Kittrell, F. , Medina, D. , Berget, S. M. (1999) Stage-specific changes in SR splicing factors and alternative splicing in mammary tumorigenesis. Oncogene, 18, 3574-82.

Stormo, G. D. (2000) Gene-finding approaches for eukaryotes. Genome Res, 10, 394-7.

Sureau, A. , Gattoni, R. , Dooghe Y. , Stevenin J. , Soret J. (2001) SC35 autoregulates its expression by promoting splicing events that destabilize its mRNAs. EMBO J 20,1785-96.

Thanaraj, T. A. (1999) A clean data set of EST-confirmed splice sites from Homo sapiens and standards for clean-up procedures. Nucleic Acids Res, 27, 2627-37.

Uberbacher, E. C. , Mural, R. J. (1991) Locating protein-coding regions in human DNA sequences by a multiple sensor-neural network approach. Proc Natl Acad Sci U S A, 88, 11261-5.

Uberbacher, E. C. , Xu, Y. , Mural, R. J. (1996) Discovering and understanding genes in human DNA sequence using GRAIL. Methods Enzymol, 266, 259-81.

Valve, E. M. , Nevalainen, M. T. , Nurmi, M. J. , Laato, M. K. , Martikainen, P. M. , Harkonen, P. L. (2001) Increased expression of FGF-8 isoforms and FGF receptors in human premalignant prostatic intraepithelial neoplasia lesions and prostate cancer. Lab Invest, 81, 815-26.

Varley, J. M. , McGown, G. , Thorncroft, M. , Santibanez-Koref, M. F. , Kelsey, A. M. , Tricker, K. J. , Evans, D. G. , Birch, J. M. (1997) Germ-line mutations of TP53 in Li-Fraumeni families: an extended study of 39 families. Cancer Res, 57, 3245-52.

Vogan, K. J. , Underhill, D. A. , Gros, P. (1996) An alternative splicing event in the Pax-3 paired domain identifies the linker region as a key determinant of paired domain DNA-binding activity. Mol Cell Biol, 16, 6677-86.

Wagner, E. J. , Garcia-Blanco, M. A. (2001) Polypyrimidine tract binding protein antagonizes exon definition. Mol Cell Biol, 21, 3281-8.

Woerfel, G. , Bindereif, A. (2001) In vitro selection of exonic splicing enhancer sequences: identification of novel CD44 enhancers. Nucleic Acids Res, 29, 3204-11.

Wollerton, M. C. , Gooding, C. , Robinson, F. , Brown, E. C. , Jackson, R. J. , Smith, C. W. (2001) Differential alternative splicing activity of isoforms of polypyrimidine tract binding protein (PTB). RNA, 7, 819-32.

Xerri, L. , Hassoun, J. , Devilard, E. , Birnbaum, D. , Birg, F. (1998) BCL-X and the apoptotic machinery of lymphoma cells. Leuk Lymphoma, 28, 451-8.

Zhao, H. , Fang, J. , Du, G. (1998) [Expression of standard and variant CD44 in human lung cancer] Zhonghua Jie He He Hu Xi Za Zhi, 21, 541-3.

Chapter 17

SINGLE-NUCLEOTIDE POLYMORPHISMS, DNA REPAIR, AND CANCER

Qingyi Wei[1], Erich M. Sturgis[1], Margaret R. Spitz[1],
Harvey W. Mohrenweiser[3], Ilya Shmulevich[2], Shouming Kong[2], David Cogdell[2], Qing Mi[2], and Wei Zhang[2]

[1] *Department of Epidemiology*

[2] *Department of Pathology*

University of Texas M. D. Anderson Cancer Center, Houston, Texas, USA

[3] *Biology and Biotechnology Research Program, Lawrence Livermore National Laboratory, Livermore, California, USA*

1. Introduction

Cancer develops in only a small fraction of individuals exposed to environmental carcinogens, suggesting that there are differences in individual susceptibility to carcinogenesis and the impact of gene-environment interactions. Following exposure to carcinogens, the agents undergo a series of metabolic activation and detoxification steps that determine the internal dose of exposure and the amount of DNA damage incurred. However, both endogenous and exogenous carcinogens or genotoxic agents also cause cell-cycle delays that allow cells to repair DNA damage, and this cellular DNA repair capacity (DRC) is central to maintaining genomic integrity and normal cellular functions. Studies have shown that polymorphisms of genes, including the single-nucleotide polymorphism (SNP), that control the metabolism of various chemical agents and DNA repair may contribute to variations in carcinogenesis in the general population. Thus, identification of the SNPs, singly or in combination, that are associated with DNA repair would have an important value in cancer risk prediction, prevention, and therapy. In this chapter, we will review the DNA repair system and describe some known examples of SNP involvement in cancer. We will also describe the complexity of the SNP data and the resultant need for a SNP database and computational models for SNP analysis.

2.   Overview of DNA repair systems/pathways

2.1   DNA Damage

DNA is routinely assaulted by intracellular metabolites and environmental agents (Bartsch *et al.*, 2000; Nair *et al.*, 2001) that cause a wide range of damage (Friedberg *et al.*, 1995; Garner, 1998; Povey, 2000). Reactive oxygen species (e.g., superoxide and hydroxyl radicals) are one such agent. They are the by-products of normal cellular metabolism, but they constantly attack DNA, producing modified bases, oxidized apurinic sites, and DNA strand breaks. It is estimated that several thousand lesions are generated per cell per day, even in the absence of exposure to exogenous mutagens (Lindahl, 1993, Collins, 1999). A number of anticancer agents, radiomimetic antibiotics, and environmental mutagens also generate free radicals and induce this class of DNA damage (Collins, 1999). Free radical-induced damage is the most common class of DNA damage (Collins, 1999; Marnatt, 2000).

The damage induced by ionizing radiation (IR) is particularly complex. The direct action of IR results in the formation of single-strand breaks (SSB), double-strand breaks (DSB), and DNA-protein cross-links. In particular, one Sievert of radiation exposure induces approximately 1000 SSBs and 40 DSBs per cell (Ward, 1995). IR exposure also increases the cellular level of reactive oxygen species, which are formed mainly through the radiolysis of water, and this elevates oxidative damage. Exposure to ultraviolet (UV) light produces photoproducts and cyclobutane pyrimidine dimers (Cleaver *et al.*, 2001; Friedberg, 2001). Recent evidence suggests that UV exposure also generates reactive oxygen species in cells and increases the levels of oxidative damage in the DNA (Cooke *et al.*, 2000). Similarly, a range of covalently attached DNA adducts are found in cells exposed to different chemical mutagens including environmental and lifestyle exposures and pharmaceutical agents (Garner, 1998; Airoldi *et al.*, 1998). Many of these DNA adducts can also be found in normal tissues from ostensibly unexposed individuals (Gupta and Lutz, 1999; Povey, 2000).

Even in the absence of obvious exposures to mutagens, cells must constantly deal with the existence of a wide range of premutagenic lesions in their DNA. For examples, potential mutagenic events can also occur during replication of the DNA in preparation for cell division. Different polymerases replicate DNA with different degrees of fidelity (Friedberg *et al.*, 2001; Pavlov *et al.*, 2001), with the result that all polymerases incorporate some level of inappropriate nucleotides. Another source of potentially harmful damage is DNA strand breaks that arise spontaneously during DNA replication. Damage that is not properly repaired prior to cell division can be either cytotoxic, resulting in cell death, or mutagenic, resulting in DNA sequence changes in the daughter cell (Evan and Vousden, 2001). Sequence changes or chromosomal aberrations transmitted to daughter cells can result in the generation of a precancerous somatic cell. A number of genes monitor the DNA for damage and ensure that DNA replication is delayed until the DNA repair proteins can restore the integrity of the sequence and the genome (Weinert, 1997; Orr-Weaver and Weinberg, 1998; Fearon and Dang, 1999; Pfieffer *et al.*, 2000; Khanna and Jackson, 2001).

2.2   DNA Repair

Organisms have developed several generally nonredundant pathways (Table 17.1) for repairing the different classes of DNA lesions resulting from exposures to mutagens, thereby prevent the subsequent threats to genomic integrity (Sancer, 1995). These pathways involve the interaction and sequential activity of multiple proteins (Wood *et al.*, 2001; Romen and Glickman, 2001; Hoeijmakers, 2001). The DNA repair capacity has even been developed so that it can counter the damaging effects of the natural environment. An example is the human skin repair capacity. However, because it can just meet the repair demand posed by sunlight exposure at midday, overloaded DNA damage can lead to either cell death or mutants such as cancerous cells that escape the repair systems. Of further concern in this regard, emerging data, as recently reviewed by Berwick and Vineis (2000), demonstrate that a reduced DNA repair capacity in the general population is associated with a greater risk of cancer including the skin (Wei *et al.*, 1993; 1996; Cheng *et al.*, 1998; Wei *et al.*, 2000).

3.   SNPs of DNA repair genes

Knowledge of the sequence of the human genome has made it increasingly feasible to screen for DNA sequence variations among individuals in the human population (Wang *et al.*, 1998). The largest effort to identify SNPs is the SNP Consortium initiative (Sachidanandam *et al.*, 2001), which has identified 1.4 million SNPs, most from resequencing a large number of genomic regions in a set of 24 DNA samples. These data are very useful for constructing the very high density genetic map required to construct haplotypes and also for mapping genes with low penetrance and complex diseases.

Type	Genes Involved
Base-excision repair	At least 25 genes: DNA ligase (*LIG1, LIG3*), DNA glycosylase (*hOGG1, hMYH, hNTG1, AAG, TDG, UDG1*), *NTH1, APE1, POLB, XRCC1, FEN1, RPA1-3, PCNA, PARP, POLE1-5, POLD1, POLD2, NUDT1, NTHL1, PCNA, RFC*
Nucleotide-excision repair	>30 genes: *XPA, XPC, ERCC1, ERCC2, ERCC3, ERCC4, ERCC5, CSB/ERCC6, RPA1-3, RFA1-5, DDB1, DDB2, CNK1, CNK2, REV1, REV7, REV3L, HHR23A, HHR23B, hSSB, p34, p38, p41, p44, p62, hssL1, PCNA,* DNA polymerase (*POLD1, POLD2, POLE1-4, POLH*), DNA ligase (*LIG1*)
Mismatch repair	At least 6 genes: *hMLH1, hPMS1, hPMS2, hMSH2, hMSH3, hMSH6/hGTBP*
Double-strand break repair:	
Nonhomologous end-joining	>15 Genes: *DNA-PKcs, KU70 (XRCC6), KU86 (XRCC5), MRE11, RAD50, NBS1, XRCC4, LIG4*
Homologous recombination	>15 Genes: *HsRAD51, HsRAD52, HsRAD54, ERCC1, XPF, XRCC1, XRCC2, XRCC3, XRCC4, XRCC7, XRCC8, RAD51B, RAD51C, RAD51D*
Damage-recognition checkpoint	>30 genes: *RAD1, RAD9, RAD17, RAD24, XRCC9, HUS1, ATM, ATR, BRCA1, BRCA2, P53, MDM2*

Table 17.1.   Genes in Human DNA Repair Pathways

3.1   SNP Detection

Variation within or adjacent to the coding regions or exons of genes is expected to have a higher potential to cause functionally relevant sequence variants than in other regions, such as introns and intergenic regions. Therefore, several groups have screened the two major publicly available sequence databases in a search for sequence variants, described as *in silico* efforts. The Expressed Sequence Tag databases have been similarly screened in an effort to identify amino acid substitution variants (Buetow *et al.*, 1999; Clifford *et al.*, 2000; Picoult-Newberg *et al.*, 1999; Garg *et al.*, 1999). These data-mining strategies were also employed in the SNP Consortium effort (Sachidanandam *et al.*, 2001). Most of the DNA repair genes are expressed in a low copy number and the repair genes are not well represented in these databases. Only the most common SNPs are being identified and only a limited number of SNPs result in amino acid substitutions.

Methods used to search for SNPs in the genome range from experimental to informatic. Experimental protocols are targeted at specific stretches of DNA. Conformation-based techniques, coupled to chemical or enzymatic methods, and high-performance liquid chromatography (HPLC) detect the heterozygosity of the samples through the resulting structural change. These techniques, which are designed to detect SNPs, are followed by sequencing to confirm the specific location of the SNP (Shi, 2001).

To discriminate alleles at known SNP positions generally requires amplification of the region of interest. After this step, analysis by restriction fragment length polymorphism (RFLP) analysis, fluorescent probe assays, direct sequencing, chip technologies, or primer extension techniques can score the allele (Kwok, 2000; Shi, 2001; Halushka, 1999). However, the need to amplify the target region in the genome has been problematic due to time, cost, and sample limitations. More directly, novel restriction endonucleases coupled to fluorescent detection systems, known as the Invader Assay, have allowed the scoring of SNP positions in unamplified samples (Loyamichev, 1999). Speeding up the process are chip-based techniques that can score hundreds of positions simultaneously while samples from affected individuals are being pooled, which allows hundreds of samples to be screened in parallel. This can compare patients with cancer to unaffected individuals, leading to rapid identification of SNPs related to desired phenotypes (Chicurel, 2001).

A number of efforts have focused on direct resequencing of the exons of genes, usually exons of genes expected to be relevant to diseases, such as cardiovascular disease, asthma, hypertension, rheumatoid arthritis, for SNPs resulting in amino acid substitutions (Cambien *et al.*, 1999; Halushka *et al.*, 1999; Cargill *et al.*, 1999; Ohnishi *et al.*, 2000; Yamada *et al.*, 2000). The data from these screens have provided a catalog of variants that can serve as reagents for biochemical and molecular epidemiology studies that directly examine the relationship between specific genetic sequences and cancer risk. This has been described as a "genotype to phenotype" approach to molecular epidemiology studies (Mohrenweiser and Jones, 1998).

Of specific interest for this chapter have been the screens for variations in the DNA repair genes. The laboratories of H. Mohrenweiser at Lawrence Livermore National Laboratory (Shen *et al.* 1998) and M. Olson at The University of Washington (personal communication) have initiated systematic screens for common SNPs in many DNA repair genes and repair-related pathways. The strategy for identifying variants in both of these efforts has been to directly sequence the polymerase chain reaction (PCR) fragments following amplification of the exons and adjacent introns or untranslated regions of a gene from genomic DNA. The Mohrenweiser laboratory has screened over 35 DNA

Pathway	No. of Genes in Pathway*	No. of Genes Screened	No. of Variants Identified	No. of Alleles with >2% frequency
Base-excision repair	25	20	73	16
Nucleotide-excision repair	33	16	51	17
Double-strand break repair	∼ 30	14	34	10
Mismatch repair	7	6	41	13
Damage recognition and cell-cycle checkpoint	∼ 30	11	65	12
Total	> 110	60	235	63

Table 17.2. Amino Acid Substitution Variants in Genes in Different DNA Repair Pathways (*Genes with roles in more than one pathway are counted in all relevant pathways. The total consists of unique genes and variants and is not the sum of the column.)

repair or repair-related genes thus far, most in a set of 92 samples from the DNA Polymorphism Discovery Resource available from the Cornell Institute for Medical Research. The samples in this NIH resource (Collins *et al.*, 1998) are from U. S. residents selected to represent the major ethnic groups of the population. The individuals in this sample set are from population groups in the following ratio: European-American 2; African-American 2; Mexican-American 1; Native-American 1; Asian-American 2. Data regarding the variants identified are available at <http://greengenes.llnl.gov/dpublic/secure/reseq /reseq_home_page.html>.

In a similar effort initiated by the laboratory of Dr. Maynard Olsen at The University of Washington, some 25 repair and repair-related genes have been screened for variations, including 15 genes not screened by the Lawrence Livermore group in 45 to 450 individuals from the same sample set. Their data are available at <http://www.genome.utah.edu/genesnps>. Others have reported results from screens of these and additional repair genes in other sample sets (Broughton *et al.*, 1996; Fan *et al.*, 1999; Bell *et al.*, 2000; Kato *et al.*, 2000; Ma *et al.*, 2000, Passarino *et al.*, 2001; Wagner *et al.*, 1999; Thorstenson *et al.*, 2001). The amino acid substitution data obtained in these screens are organized by repair pathway in Table 17.2. The data from gene-specific databases, generated in the course of characterizing cancer genes and cancer families, have not been included in this summary as it is often difficult to ascertain whether the variant screens in these data are complete. The degree to which the individuals screened are representative of the general population in many of these latter databases is also often difficult to discern.

The objective of these SNP discovery projects is just to create a catalog of SNPs. It should be remembered that the estimated allele frequencies are based on data from only a small number of individuals. It is also not possible to account for potential differences in allele frequencies among ethnic groups when screening the samples from the DNA Polymorphism Discovery Resource. The frequency of some of the variants could be high within specific subpopulations. Thus, the estimated allele frequencies should never be used as a reference for comparing allele frequencies among groups.

Individuals with homozygous variant alleles and individuals with multiple amino acid substitutions in a gene are also observed, as are individuals with variants in multiple genes in a repair pathway. However, extrapolating from the current estimate of the number of different variants per gene and the average variant allele frequency (3-4%), the "typical" individual should have 5 to 6 variants among the 25 genes of the BER pathways and 6 to 7 variants among the genes of the DSBR and damage recognition (DRC) pathways. Most individuals in the population are expected to have variant subunits for several proteins that are components of multimeric complexes or sequential steps of each repair pathway. Thus, the genotype for each individual is very complex, considering the genes for a repair pathway or the genes responsible for repairing a class of DNA damage. The challenge therefore is to develop metrics of repair pathway genotypes that emulate repair capacity phenotypes, which are an integrated measure or sum of the impact of the variants in the genes of a pathway.

Although the significance of these variants is largely unknown, the implication is that variants that cause amino acid substitutions affect the function of the proteins and therefore the efficiency of DNA repair. However, those variants that do not cause an amino acid change may also affect DNA repair function because they may lie in introns that regulate splicing, may cause mRNA instability, or may be linked to genetic changes in other unknown genes. Therefore, it is important to understand the impact of these polymorphisms on disease to further understand their functional relevance.

3.2 Selected SNPs of DNA repair genes associated with cancer

XRCC1. The x-ray repair cross-complementing group 1 (*XRCC1*) gene is located on chromosome 19q13.2 (Shannon and Stubbs, 1998), and several polymorphisms of *XRCC1* were recently identified (Shen *et al.*, 1998). These include polymorphisms resulting in a non conservative amino acid substitution at C26304T of codon 194 (Arg194Trp) in exon 6, G27466A of codon 280 (Arg280His) in exon 9, and G28152A of codon 399 (Arg399Gln) in exon 10. Although the functional relevance of these variants is unknown, codon 399 is within the *XRCC1* BRCT domain (codons 314-402) (Zhang *et al.*, 1998), which is highly homologous to *BRCA1* (a gene also involved in DNA repair), containing a binding site for poly(ADP-ribose) polymerase (Masson *et al.*, 1998). Because the role of XRCC1 in BER is to bring together DNA polymerase β, DNA ligase III (LIG III), and poly(ADP-ribose) polymerase at the site of DNA damage (Caldecott *et al.*, 1994; Kubota *et al.*, 1996; Cappelli *et al.*, 1997), the codon 399 variant could have an altered repair activity. The codon 194 polymorphism resides in the linker regions of the XRCC1 N-terminal domain separating the helix 3 and DNA polymerase β involved in binding a single-nucleotide gap DNA substrate (Marintchev *et al.*, 1999). Therefore, it is less likely to cause a significant change in the repair function. Lunn *et al.* (1999) reported that the codon 399 variant was associated with higher levels of both aflatoxin B1-DNA adducts and glycophorin A variants in a normal population, suggesting that this variant may be an adverse genotype. The *XRCC*1 399Gln was also reportedly associated with higher DNA adduct levels only in never-smokers (Matullo *et al.*, 2001) and in patients with ionizing radiation sensitivity (Hu *et al.*, 2001).

Several molecular epidemiological studies have provided evidence for an association between polymorphisms of the DNA repair gene *XRCC1* and the risk of cancers of the esophagus, stomach, and lung in the Asian population (Lee *et al.*, 2001; Shen *et al.*, 2000; Ratnasinghe *et al.*, 2001); the breast in African Americans (Duell *et al.*, 2001); the colon in Egyptians (Abdel-Rahman *et al.*, 2000), and the head and neck in Caucasians (Sturgis *et al.*, 1999). However, others have failed to identify these polymorphisms as risk factors for cancers of the lung (Divine *et al.*, 2001; Bulkiewicz *et al.*, 2001), breast (Duell *et al.*, 2001), and bladder (Stern *et al.*, 2001; Matullo *et al.*, 2001).

XPC. The xeroderma pigmentosum complementing group C (XPC) gene contains 15 exons and codes for a 940-amino acid protein that plays a role in DNA damage recognition in the process of NER (Leg-erski *et al.*, 1992). Cells with deficient XPC are UV sensitive. Khan *et al.* (2000) identified a common biallelic polymorphism (PAT) in the XPC gene. This polymorphism consists of an insertion of 83 bases of A and T (poly(AT)) and a 5-base deletion within intron 9 and is in linkage disequilibrium with an SNP in exon 15 of the *XPC* gene (A2920C, Lys939→Gln). In a hospital-based case-control study of 287 non-Hispanic white patients with newly diagnosed squamous cell carci-nomas of the head and neck cancer and 311 control subjects matched by age, sex, ethnicity, and smoking status, Shen *et al.* (2001) found that the *XPC*-PAT+ allele was more frequent in the cases (0.409) than in the controls (0.333; P = 0.007). Specifically, 50 cases (17.4%) and 37 controls (11.9%) were *XPC*-PAT+/+, and 135 (47.0%) cases and 133 controls (42.8%) were *XPC*-PAT+/-. Subjects who were *XPC*-PAT+/- and *XPC*-PAT+/+ were at significantly increased risk for squamous cell carcinomas of the head and neck (adjusted odds ratios = 1.44 and 1.85, respectively (95% confidence intervals, 1.01-2.05 and 1.12-3.05, re-spectively; trend test, P = 0.007)). There was no ethnic difference in the frequency of the *XPC*-PAT+ allele among the four groups of non-Hispanic whites, African-Americans, Hispanic-Americans, and na-tive Chinese. The findings appeared to support the hypothesis that the *XPC*-PAT+ allele contributes to the risk of squamous cell carcinomas of the head and neck.

XPD/ERCC2. The XPD protein is an evolutionarily conserved helicase, a subunit of transcription factor IIH (TFIIH) that is essential for transcription and NER (Coin *et al.*, 1998). The gene is located on chromosome 19q13.2 (Lamerdin *et al.*, 1996). Mutations in *XPD* prevent its protein from interacting with *p44*, another subunit of TFIIH (Reardon *et al.*, 1996) and cause decreased helicase activity, resulting in a defect in NER. Mutations at different sites result in distinct clinical phenotypes (Taylor *et al.*, 1997).

Several studies have identified an association between polymorphisms of the DNA repair gene *XPD/ERCC2* and cancers of the skin (Tomescu *et al.*, 2001) and head and neck (Sturgis *et al.*, 2000), as well as cancers of the lung and brain in subgroups of Caucasians (Bulkiewicz *et al.*, 2001; Caggana *et al.*, 2001). However, other studies did not find evidence for a statistically significant main effect on basal cell carcinomas (Vogel *et al.*, 2001; Dybdahl *et al.*, 2000) and lung cancer (Spitz *et al.*, 2001).

XRCC3. The *XRCC3* gene, which is located on chromosome 14q32.3 and structurally related to *Rad51*, participates in homologous recombinational repair of DNA DSBs and cross-links (Tebbs *et al.*, 1995; Liu *et al.*, 1998). The XRCC3 protein is required for the assembly and stabilization of Rad51. Therefore, XRCC3-deficient cells do not form Rad51 foci after radiation damage and exhibit genetic instability and increased sensitivity to DNA-damaging agents such as UV light (Masson *et al.*, 2001). *XRCC3* is polymorphic for a C→T substitution at position 18067 in exon 7 that results in an amino acid change (Thr241Met) (Shen *et al.*, 1998), that is thought to affect the functions of the enzyme and/or its interaction with other proteins involved in DNA damage and repair.

Recently, Winsey *et al.* (2000) determined in the United Kingdom the variant allele frequency in 125 patients with melanoma and 211 controls consisting of cadaveric renal transplant donors and found that the *XRCC3*-Thr241Met polymorphism was significantly associated with an increased risk of melanoma (OR=2.36, 95% CI=1.44-3.86). In another case-control study of 124 bladder cancer patients and 85 hospital controls (urological and non urological), Matullo *et al.* (2001) found that the *XRCC3*-Thr241Met polymorphism was associated with an increased risk of bladder cancer, with the effect of XRCC3 more evident in non smokers. They also found that *XRCC3* variant apparently interacted with the N-acetyltransferase type 2 (*NAT2*) genotype, suggesting that *XRCC3* is involved in a common repair pathway of bulky DNA adducts. In addition, the likelihood of DNA adduct levels above the median was greater in NAT-2 slow acetylators, who are homozygotes for the *XRCC3*-241Met variant allele. However, this *XRCC3*-Thr241Met polymorphism was not associated with a greater risk of lung cancer in a hospital-based case-control study conducted in a Polish population (Butkiewicz *et al.*, 2001) and in another population-based case-control study (178 lung cancer cases and 453 controls) of U.S. Caucasians (David-Beabes *et al.*, 2001).

4.   SNP databases

The Human Genome Projects revealed that SNPs are the most common genetic variations in the genome and occur about once every 1,200 bp (Sherry *et al.*, 2001). Thus, there are more than 10 million SNPs in the human genome. The challenge is to identify the SNPs that are informative regarding human diseases. It is very important to compile information on SNPs for large-scale association genetics studies, population genetics studies, evolutionary biology research, and positional cloning and physical mapping projects. Several SNP databases have been developed. The two most comprehensive ones are Celera's Human

RefSNP database and The National Center for Biotechnology Information (NCBI) dbSNP database.

4.1   Human RefSNP

The Celera Human RefSNP database (http://www.celera.com/) is part of the Celera Discovery System (CDS) and contains a total of 3.6 million unique SNPs determined from a comprehensive database of computed and validated variations in the human genome framework from both Celera and public data sources. These sources are summarized in Table 17.3.

Source	Number of SNPs
Celera genomic sequences	2,459,317
DbSNP	1,496047
HGMD	14,918
HGBase	486,590

Table 17.3.   Celera Human RefSNP Sources

The SNP data in the CDS are fully integrated with the genome, gene, and protein structure data. There are two types of text search: basic and advanced. Advanced searches offer many options that meet specific requirements, from finding all the SNPs in the HGBase and selecting a source box with an HGBase item to viewing all the SNPs associated with a missense mutation, by selecting the missense mutation item in the gene-structure box. The search output can be displayed in terms of SNP frequencies, population information, gene structure, and effects on protein. SNPs are also linked to disease associations as defined by Online Mendelian Inheritance in Man (OMIM). Each SNP sequence plus 300 5' bases and 300 3' bases is linked to the output. Three formats of sequence data that can be selected are FASTA, which yields a flank sequence; Primer express, which indicates the search SNP such as (A/G) and whether there are other SNPs in this sequence signed as N; and Primer+, which lists every SNP in detail such as (A/G). The information regarding the sequence around an SNP is useful for amplicon and SNP primer design. The Human RefSNP database is well organized. However, the data on only 8405 SNPs are validated; most of the other SNP data are found by computer data-mining procedures, and most of the SNP frequency data are based on a small sample size and needs to be further characterized.

4.2   dbSNP

NCBI in collaboration with the National Human Genome Research Institute (NHGRI), has established the dbSNP database (http://www.ncbi.nlm.nih.gov/SNP/) to serve as a central repository for sequence variations, which include single-base changes (99.77%) and short insertions and deletions (Sherry *et al.*, 1999). In November 2001, dbSNP had submissions for 3,053,511 SNPs from 168 registered groups. These three million SNPs can be grouped into 1,805,951 non redundant SNPs. The NCBI assigns a dbSNP accessioning ss# to each submitted variation and an rs# to each reference identifier. As a public database, submissions are welcomed from all sources, public and private. The various sources of data in the dbSNP are summarized in Table 17.4.

Source	Percentage of Total Submissions
Human genome project	65%
Private investigator/corporate	28%
EST database mining	6%
NHGRI	1%

Table 17.4.   dbSNP Sources

Although the dbSNP has been designed to collect the information that includes the observed alleles at a particular locus, the degree of heterozygosity, the flanking sequence that surrounds the mutation, the assay methods and conditions, and a pointer to a companion STS or GenBank record, many SNPs in the dbSNP do not have allele frequency information, and the length of sequence surrounding the SNP provided by different submitters varies .

NCBI integrates dbSNP as part of its discovery space, so dbSNP can be a "first point of entry" into the resource network for the query and retrieval of specific SNP records, or it can serve as an information server for searches that start in other resources, such as GenBank, PubMed, LocusLink, OMIM, and Unigene. dbSNP can be searched with the BLAST algorithm, which compares a user-submitted sequence against all flanking sequence records in dbSNP. dbSNP can also be queried by other NCBI resources; for example, using LocusLink, queries can be done by gene name or nomenclature association.

Following are the other common SNP database Web sites:

- The SNP Consortium LTD (TSC)
  http://snp.cshl.org/

- Human Genic Bi-Allelic Sequences (HGBase)
  http://hgbase.interactiva.de/

- Human SNP Database
  http://www-genome.wi.mit.edu/SNP/human/index.html

- Japanese Single-Nucleotide Polymorphisms (JSNP)
  http://snp.ims.u-tokyo.ac.jp/

One caveat about the current SNP databases is that the inclusion of SNPs is based on a sequence comparison in a limited number of individuals. Therefore, some of the listed SNPs could be the result of a sequencing mistake. Thus it is important that the candidate SNPs in the databases be validated in a larger population. According to a recent research study of 1,200 SNPs in three ethnic groups, more than 80% of the candidate SNPs from public SNP databases were found to be polymorphic, and only about 50% were common ones in any given population (Marth *et al.*, 2001).

5. Analysis of SNP data

A central goal in SNP data analysis is to find informative SNPs and combinations that are associated with certain characteristics of interest, such as phenotypes, sensitivity to DNA damage and toxicity response, or generally, their belongingness to one of a number of *classes*. It is expected that the relationship between these higher-level macroscopic phenomena and SNPs will be subtle in the sense that there will be multivariate effects. For example, the joint effects of two SNPs acting in combination have already been reported in the literature (Skibola *et al.*, 1999; van der Put *et al.* 1998; Weisberg *et al.*, 1998). Uncovering such multivariate relationships directly from the data requires robust inference techniques.

The task of uncovering the possible relationships between SNP variations and class membership falls naturally within the framework of computational learning theory. The inference (learning) of unknown functions from examples of their input-output behavior constitutes the fundamental problem in scientific investigation. The central aim of computational learning theory is to define and study useful models of learning from an algorithmic point of view (Anthony and Biggs, 1992). A major focus has been on the construction of algorithms for the efficient determination of Boolean formulae from examples. This type of induction of Boolean logic or the design of Boolean classifiers forms the core of many data-mining and knowledge-discovery algorithms (Valiant, 1984;

Crama *et al.*, 1988; Quinlan, 1986). Specifically, we are given a set of data containing "positive examples" and "negative examples," and our task is to establish a Boolean function f, possibly with some specified properties, so that f is true on every positive example and false on every negative example. The Boolean framework is especially useful in applications where the feature space, from which we draw our examples, is discrete valued (Boros *et al.* 1996). Although in SNP analyses, we will typically be interested in two-class classification problems (i.e., dichotomies, such as toxic verses non toxic response to radiation), it is possible to apply Boolean methods to problems with an arbitrary number of classes (Moreira and Mayoraz, 1998).

For example suppose that each example (patient sample) we collect consists of a number of SNP sites of interest. Each site can take on only a small number of values encoding the exact nucleotide variations. Specifically, if we assume that in each site there are only two possible variations, which is the most common case, then there are only three possible combinations of the two alleles: (1) both copies are Type 1; (2) both copies are Type 2; (3) one copy is Type 1 and the other copy is Type 2 (heterogeneous). Thus, each SNP site can be coded with two bits of information. The inference problem is then posed as follows: given a set of examples (samples with observed SNP values together with the corresponding known class membership), establish a rule or an *extension* f that is consistent with the given data. This amounts to finding a logical (diagnostic) explanation of these data and is known as the *Consistency Problem* (Pitt and Valiant, 1988; Shmulevich *et al.*, 2000). The Consistency Problem is closely related to the important *Probably Approximately Correct* (PAC) learning model (Valiant, 1984). The central idea behind PAC is that for an algorithm to successfully learn an unknown *concept* (Boolean formula), it should produce, with high probability, a *hypothesis* that is a good approximation of the concept.

In reality, a consistent extension may not exist either due to errors, or more likely, due to a number of underlying latent factors not utilized by the learning algorithm. Thus, the observed examples may be inconsistent. In this case, we may have to give up our goal of establishing a consistent extension and settle for a Boolean formula that minimizes the number of misclassifications. This problem is known as the *Best-Fit Extension Problem* (Boros *et al.*, 1998; see also Chapter 11).

Finally, there exists a problem of *overfitting* the data. A learning procedure may output many Boolean formulae or classifiers for a particular data set. While each of these classifiers may seem to perform equally well on the given data set, not all of them are likely to perform equally well on future examples. Thus, the goal should be to find the best (optimal) classifier from the point of view of minimizing the error made by it in future, unknown cases, thus capturing the underlying regularity, not the noise. A standard approach to achieve this goal is to penalize the complexity of the model, or in our case, the extension. Thus, a good extension must tread a fine line: it must be sufficiently complex to describe the data accurately, but without losing generalizability. The Minimal Description Length principle (Rissanen, 1978) provides the means to achieve this and extends Occam's Razor Principle: Among competing explanations, all of which are equally good, choose the simplest (see also Chapter 10). Another promising approach to this problem is to use recently proposed theories of data generation (Franco, 2000). Such models would allow us to find optimal Boolean classifiers in the sense of minimizing the probability of error on future examples.

6. Future outlook

One of the goals of the Human Genome Project is to identify sequence variation in human populations. Although the screening for SNPs in human genes is incomplete, the number of genes that have been systematically screened is growing. It is now becoming increasingly important to perform genotype-related phenotyping assays to screen for functional variants. Results from such phenotyping assays can indicate whether a marginal reduction in repair capacity is a polymorphic trait and associated with elevated risk of diseases including cancer. Judging from the large number of amino acid SNPs observed, we can expect to identify 60 to 90 different amino acid substitutions in each DNA repair pathway. As genotyping costs decline, it will also be feasible to increase the scale of the genotyping effort. Most individuals will have four to seven variant genes within the 15-25 genes of a repair pathway, and thus have very complex genotypes, a realization that has changed the original perception of "wild-type." This variation in genetic background raises the issue of a proper control population. In particular, the reported discrepancies in the association between DNA repair gene polymorphisms and risk of cancer may be largely due to differences in the genetic background of the control populations, particularly in various ethnic groups. There is also a possibility of selection bias stemming from survival that may be related to certain variants. Furthermore, multiple tests for numerous genotypes in a relatively small number of subjects adds complexity to the data analysis. Regardless, the immediate challenge is to use genotyping data to identify individuals with an elevated risk of cancer so that programs to reduce the incidence of cancer can be implemented. Therefore, there is an increasing need for large, well-characterized cohorts with good estimates of exposures and significantly elevated cancer incidence. With improvements in our ability to analyze genotype data, such cohort studies will provide an accurate picture of the role of genetic polymorphisms of DNA repair genes in the etiology of cancer.

Acknowledgments

This work was supported in part by a National Institute of Environmental Health of Sciences grant R01 ES 11740 (Q. W.), and in part by the Tobacco Settlement Fund as appropriated to the University of Texas M. D. Anderson Cancer Center by the Texas Legislature (Genomics program; S. H and W. Z), and a portion of the work was performed under the auspices of the U.S. Department of Energy by the University of California, Lawrence Livermore National Laboratory under contract No. W-7405-Eng-48 and supported by an Interagency Agreement with NIEHS (Y1-ES-8054-05) (H.W.M).

References

Abdel-Rahman, S. Z., Soliman, A. S., Bondy, M. L., Omar, S., El-Badawy, S. A., Khaled, H. M., Seifeldin, I. A., Levin, B. (2000) Inheritance of the 194Trp and the 399Gln variant alleles of the DNA repair gene XRCC1 are associated with increased risk of early-onset colorectal carcinoma in Egypt. *Cancer Lett.*, 159, 79-86.

Airoldi, L., Pastorelli, R., Magagnotti, C. and Fanelli, R. (1999) Carcinogen-DNA adducts as tools in risk assessment. *Adv. Exp. Med. Biol.*, 472, 231-240.

Anthony, M. and Biggs, N. (1992) Computational Learning Theory, *Cambridge, UK: Cambridge University Press.*

Bartsch, H., Nair, U., Risch, A., Rojas, M., Wikman, H., Alexandrov, K. (2000) Genetic polymorphism of CYP genes, alone or in combination, as a risk modifier of tobacco-related cancers. *Cancer Epidemiol Biomarkers Prev.*, 9, 3-28.

Bell D. W., Wahrer, D. C., Kang, D. H., MacMahon, M. S., FitzGerald, M. G., Ishioka, C., Isselbacher, K. J., Krainer, M. and Haber, D. A. (1999) Common nonsense mutations in RAD52. *Cancer Res.*, 59, 3883-3888.

Berwick M., Vineis P. (2000) Markers of DNA repair and susceptibility to cancer in humans: an epidemiologic review. *J. Natl. Cancer Inst.*, 92, 874-97.

Boros, E., Hammer, P. L., Ibaraki, T., Kogan, A., Mayoraz, E. and Muchnik, I. (1996). An Implementation of Logical Analysis of Data. *Rutcor Research Report* 22-96 RUTCOR, Rutgers University.

Boros, E., Ibaraki, T., Makino, K. (1998) Error-free and best-fit extensions of partially defined Boolean functions. *Information and Computation*, 140, 254-283.

Broughton, B. C., Steingrimsdottir, H., Lehmann, A. R. (1996) Five polymorphisms in the coding sequence of the xeroderma pigmentosum group D gene. *Mutat. Res.*, 362, 209-211.

Buetow, K. H., Edmonson, M. N., Cassidy, A. B. (1999) Reliable identification of large numbers of candidate SNPs from public EST data. *Nature Genet.*, 21, 323-325.

Butkiewicz, D., Rusin, M., Enewold, L., Shields, P. G., Chorazy, M., Harris, C. C. (2001) Genetic polymorphisms in DNA repair genes and risk of lung cancer. *Carcinogenesis*, 22, 593-7.

Caggana, M., Kilgallen, J., Conroy, J. M., Wiencke, J. K., Kelsey, K.T., Miike, R., Chen, P., Wrensch, M. R. (2001) Associations between ERCC2 polymorphisms and gliomas. *Cancer Epidemiol Biomarkers Prev.*, 10, 355-60.

Caldecott, K. W., McKeown, C. K., Tucker, J. D., Ljungquist, S., Thompson, L. H. (1994) An interaction between the mammalian DNA repair protein XRCC1 and DNA ligase III. *Mol Cell Biol.*,14, 68-76.

Cappelli, E., Taylor, R., Cevasco, M., Abbondandolo, A., Caldecott, K., Forsina, G. (1997) Involvement of XRCC1 and DNA ligase gene products in DNA base excision repair. *J Biol. Chem.*, 272, 23970-75.

Cargill, M., Altshuler, D., Ireland, J., Sklar, P., Ardlie, K., Patil, N., Shaw, N., Lane, C. R., Lim, E. P., Kalyanaraman, N., Nemesh, J., Ziaugra, L., Friedland, L., Rolfe, A., Warrington, J., Lipshutz, R., Daley, G. Q., Lander, E. S. (1999) Characterization of single-nucleotide polymorphisms in coding regions of human genes. *Nature Genet.*, 22, 231-238.

Cheng, L., Eicher, S. A., Guo, Z., Hong, W. K., Spitz, M. R., Wei, Q. (1998) Reduced DNA repair capacity in head and neck cancer patients. *Cancer Epidemiol Biomarkers Prev.*, 7, 465-8.

Chicurel, M. (2001) Faster, better, cheaper genotyping, *Nature*, 412, 580-2.

Cleaver, J. E., Karplus, K., Kashani-Sabet, M., Limoli, C. L. (2001) Nucleotide excision repair, "a legacy of creativity." *Mutat. Res.*, 485, 23-36

Clifford, R., Edmonson, M., Hu, Y., Nguyen, C., Scherpbier, T., Buetow, K. H. (2000) Expression-based genetic/physical maps of single-nucleotide polymorphisms identified by the cancer genome anatomy project. *Genome Res.*,10, 259-1265.

Coin, F., Marinoni, J. C., Rodolfo, C., Fribourg, S., Pedrini, A. M., Egly, J. M. (1998) Mutations in the XPD helicase gene result in XP and

TTD phenotypes, preventing interaction between XPD and the p44 subunit of TFIIH. *Nat. Genet.*, 20, 184-8.

Collins, F. S., Brooks, L. D., Chakravarti, A. (1998) A DNA polymorphism discovery resource for research on human genetic variation. *Genome Res.*, 8, 1229-1231.

Collins, A. R. (1999) Oxidative DNA damage, antioxidants, and cancer. *BioEssays*, 21, 238-246.

Cooke, M. S., Mistry, N., Ladapo, A., Herbert, K. E. and Lunec, J. (2000) Immunochemical quantitation of UV-induced oxidative and dimeric DNA damage to human keratinocytes. *Free Radical Res.*, 33, 369-381.

Crama, Y., Hammer, P., and Ibaraki, T. (1988) Cause-Effect Relationships and Partially Defined Boolean Functions. *Annals Operations Res.*, 16, 299-326.

David-Beabes, G. L., Lunn, R. M., London, S. J. (2001) No Association between the XPD (Lys751Gln) Polymorphism or the XRCC3 (Thr241Met) Polymorphism and Lung Cancer Risk. *Cancer Epidemiol Biomarkers Prev.*,10, 911-2.

Divine, K. K., Gilliland, F. D., Crowell, R. E., Stidley, C. A., Bocklage, T. J., Cook, D. L., Belinsky, S. A. (2001) The XRCC1 399 glutamine allele is a risk factor for adenocarcinoma of the lung. *Mutat Res.*, 461, 273-8.

Duell, E. J., Millikan, R. C., Pittman, G. S., Winkel, S., Lunn, R. M., Tse, C. K., Eaton, A., Mohrenweiser, H. W., Newman, B., Bell, D. A. (2001) Polymorphisms in the DNA repair gene XRCC1 and breast cancer. *Cancer Epidemiol Biomarkers Prev.*, 10, 217-22.

Dybdahl, M., Vogel, U., Frentz, G., Wallin, H., Nexo, B. (1999) Polymorphisms in the DNA repair gene *XPD*: Correlations with risk and age at onset of basal cell carcinoma. *Cancer Epidemiol Biomarkers Prev.*, 8, 77-81.

Evan, G. I., Vousden, K. H. (2001) Proliferation, cell cycle and apoptosis in cancer. *Nature,* 411, 342-348.

Fan, F., Liu, C., Tavare, S., Arnheim, N. (1999) Polymorphisms in the human DNA repair gene XPF. *Mutat Res.*, 406, 115-20.

Fearon, E. R., Dang, C. V. (1999) Cancer genetics: Tumor suppressor meets oncogene. *Curr. Biol.*, 9, R62-R65.

Franco, J. (2000) Finding optimal Boolean classifiers. In Approximation and Complexity in Numerical Optimization: Continuous and Discrete Problems. (*Nonconvex Optimization and its Applications*, Vol. 42), Pardalos (Ed.), Kluwer.

Friedberg, E. C., Walker, G. C., Siede, W. (1995) DNA Repair and Mutagenesis. *Washington, DC: ASM Press.*

Friedberg, E. C. (2001) How nucleotide excision repair protects against cancer. *Nature Reviews*, 1, 22-33.

Friedberg, E. C., Fischhaber, P. L. and Kisker, C. (2001) Error-prone DNA polymerases: novel structures and the benefits of infidelity. *Cell*, 107, 9-12.

Garg, K., Green, P. and Nickerson, D. A. (1999) Identification of candidate coding region single nucleotide polymorphisms in 165 human genes using assembled expressed sequence tags. *Genome Res.*, 9, 1087-1092.

Garner, R. C., (1998) The role of DNA adducts in chemical carcinogenesis. *Mutat. Res.*, 402, 67-75.

Gupta, R. C. Lutz, W. K. (1999) Background DNA damage for endogenous and unavoidable exogenous carcinogens: a basis for spontaneous cancer incidence? *Mutat. Res.*, 424, 1-8.

Halushka, M. K., Fan, J. B., Bentley, K., Hsie, L., Shen, N., Weder, A., Cooper, R., Lipshutz, R., Chakravarti, A. (1999) Patterns of single-nucleotide polymorphisms in candidate genes for blood-pressure homeostasis. *Nature Genet.*, 22, 239-247.

Hoeijmakers, J. H. (2001) Genome maintenance mechanisms for preventing cancer. *Nature* 411, 366-374.

Hu, J. J., Smith, T. R., Miller, M. S., Mohrenweiser, H. W., Golden, A., Case, L. D. (2001) Amino acid substitution variants of APE1 and XRCC1 genes associated with ionizing radiation sensitivity. Carcinogenesis, 22, 917-22.

Kato, M., Yano, K., Matsuo, F., Salto, H., Katagiri, T., Kurumizaka, H., Yoshimoto, M., Kasumi, F., Akiyama, F., Sakamoto, G., Nagawa, H., Nakamura, Y., Miki, Y. (2000) Identification of RAD51 alteration in patients with bilateral breast cancer. *J. Hum. Genet.*, 45, 133-137

Khan, S. G., Metter, E. J., Tarone, R. E., *et al.* (2000) A new xeroderma pigmentosum group C poly(AT) insertion/deletion polymorphism. *Carcinogenesis*, 21, 1821-5.

Khanna, K. K., Jackson, S. P. (2001) DNA double-strand breaks: signaling, repair and the cancer connection. *Nature Genet.* 27, 247-254.

Kubota, Y., Nash. R. A., Klungland, A., Schar, P., Barnes, D. E., Lindahl, T. (1996) Reconstitution of DNA base excision-repair with purified human proteins: interaction between DNA polymerase beta and the XRCC1 protein. *EMBO J.*, 15, 6662-70.

Kwok, P. Y. Approaches to allele frequency determination. (2000) *Pharmacogenetics*, 1, 231-5.

Lamerdin, J. E., Stilwagen, S. A., Ramirez, M. H., Stubbs, L., Carrano, A. V. (1996) Sequence analysis of the ERCC2 gene regions in human, mouse, and hamster reveals three linked genes. *Genomics*, 34, 399-409.

Lee, J. M., Lee, Y. C., Yang, S. Y., Yang, P. W., Luh, S. P., Lee, C. J., Chen, C. J., Wu, M. T. (2001) Genetic polymorphisms of XRCC1 and risk of the esophageal cancer. *Int J Cancer*, 95, 240-6.

Legerski, R., Peterson, C. (1992) Expression cloning of a human DNA repair gene involved in xeroderma pigmentosum group C. *Nature*, 359, 70–73.

Lindahl, T. (1993) Instability and decay of the primary structure of DNA. *Nature*,362,709-715.

Loyamichev, *et al.* (1999) Third wave technologies, polymorphism identification and quantitative detection of genomic DNA by invasive cleavage of oligonucleotide probes. *Nature Biotech,* 17, 292-6.

Lunn, R. M., Langlois, R. G., Hsieh, L. L., Thompson, C. L., Bell, D. A. (1999) XRCC1 polymorphisms: effects on aflatoxin B1-DNA adducts and glycophorin A variant frequency. *Cancer Res.*,59, 2557-61.

Ma, X., Jin, Q., Forsti, A., Hemminki, K., Kumar, R. (2000) Single nucleotide polymorphism analyses of the human proliferating cell nuclear antigen (PCNA) and flap endonuclease (FEN1) genes. *Int. J. Cancer*, 88, 938-942.

Marintchev A, Mullen MA, Maciejewski MW, Pan B, Gryk MR, Mullen GP. (1999) Solution structure of the single-strand break repair protein XRCC1 N-terminal domain. *Nat. Struct. Biol.*, 6, 884-93.

Marth, G., Yeh, R., Minton, M., Donaldson, R., Li, Q., Duan, S., Davenport, R., Miller R. D., Kwok P. Y. (2001) Single-nucleotide polymorphisms in the public domain: how useful are they? *Nature Genet.*, 27, 371-372.

Masson, J.Y., Stasiak, A. Z., Stasiak, A. *et al.* (2001) Complex formation by the human RAD51C and XRCC3 recombination repair proteins. *Proc Natl Acad Sci U S A*, 98, 8440-6.

Masson, M., Niedergang, C., Schreiber, V., Muller, S., Demarcia, J. M., Demurcia, G. (1998) XRCC1 is specifically associated with PARP polymerase and negatively regulates its activity following DNA damage. *Mol. Cell Biol.*, 18, 3563-71.

Matullo, G., Guarrera, S., Carturan, S., Peluso, M., Malaveille, C., Davico, L., Piazza, A., Vineis, P. (2001) DNA repair gene polymorphisms, bulky DNA adducts in white blood cells and bladder cancer in a case-control study. *Int. J. Cancer*, 92, 562-7

Matullo, G., Palli, D., Peluso, M., Guarrera, S., Carturan, S., Celentano, E., Krogh, V., Munnia, A., Tumino, R., Polidoro, S., Piazza, A., Vineis, P. (2001) XRCC1, XRCC3, XPD gene polymorphisms, smoking and (32)P-DNA adducts in a sample of healthy subjects. *Carcinogenesis*, 22, 1437-1445.

Mohrenweiser, H. W., Jones, I. M. (1998) Variation in DNA repair is a factor in cancer susceptibility: a paradigm for the promises and perils of individual and population risk estimation? *Mutat. Res.*, 400, 5-24.

Moreira, M., Mayoraz, E. (1998). "Improved Pairwise Coupling Classification with Correcting Classifiers", *Proceedings of the 10th European Conference on Machine Learning, Chemnitz, Germany*, April, 160-171.

Nair, U., Bartsch, H. (2001) Metabolic polymorphisms as susceptibility markers for lung and oral cavity cancer. *IARC Sci Publ.*,154, 271-90.

Ohnishi, Y., Tanaka, T., Yamada, R., Suematsu, K., Minami, M., Fujii, K., Hoki, N., Kodama, K., Nagata, S., Hayashi, T., Kinoshita, N., Sato, H., Kuzuya, T., Takeda, H., Hori, M.,Nakamura, Y. (2000) Identification of 187 single nucleotide polymorphisms (SNPs) among 41 candidate genes for ischemic heart disease in the Japanese population. *Hum. Genet.*,106,288-292

Orr-Weaver T. L., Weinberg, R. A. (1998) A checkpoint on the road to cancer. *Nature*, 392, 223-224.

Passarino, G., Shen, P., Van Kirk, J. B., Lin, A. A., De Benedicts, G., Cavalli Sforza, L. L., Oefner, P. J. and Underhill, P. A. (2001) The Werner syndrome gene and global sequence variation. *Genomics*, 71, 118-122.

Pavlov, Y. I., Shcherbakova, P. V., Kunkel, T. A. (2001) In vivo consequences of putative active site mutations in yeast DNA polymerases a, e, g, d. *Genetics*,159, 47-64.

Petit, C., Sancar, A. (1999) Nucleotide excision repair: from E. coli to man. *Biochimie* 81,15-25.

Picoult-Newberg L., Ideker, T. E., Pohl, M.G., Taylor, S.L., Donaldson, M.A., Nickerson, D.A., Boyce-Jacino, M. (1999) Mining SNPs from EST databases. Genome Res. 9(2), 167-174.

Pitt, L., Valiant. L.G. (1988) Computational limitations on learning from examples. *ACM*, 35, 965-984.

Povey, A. C. (2000) DNA adducts: endogenous and induced. *Toxicol. Pathol.*,28, 405-414.

Ratnasinghe, D., Yao, S. X., Tangrea, J. A., Qiao, Y. L., Andersen, M. R., Barrett, M. J., Giffen, C. A., Erozan, Y., Tockman, M. S., Taylor, P. R. (2001) Polymorphisms of the DNA repair gene XRCC1 and lung cancer risk. *Cancer Epidemiol Biomarkers Prev.*,10, 119-23.

Reardon, J. T., Ge, H., Gibbs, E., Sancar, A., Hurwitz, J., Pan, Z. Q. (1996) Isolation and characterization of two human transcription factor IIH (TFIIH)-related complexes: ERCC2/CAK and TFIIH. *Proc Natl Acad Sci USA*, 93, 6482-7.

Rissanen J. (1978) Modeling by shortest data description. *Automatica*, 14, 465-471.

Romen A., Glickman,B. W. (2001) Human DNA repair genes. *Environ. Mol. Mutag.*, 37, 241-283.

Sachidanandam R., Weissman, D., Schmidt, S. C., Kakol, J. M., Stein, L. D., Marth, G., Sherry, S., Mullikin, J. C., Mortimore, B. J., Willey, D. L., Hunt, S. E., Cole, C. G., Coggill, P. C., Rice, C. M., Ning, Z., Rogers, J., Bentley, D. R., Kwok, P. Y., Mardis, E. R., Yeh, R. T., Schultz, B., Cook, L., Davenport, R., Dante, M.. Fulton, L., Hillier, L., Waterston, R. H., McPherson, J. D., Gilman, B., Schaffner, S., Van Etten, W. J., Reich, D., Higgins, J., Daly, M. J., Blumenstiel, B., Baldwin, J., Stange-Thomann, N., Zody, M. C., Linton, L., Lander, E. S., Attshuler, D. (2001) A map of human genome sequence variation containing 1.42 million single nucleotide polymorphisms. *Nature* 409, 928-933.

Sancer, A. (1995) DNA repair in humans. *Annu Rev Genet*, 29, 69-105.

Shannon, M., Stubbs, L. (1998) Analysis of homologous XRCC1-linked zinc-finger gene families in human and mouse: evidence for orthologous genes. *Genomics*, 49,112-21.

Shen, H., Sturgis, E. M., Khan, S. G., Qiao, Y., Shahlavi, T., Eicher, S. A., Xu, Y., Wang, X., Strom, S. S., Spitz, M. R., Kraemer, K. H., Wei, Q. (2001) An intronic poly (AT) polymorphism of the DNA repair gene XPC and risk of squamous cell carcinoma of the head and neck: a case-control study. *Cancer Res.*, 6, 3321-5.

Shen, H., Xu, Y., Qian, Y., Yu, R., Qin, Y., Zhou, L., Wang, X., Spitz, M. R., Wei, Q. (2000) Polymorphisms of the DNA repair gene XRCC1 and risk of gastric cancer in a Chinese population. *Int J Cancer.*, 88, 601-6.

Shen, M. R., Jones, I. M., Mohrenweiser, H. (1998) Nonconservative amino acid substitution variants exist at polymorphic frequency in DNA repair genes in healthy humans. *Cancer Res.*, 58, 604-8.

Sherry, S. T., Ward, M. H., Kholodov, M., Baker, J., Phan, L., Smigielski, E. M. and Sirotkin, K. (2001) dbSNP: the NCBI database of genetic variation. *Nucleic Acids Res.*, 29,308-311.

Shi, M. M. (2001) Enabling large-scale pharmacogenetic studies by high-throughput mutation detection and genotyping technologies. *Clin. Chemi.*, 47,164-172.

Shmulevich, I., Gabbouj, M., and Astola, J. (2001) Complexity of the Consistency Problem for Certain Post Classes. *IEEE Transactions on Systems, Man, and Cybernetics*, Part B, 31, 251-253.

Skibola, C. F., Smith, M.T., Kane, E., Roman, E., Rollinson, S., Cartwright, R. A., Morgan, G. (1999) Polymorphisms in the methylenetetrahydrofolate reductase gene are associated with susceptibility to acute leukemia in adults *Proc Natl Acad Sci USA*, 96, 12810-12815.

Spitz, M. R., Wu, X., Wang, Y., Wang, L. E., Shete, S., Amos, C. I., Guo, Z., Lei, L. Mohrenweiser, H., Wei, Q. (2001) Modulation of nucleotide excision repair capacity by XPD polymorphisms in lung cancer patients. *Cancer Res.*, 61, 1354-7.

Stern, M. C., Umbach, D. M., van Gils, C. H., Lunn, R. M., Taylor, J. A. (2001) DNA repair gene XRCC1 polymorphisms, smoking, and bladder cancer risk. *Cancer Epidemiol Biomarkers Prev.*, 10, 125-31.

Sturgis, E. M., Castillo, E. J., Li, L. *et al.* (1999) Polymorphisms of DNA repair gene XRCC1 in squamous cell carcinoma of the head and neck. *Carcinogenesis*, 20, 2125-9.

Sturgis, E. M., Zheng, R., Li, L., Castillo, E. J., Eicher, S. A., Chen, M., Strom, S. S., Spitz, M. R., Wei, Q. (2000) XPD/ERCC2 polymorphisms and risk of head and neck cancer: a case-control analysis. *Carcinogenesis*, 21, 2219-23.

Taylor, E. M., Broughton, B. C., Botta, E. *et al.* (1997) Xeroderma pigmentosum and trichothiodystrophy are associated with different mutations in the XPD (ERCC2) repair/transcription gene. *Proc Natl Acad Sci USA*, 94, 8658-63.

Tebbs, R. S., Zhao, Y., Tucker, J. D. *et al.* (1995) Correction of chromosomal instability and sensitivity to diverse mutagens by a cloned cDNA of the XRCC3 DNA repair gene. *Proc Natl Acad Sci U S A*, 92, 6354-8.

Thorstenson, Y. R., Shen, P., Tusher, V. G., Wayne, T. L., Davis, R. W., Chu, G., Oefner, P. J. (2001) Global analysis of ATM polymorphism reveals significant functional constraint. *Am. J. Hum. Genet.*, 69, 396-412.

Tomescu, D., Kavanagh, G., Ha, T., Campbell, H., Melton, D. W. (2001) Nucleotide excision repair gene XPD polymorphisms and genetic predisposition to melanoma. *Carcinogenesis*, 22, 403-8.

van der Put, N. M., Gabreels, F., Stevens, E. M., Smeitink, J. A., Trijbels, F. J., Eskes, T. K., van den Heuvel, L. P., Blom, H. J. (1998) A second common mutation in the methylenetetrahydrofolate reductase gene: an additional risk factor for neural-tube defects? *Am. J. Hum. Genet.*, 62, 1044-51.

Vogel, U., Hedayati, M., Dybdahl, M., Grossman, L., Nexo, B. A. (2001) Polymorphisms of the DNA repair gene XPD: correlations with risk of basal cell carcinoma revisited. *Carcinogenesis*, 22, 899-904.

Wagner, T. M. U., Hirteniehner, K., Shen, P., Moeslinger, R., Muhr, D., Fleischmann, E., Concin, H., Doeller, W., Haid, A., Lang, A. H.,

Mayer, P., Petru, E., Ropp, E., Langbauer, G., Kubista, E., Scheiner, O., Underhill, P., Mountain, J., Stierer, M., Zielinski, C., Oefner, P. (1999) Global sequence diversity of BRCA2: analysis of 71 breast cancer families and 95 control individuals of worldwide populations. *Human Mol. Genet.*, 8, 413-423.

Wang, D. G., Fan, J. B., Siao, C. J., Berno, A., Young, P., Sapolsky, R., Ghandour, G., Perkins, N., Winchester, E., Spencer, J., Kruglyak, L., Stein, L., Hsie, L., Topaloglou, T., Hubbell, E., Robinson, E., Mittmann, M., Morris, M. S., Shen, N., Kilburn, D., Rioux, J., Nusbaum, C., Rozen, S., Hudson, T. J., Lipschutz, R., Chee, M., Lander, E. S. (1998) Large-scale identification, mapping, and genotyping of single-nucleotide polymorphisms in the human genome. *Science*, 280,1077-1082.

Ward, J. F. (1995) Radiation mutagenesis: the initial DNA lesions responsible. *Radiat. Res.*, 142, 362-368.

Wei, Q., Cheng, L., Amos, C. I. *et al.* (2000) Repair of tobacco carcinogen-induced DNA adducts and lung cancer risk: A molecular epidemiological study. *J. Natl. Cancer Inst.*, 92, 1764-72.

Wei, Q., Cheng, L., Hong, W. K. *et al.* (1996) Reduced DNA repair capacity in lung cancer patients. *Cancer Res,* 56, 4103-7.

Wei, Q., Matanoski, G. M., Farmer, E. R., Hedayati, M. A., Grossman, L. (1993) DNA repair and aging in basal cell carcinoma: a molecular epidemiology study. *Proc Natl Acad Sci USA,* 90, 1614-8.

Weinert, T. (1997) A DNA damage checkpoint meets the cell cycle engine. *Science*, 277,1450-1505.

Weisberg, I., Tran, P., Christensen, B., Sibani, S., Rozen, R. (1998). "A Second Genetic Polymorphism in Methylenetetrahydrofolate Reductase (MTHFR) Associated with Decreased Enzyme Activity" *Molecular Genetics and Metabolism* 64,169-172.

Winsey, S. L., Haldar, N. A., Marsh, H. P., Bunce, M., Marshall, S. E., Harris, A. L., Wojnarowska, F., Welsh, K. I. (2000) A variant within the DNA repair gene XRCC3 is associated with the development of melanoma skin cancer. *Cancer Res.,* 60, 5612-6.

Wood, R. D., Mitchell, M., Sgouros, J., Lindahl, T. (2001) Human DNA repair genes. *Science*, 291,1284-1289.

Yamada, R., Tanaka, T., Ohnishi, Y., Suematsu, K., Minami, M., Seki, T., Yukioka, M., Maeda, A., Murata, N., Saiki, O., Teshima, R., Kudo, O., Ishikawa, K., Ueyosi, A., Tateishi, H., Inaba, M., Goto, H., Nishizawa, Y., Tohma, S., Ochi, T., Yamamoto, K., Nakamura, Y. (2000) Identification of 142 single nucleotide polymorphisms in 41 candidate genes for rheumatoid arthritis in the Japanese population. *Hum. Genet.,*106, 293-297.

Zhang, X., Morera, S., Bates, P. A., Whitehead, P. C., Coffer, A. I., Hainbucher, K., Nash, R. A., Sternberg, M. J., Lindahl, T., Freemont, P. S. (1998) Structure of an XRCC1 BRCT domain: a new protein-protein interaction module. *EMBO J.*,17, 6404-11.

Index